OXFORD MEDICAL PUBLICATIONS

Geriatric Medicine:
a case-based manual

D1243754

Geriatric Medicine:
a case-based manual

Edited by

Jeanne Y. Wei

Harvard Medical School and Beth Israel Hospital

and

Myles N. Sheehan

Stritch School of Medicine and Loyola University
Medical Center

Oxford   New York   Tokyo
OXFORD UNIVERSITY PRESS
1997

Oxford University Press, Great Clarendon Street, Oxford OX2 6DP

Oxford   New York

Athens   Auckland   Bangkok   Bogota   Bombay   Buenos Aires
Calcutta   Cape Town   Dar es Salaam   Delhi   Florence   Hong Kong
Istanbul   Karachi   Kuala Lumpur   Madras   Madrid   Melbourne
Mexico City   Nairobi   Paris   Singapore   Taipei   Tokyo   Toronto
and associated companies in
Berlin   Ibadan

Oxford is a trade mark of Oxford University Press

Published in the United States
by Oxford University Press Inc., New York

A catalogue record for this book is available from the British Library

Library of Congress Cataloging in Publication Data
Geriatric medicine : a case-based manual / edited by Jeanne Y. Wei
and Myles N. Sheehan.
p.   cm.
Includes index.
ISBN 0 19 262576 4 (pbk).
1. Geriatrics–Case studies.   I. Wei, Jeanne Y.   II. Sheehan, Myles N.
[DNLM: 1. Geriatrics–case studies. WT 100 G366354 1997]
RC952.7.G467   1997
618.97–dc20
DNLM/DLC
for Library of Congress   96–8590 CIP

Typeset by Palimpsest Book Production Limited,
Polmont, Stirlingshire

Printed in Great Britain by
The Bath Press, Bath

I dedicate this book to my parents, John and Elizabeth Sheehan. It is a great sadness to me that my father's old age has been a time of illness and decline. Both he and my mother, however, have taught me much about living life successfully and have given me the upbringing that recognizes success as more than long life, academic tributes, or a good salary.

M.N.S.

I dedicate this book to the memory of my beloved parents, Alice and George Wei, and to my wonderful husband Ken, and our precious children Michael and David, for all their encouragement, love, and support.

J.Y.W.

We also would like to dedicate this book to all of our patients.

Preface

'Why is it that the important problems of older persons are often not the ones that we know how to help?' This lament from a tired, young physician neatly summarizes the quandary that many physicians and medical students face in caring for older persons: how to deal effectively with the medical problems that so commonly arise in patients of this age group. Effective management of many of these problems is possible. It requires knowledge and a recognition of factors that might not have been previously emphasized in medical school or postgraduate residency/fellowship training. These include physiologic changes with aging, functional assessment, common geriatric syndromes, and an appreciation that medical problems are often deeply enmeshed with social and psychological issues.

The purpose of this book is to help to remedy the complaint of the physician caring for older persons. Our goal is to provide physicians and medical students with a solid foundation for their future practice in caring for an expanding elderly population. This book is aimed at complementing a general medical education. The reading of this book might not provide a solution for every problem that arises during the care of older persons. It is likely to provide the information necessary for intelligent management of many common problems. Our hope is that the reader will, by learning from the approach shown in this book, have the knowledge as to how to proceed in finding the best course of action with more unusual conditions.

We believe that this book will be useful for most clinicians and will be especially useful for four groups of individuals: 1) postgraduate residents and fellows in a variety of disciplines who care for a large and growing number of older patients; 2) practising clinicians who are interested in enhancing their geriatrics knowledge; 3) medical students involved in pre-clinical study; and 4) medical students during their clinical years. They are likely to find the cases to be an interesting way to review the pathophysiology of aging.

This book is the outcome of the rich experience of the contributors. The content and format have been shaped by their expertise in research, patient care, and teaching. An especially important influence has been the New Pathway curriculum at Harvard Medical School. This curriculum is designed for the education of generalists and seeks to provide a core knowledge of information that a physician should know regarding the common problems that physicians regularly encounter. The teaching takes advantage of a case-based format with small group tutorials. The aim is for the students, with the assistance of expert tutors, to learn as adults. In adult learning, each student takes responsibility for his or her own learning.

This book attempts to capture the spirit of Harvard's New Pathway in general medical education. Each chapter begins with a case involving an older person. The chapter then proceeds with a series of issues or questions followed by discussion of the pertinent points. A particularly attractive aspect of this case-based approach is that it allows for an integrated perspective to be given that combines the scientific, clinical, social, and psychological aspects of each case. In a field such as Geriatrics, which emphasizes a multi-disciplinary approach to the care of patients, the case-based method is an excellent format. We hope that this model of case-based learning will serve as an attractive resource for hospitals, offices, long-term care facilities, and medical schools.

The scope of this book is designed to provide a thorough, but not necessarily encyclopedic, approach to the care of older persons. There are currently a number of excellent textbooks of geriatric medicine. The case-based method used in this book is meant to foster adult learning, that is, the recognition that not everything can be memorized but that it is important to know how to learn about medicine with all its ongoing changes. Our goal is to discuss the major issues in geriatrics, recognizing that some common problems might not be covered fully, and to intrigue the reader into learning more about the topics. In considering, for example, the topic of heart disease in an older person, we do not make an effort to cover exhaustively every possible cardiac condition. Instead, we discuss a common problem, review issues in physiology that are pertinent, especially the age-related changes, and proceed to analyze the case so that the questions raised and the concepts considered may provide guidance in situations that are not specifically discussed.

There are many individuals whom we wish to acknowledge for their assistance in this work. Dr. Daniel Federman, Dean for Medical Education at Harvard Medical School, has provided guidance and encouragement not only with this project but with the efforts of the Division on Aging to further enhance geriatrics content in the medical school curriculum. Our colleagues in geriatrics and gerontology, who have contributed directly and indirectly to this book, have been invaluable teachers and have given us new insights into aging and education. Our fellows, residents, and medical students have been a source of inspiration. They have also challenged us to improve our teaching and broaden our knowledge. Particular appreciation goes to David Knauss for his expert secretarial assistance in the final stages of manuscript preparation. We also thank the staff at Oxford University Press for their help and patience.

A particular note of gratitude goes to the Brookdale Foundation for their support of Dr. Sheehan and his efforts in curriculum development in geriatrics. His contributions to this book are in part due to the generosity of the Brookdale Foundation.

Boston   J.Y.W
May 1996   M.N.S

Contents

Contents

Contributors

Juergen H. Bludau Instructor in Medicine, Harvard Medical School; Associate in Medicine, Beth Israel Hospital, USA.

Catherine E. DuBeau, Instructor in Medicine, Harvard Medical School and Gerontology Division, Brigham and Women's Hospital; Attending Staff Physician, Brigham and Women's Hospital; Research Scientist, Urology Section, Brockton/West Roxbury VAMC and Hebrew Rehabilitation Center for Aged, USA.

J. Grimley Evans, Professor of Clinical Geratology, University of Oxford, Radcliffe Infirmary, Oxford, UK.

Daniel E. Forman, Assistant Professor of Medicine, Brown University School of Medicine; Assistant Professor of Medicine, Division of Cardiology, The Miriam Hospital, Providence, Rhode Island, USA.

Tobin N. Gerhart, Clinical Assistant Professor of Orthopedic Surgery, Harvard Medical School; Director, Clinical Research, and Orthopedic Surgeon, Beth Israel Hospital, USA.

Claus Hamann, Instructor in Medicine, Harvard Medical School; Associate Director for Education and Evaluation, Brockton/West Roxbury VAMC Division of the Boston Geriatric Research, Education and Clinical Center; Staff Physician, Gerontology Division, Beth Israel Hospital, USA.

Catherine L. Kelleher, Instructor in Medicine, Harvard Medical School and Gerontology Division, Beth Israel Hospital; Associate in Medicine, Beth Israel Hospital, USA.

Lewis A. Lipsitz, Associate Professor of Medicine and Director, Geriatric Fellowship Program, Harvard Medical School; The Irving and Edyth S. Usen Director of Clinical Research, Hebrew Rehabilitation Center for Aged, USA.

Kenneth L. Minaker, Associate Professor of Medicine, Harvard Medical School; Associate Chief of Staff, Geriatric and Extended Care, Brockton/West Roxbury VAMC; Director of the Brockton/West Roxbury VAMC Division of the Boston GRECC; Associate Program Director of Beth Israel Hospital Clinical Research Center, and Associate Physician, Divisions of Gerontology, Departments of Medicine, Beth Israel and Brigham and Women's Hospitals, USA.

Mark Monane, Assistant Professor of Medicine, Harvard Medical School; Associate Physician, Gerontology Division, Brigham and Women's Hospital, USA.

Linda A. Morrow, Assistant Professor of Medicine, Harvard Medical School; Medical Director, Alexian Brothers Senior Health Center, San Jose, California, USA.

Germaine L. Odenheimer, Associate Professor in Neuropsychiatry, University of South Carolina School of Medicine; Director of Geriatric Neurology, James F. Byrnes Center for Geriatric Medicine, Education, and Research, Columbia, South Carolina, USA.

Rivka Dresner Pollak, Assistant Professor in Medicine, Hadassah Medical School, Hebrew University, Jerusalem, Israel.

David F. Polakoff, Instructor in Medicine, Harvard Medical School; Clinical Director, Brockton/West Roxbury VAMC Division of the Boston GRECC, Associate Physician, Beth Israel Hospital, USA.

Harold N. Rosen, Assistant Professor of Medicine, Harvard Medical School; Staff Physician, Divisions of Gerontology, Bone and Mineral Metabolism, and Endocrinology, Beth Israel Hospital, USA.

Andrew Satlin, Assistant Professor of Psychiatry, Harvard Medical School; Director of Geriatric Psychiatry, McLean Hospital, USA.

Lidia Schapira, Instructor in Medicine, Harvard Medical School; Staff Physician, Hematology-Oncology Division, Beth Israel Hospital, USA.

K. Lea Sewell Clinical Instructor in Medicine, Harvard Medical School, Medical Director, Clinical Trials Unit; Associate Physician in Gerontology and Rheumatology, Beth Israel Hospital, USA.

Myles N. Sheehan, Assistant Professor of Medicine, Stritch School of Medicine, Loyola University Medical Center, Maywood, Illinois. Lectures on Medicine, Harvard Medical School.

Jeanne Y. Wei, Director, Division on Aging and Associate Professor of Medicine, Harvard Medical School; Director, Claude D. Pepper Older Americans Independence Center, Harvard Medical School; Chief, Gerontology Division, Beth Israel Hospital, USA.

Approach to the older patient

Myles N. Sheehan

After finishing his training in internal medicine at a university-associated hospital two years ago, Dr. Thomas Lee joined a group practice with several internists. Thomas had married his wife while he was in his second year of medical residency and she was finishing her senior year of medical school. They had had one child, Jim, a year ago. Dr. Lee's wife, Dr. Joan Stuart, had gone back to work to complete her training. It took a lot of juggling between the two to meet their professional demands, care for their child, and find some time for each other. Dr. Lee found himself frequently challenged by conflicting demands as husband, father, and physician. He enjoyed his work as a primary care physician, although sometimes admitted that he felt weak in his skills in caring for some patients. When he could, which was not as often as he wished, he tried to spend some time in the evening reviewing the day and doing some professional reading.

It was 9 p.m. At last it was quiet. Tom sat down in the living room of the apartment. Joan was on call at the hospital tonight. It meant an especially hectic day: around at the hospital early, see patients at the office, get out in time to pick up the baby at daycare, shop for errands, cook a meal, give Jim a bath and put him to bed. Thankfully, bedtime came without too much of a struggle. And now there was a little time before it got too late to think about what happened today, make some plans for tomorrow, and read a bit.

It was hard to forget the two new patient visits of the morning. Mr. Haas and Mr. Abboud, both 80 years old, each very different from the other. Mr. Haas came because he had recently moved to the area, needed a doctor, and wanted his medications reviewed and renewed. Mr. Abboud came with his daughter. His doctor had recently retired and now he needed someone to care for him. He was sick. His steps were labored, his breathing heavy, his exam remarkable for all the signs of congestive heart failure: elevated neck veins, râles half way up, soft heart tones, an S3, and pitting edema to his knees.

Mr. Haas seemed pretty young for 80 years of age. He carried an impressive problem list: prostate cancer, hypertension, osteoarthritis. But none of the problems seemed to interfere with his activity or enjoyment of life to any great degree. He had moved to the area to be closer to his children after the death of his wife two years previously.

Even allowing for the seriousness of the congestive heart failure, Mr. Abboud seemed old. His daughter said that at those times when her father's breathing was better and his legs less swollen, he did not seem, overall, much improved. His memory was poor. Sometimes Mr. Abboud did not remember if his daughter had come by when she had been there only that morning. From time to time, it seemed he did not make it to the toilet on time, he sometimes smelled of urine and his clothing could be stained. It was hard for him to get out. Another of his children, a son, had taken over doing the banking and paying Mr. Abboud's bills.

Dr. Lee mused over the events of the day. What happens to people with age? Why is it so different in different people? How could I better take care of my older patients? What sort of things should I be alert to and expect?

Approaching the older patient

The vast differences between the two 80-year-old men, Mr. Haas and Mr. Abboud, highlight that there is no such thing as a standard older patient. Before becoming patients, Mr. Haas, Mr. Abboud, and anyone who approaches a physician for care, are persons first and remain persons, not things to be objectified. This is true no matter what the age of the individual. Older persons, however, may be particularly vulnerable to such objectification. An advanced chronological age carries with it a number of connotations, only some of which are accurate. Not all old people are wise, nor are all feeble. Some old people are filled with life, others are depressed and difficult. Physicians who approach older persons with a set of expectations based on age may find themselves confused and confounded by the experience of the elderly. The variety and difference between older people makes any type of blanket statement regarding what to expect, except diversity, overly simple and filled with error. In considering the experience of aging and how physicians can provide good care for older persons, one must recognize that although there will be an emphasis on illness, frailty, and the need for assistance of various sorts, most older people are healthy, independent, and lead fulfilled lives.

Dr. Lee's concern about providing better care is not only a laudable sentiment, but an imperative created by a rapidly expanding aging population. There are three major forces leading to this change in the age distribution of population in developed countries: the aging of the post World War II baby boomers, the increase in life span, and the decline in fertility rates. Put another way, there are a large number of people approaching middle age who, in coming decades, will be old, they will likely live to a later age than their parents did, and both these baby boomers and their children are opting for smaller families (Suzman *et al.* 1992). In 1989, 12.5% of the population of the United States was 65 years of age and over. In 2030, that percentage is expected to swell to 22% (Gaylord 1991). In the United States, the fast-growing segment of the population consists of individuals 85 years of age and over. Estimates of the percentage of the population 65 years and older show a massive demographic shift in developed countries in the next few decades. Women have a longer life expectancy than men, with a baby girl having a projected life expectancy of 79 years and a baby boy 72 years. Life expectancy for individuals who reach 65 and over shows a considerable number of average years of remaining life expected: for women nearly 20 years, for men 15 years (Cassel and Brody 1990). These remaining years are, for most people, years of independent living. Although a large number of people over 85 will spend some portion of time in a nursing home in the months prior to death, most people 65 and older are living without assistance (Soldo and Manton 1988).

Advanced age serves as a marker for complexity and diversity rather than stereotypical images of aging. This complexity and diversity is a component of the older person's life in all its aspects: biological, psychological, and social. Although younger persons may well have complex and diverse experiences, the accumulation of years lends added emphasis to these concerns in the elderly. Three points illustrate this fact.

First, aging as a process is understood imperfectly. How and why people age is the subject of intense research. Despite advances in understanding, our knowledge of the aging process is fragmentary. In dealing with an older individual, it is frequently unclear what types of changes are expected with aging, what represents the toll of disease and extrinsic factors like lack of exercise, smoking, or poor nutrition, and what parts of the aging process can be modified. Our imperfect knowledge of aging, as well as the diversity of older patients has clinical consequences. Caring for older patients demands careful evaluation of an older person with an extensive history, thorough physical exam, and thoughtfully planned diagnostic testing. Ascribing changes in any individual to 'old age' is usually wrong. Greater specificity should be sought not simply out of curiosity, but in the effort to better understand and assist the patient.

Second, a long life, of necessity, implies a variety of experiences, stresses, and changes. Older people have, over time, developed styles of relating to other individuals, coping (or not) coping with challenges, and adjusting to success and disappointment. An awareness of how the person has related to the experience of life over time can be of great value for a physician in assisting an older person.

Third, and closely related to the psychological richness of older persons, is the extensive network of social relationships that have been part of the life of an individual in his or her seventies, eighties, and nineties. It is pertinent for the physician who cares for an older person to know who is important to that person, who assists the elder in trouble, whom they love, and who may be a source of disappointment and frustration. Recognizing a lack of relationships signals caregivers to a number of possibilities: multiple losses, a reclusive personality, a strong independent streak, or a habit of alienating others.

An awareness of complexity and richness in the life of older persons has ramifications for how the physician approaches an older patient. Two interrelated concerns must be addressed: first, practical aspects of meeting older patients and second, the process of history taking, physical exam, and problem solving.

Meeting older patients

For some elderly, just getting to the doctor is a considerable challenge. Older people may have difficulty receiving medical care because of problems with access to care. Effective medical care requires a number of skills: find a doctor, make the appointment, secure transportation, find a way to get into the office, disrobe for the examination, understand the physician's instructions, arrange for further tests and/or prescriptions, and, then, comply with instructions. Each of these tasks, for a variety of reasons, can be difficult for an older person. Hearing loss, visual impairment, or cognitive decline interfere with tasks like using the phone, making appointments, and understanding instructions. Older persons with difficulty in driving or walking may have an arduous struggle to get to the doctor's office. Public transportation systems are not always available or accessible. In urban areas, crime and fear may keep older persons in their homes, afraid to venture out. Even once at the address of the doctor, stairs, steps, and other barriers to easy access can make a trip to the doctor an ordeal to be avoided.

Paying for health care presents another barrier for some elderly persons, despite government attempts to cover costs. In some countries, financial problems with care for older persons are eased with universal health insurance. In the United States, the Medicare program covers many, but not all, of the medical costs of those over 65 years of age. The Medicare program is divided into two components, Medicare A and Medicare B. Medicare A will pay for most of the costs of an acute hospitalization and, in the immediate post-hospital period, some of the costs for further rehabilitation and assistance if medically necessary. Medicare B requires an additional payment on the part of those individuals who desire this coverage for doctor's visits. Other insurance must be purchased to cover the gaps in Medicare coverage as well as pay for drugs. This supplemental insurance can be expensive, especially for an older person on a fixed income. Because of problems with reimbursement, physicians in some areas in the United States occasionally refuse to take older persons as patients. The Medicare program provides substantial assistance to many elderly but it is not complete. Long-term hospital or nursing home care is not covered by Medicare.

Another barrier for older persons that may affect their care is the way older people themselves approach the medical system. Older patients often are well-informed consumers of medical care, take an active part in decision making, and demand explanations and alternatives from their physicians of any proposed therapy. Many other older persons are not so aggressive. For some, after a lifetime of quiet hard work and respect for authority, physicians represent exalted experts who should not be bothered for a minor complaint and never questioned. Unfortunately, many of these same submissive older persons may be quite fatalistic regarding their health as they age, minimize any symptom, and not seek clarification from the doctor when they are confused about her or his recommendations. The idea that many conditions commonly associated with older age have the potential for reversal may not occur to older patients. Older individuals can, like poor physicians, readily assume that many symptoms are an inevitable part of aging and thus not seek attention until extremely ill. Falls, memory loss, urinary incontinence, and other syndromes may seem to an older patient as just another sign that they are getting old (Levkoff et al. 1988).

For doctors who are concerned with providing excellent medical care, the physical, financial, and behavioral barriers potentially facing older persons lend added urgency to the need for the physician to do his or her best to assist those older patients with whom he or she comes into contact. Although the circumstances of the clinical encounter (emergency visit vs. scheduled visit; new patient encounter vs. brief follow-up; office appointment vs. emergency room or hospital evaluation) will dictate much of the interaction between doctor and patient, there are approaches to avoid and others to consider in meeting with an older patient.

Dr. Lee always tried to meet new patients in the waiting room and then direct them to the examining room. Mr. Abboud and Mr. Haas provided a study in contrasts. He had come out to meet Mr. Haas first. As he called out his patient's name, Dr. Lee was impressed that the youthful appearing man who answered was 80 years old. Mr. Haas was carefully dressed in a navy blue blazer and grey flannel slacks. His shoes were freshly shined. A perfectly knotted striped tie matched the crisp white cotton shirt. Mr. Haas rose easily from his chair. He confidently strode over to the doctor and shook his extended hand with a firm grasp. Dr. Lee introduced himself and invited Mr. Haas to one of the rooms in the suite that made up his office. As the doctor followed Mr. Haas into the room he told his new patient: 'Mr. Haas, it is good to meet you. Please be seated. Could you let me know how you think I could be of assistance?'

Two hours, and several other patients later, Dr. Lee returned to his waiting room to greet Mr. Abboud. He called out Mr. Abboud's name and was answered by a middle-aged woman who was sitting with an elderly man. The older man was dressed in a clean sweatshirt and baggy pants. Dr. Lee noted that his black shoes were scuffed at the toes. The upper portions of the shoes were weathered and in poor shape. Mr. Abboud's daughter turned to help her father out of the chair. He put each hand on the arms of the chair in the waiting room and slowly pushed himself up, using the arms for balance and support, as if they were parallel bars. Mr. Abboud walked with slow, small steps toward Dr. Lee, his daughter protectively holding one of his arms. His face seemed somewhat blank and his color a pasty grey-blue. He was breathing heavily. As Dr. Lee introduced himself to Mr. Abboud, he received only a muffled hello from his new patient. Dr. Lee brought Mr. Abboud and his daughter to the examining room. 'Mr. Abboud,' he said, 'Please take a seat. May I ask your daughter to join us for some of our conversation together? We will have time alone, also, but with your permission, I would like her to join us.'

The encounter between patient and physician always tests the observational skills of the physician. This is especially the case with the older patient. A discerning eye on the part of the doctor may detect a number of conditions that otherwise could be missed. The attention Dr. Lee paid to the clothing and shoes of his new patients is not evidence of a fetish regarding men's fashion. Rather, it is the educated gaze of a skilled clinician looking for every clue that will help her or him understand the patient. Mr. Haas' natty attire, brisk walk, and firm handshake all provide common sense information that this man is doing well: he cares about his appearance, he transfers easily from a chair, walks without a problem, understands social convention, and projects an image of confidence and health. A particularly astute clinician might raise an inward question as to the meaning of Mr. Haas' sartorial elegance: is it a sign of appropriate self-concern and respect for the physician or does it indicate some excessive concern over personal appearance and the possibility of a narcissistic personality? An older person who is very concerned about appearance may have a difficult time coping with any illness of disability that may detract from an image of good health, relative youth, and vigor.

The information gleaned from greeting and observing Mr. Abboud on the way from waiting room to office raises an extensive differential. Mr. Abboud's daughter answers when Dr. Lee calls out. Why is this? Is Mr. Abboud deaf, confused, depressed and apathetic, or is his daughter overprotective and domineering? Mr. Abboud is dressed in baggy pants. Has he lost weight recently or is the bagginess from an adult diaper worn because of urinary incontinence? Dr. Lee noted how hard it was for Mr. Lee to get out of the chair. Some of the many possible explanations came to mind: proximal muscle weakness, steroid use, lack of exercise, and atrophy of the quadriceps? The shuffling gait and blank expression raise the question of Parkinson's disease, even though Dr. Lee did not see the resting tremor that would make the classic triad of tremor, masked facial expression, and a rigid posture. Mr. Abboud's shoes show more than simply extensive use. Worn out or crusty upper portions of shoes may be due to urine, suggesting either urinary incontinence or difficulty, for whatever reason, in using the toilet. The scuffed front portion of the shoes suggests that Mr. Abboud may be tripping and falling. By the time Dr. Lee had ushered Mr. Abboud into his office, he had a number of concerns based only on a discerning eye.

More than a discerning eye is required on the part of the physician. A discerning tongue can prevent well-intentioned slips that are patronizing or offensive to older patients. Always refer to an older person, unless invited by the person, by their last name. Occasionally, individual caretakers will treat older persons like children. This can take the form of presuming to call an individual by their first name or adding epithets like 'honey' or 'dear.' Part of this infantilization may come from a desire to assist and protect a frail elder, somehow

confusing vulnerability with a child's lack of experience and need for protection. The recognition that a patient may need assistance is a valuable clinical insight. This should not be confused with comments and actions that may be considered patronizing when given by a caregiver many years the junior of the patient.

An awareness of the possibility of difficulties with vision and hearing can help the physician with an older patient. This does not mean that the doctor thrusts his or her face a few inches away from the patient and screams. A more constructive approach, after introductions and a few questions, is to ask the person if he or she has any difficulties hearing or seeing the physician. A negative answer may not be accurate. The physician, while sorting through the rest of the data, needs to consider if difficulties in obtaining a history and performing the maneuvers of physical examination could be due to poor hearing, decreased vision, and ability to follow instructions, or ascribed to a cognitive problem. Parts of the exam that are frequently neglected, such as looking at ears and looking into eyes, assume a greater significance in older patients. Assuming that an older person is deaf without looking into the ear canals is sloppy medicine on the part of the caregiver. Cerumen can render a person hard of hearing. Individuals with hearing aids may seem to be deaf despite their devices. This may be due to battery failure rather than intrinsic hearing loss. A useful trick for testing hearing aids is for the examiner to cup a hand over the hearing aid in place. If the battery and hearing aid are working, there should be the unpleasant squeal of feedback. It is best to warn the patient before doing this maneuver!

Because of the richness of relationships that characterizes a life lived over many years, the presence, or absence of a companion who comes with an older patient is a significant piece of information in a comprehensive evaluation. There are three reasons why this is true. First, the presence of a daughter, son, spouse, friend or some other person is evidence that the patient has other people for support. Second, the absence of some companion gives a warning that the person may be lacking in these relationships, either because of death or a problematic personality that drives others away. Third, an older person who comes alone may still have a number of significant relationships. An unaccompanied visit can be the sign of independence and robust health.

Although no hard and fast rule can be given, obtaining information from family members and friends of older patients is often an appropriate and necessary part of caring for the older person. In the case of older people who are unreliable sources of information, either because of acute illness or alterations in mental status, speaking with those closest to the patient may be the only way to attain an accurate history. Even in the majority of cases where the older person is articulate and knowledgeable, the insights of a friend or family member may provide a more comprehensive sense of how the patient functions. Physicians must be vigilant, however, in their concern for the privacy of all patients. It is not appropriate to converse with family members about an older patient in the way a pediatrician would speak with the parent of a toddler. The older person's permission should be obtained prior to any discussions unless illness prevents the patient from deciding. Dr. Lee's method of asking Mr. Abboud if his daughter could join them for the history could be criticized. It would have made it awkward for Mr. Abboud to say no. At the same time, respect for a patient's privacy and confidentiality need not mean that the physician assumes a suspicious and adversarial relationship with family and friends of the patient. Dr. Lee told Mr. Abboud that they would have time together privately. Such an approach strikes a balance between a recognition of the need for private doctor–patient communication and the recognition of the valuable insights that can be obtained from family and friends.

History taking, physical exam, and problem solving

As Dr. Lee thought about Mr. Haas and Mr. Abboud, he considered some of the structural problems presented by caring for older persons. A patient like Mr. Abboud made him very anxious. Dr. Lee had 50 minutes for a new patient visit. The problems Mr. Abboud was experiencing seemed far too great to deal with in that time. Dr. Lee realized that among some of his colleagues, 50 minutes was an extremely generous time allotment for a new patient encounter. How could he possibly do anything right for Mr. Abboud? What about his anxiety? Would he be driven hopelessly behind with the other patients scheduled for today? He remembered, as he sat in his living room in the evening, how he had felt a bit panicky that it might be hard to get to daycare on time to pick up his son Jim.

Obtaining historical information from older persons may require different approaches and emphases than with a younger patient. Frequently, the history of an older person, especially an established patient, is straightforward. The person will present with a complaint, appropriate questions will be asked to further elucidate the symptom and construct a provisional list of diagnoses, and then, with the insights gathered from a focused physical exam and appropriate laboratory testing (if needed), a diagnosis and course of action can

be mapped out. It may be the case, however, that this standard process of information gathering will fail.

Dr. Lee's thoughts about time pressures in caring for older persons are a very practical concern. It should be pointed out that time pressures in medical care are not caused only by older patients. The problems that lead individuals of every age to seek the attention of a physician do not easily fit into the grids of an appointment book. Frequently, however, older persons may be a special challenge for physicians in a busy practice. Considering some of the challenges and possible responses will not resolve every issue, but it allows a variety of approaches that may assist physicians in providing good care, not disrupting schedules, and avoiding the possibility of resentment against older persons as a drain on a practice.

Dr. Lee, on the advice of one of his senior partners in the practice, always speaks with his new patients before having them disrobe for examination. He recalled how his partner had said: 'Listen, Tom. You meet a person for the first time with one of those stupid examining gowns, they're half naked, feel foolish, and most probably feel so vulnerable that the last thing they want is to open up to you about what's bothering them. You want two things: let the person know you care about them and that you want to deal with their problems in a respectful but efficient way.'

Over time, Dr. Lee had developed a structure for his new patient visits. He thought how it had worked out with Mr. Abboud. After getting settled in the office, he had asked Mr. Abboud and his daughter how they hoped he could be of assistance. Mr. Abboud's daughter had answered: 'Dr. Crane, my dad's old doctor, has just retired. It has been a month since dad went for his last visit to Dr. Crane. My dad is sick. He has heart trouble and is not doing well. I know he is old but I want him to do better. He needs his medications renewed and maybe something to help his energy.' Dr. Lee asked Mr. Abboud: 'Mr. Abboud, your daughter has told me some things that I agree need attention. What do you think? Is there anything you want to add or correct?' Mr. Abboud responded: 'No, Doctor, She's got it about right. I wish I could breath better.'

Dr. Lee thought for a moment about what Mr. Abboud and his daughter had just told him, the kind of information he needed to take care of Mr. Abboud, and how he would try and structure the time. He responded: 'Well Mr. Abboud, I will try very hard to help you with your breathing as well as consider some of the other problems your daughter mentioned. For today's visit we have about fifty minutes together. In that time I want to concentrate on your breathing problem but also, as much as I can, find out about how you feel you are doing. I would like to speak with you and your daughter for about fifteen or twenty minutes, then examine you, and then come up with a plan for what to do. I may not be able to get to everything that is bothering you, but I want to know about the problems so I can assist you as much as is possible.'

Dr. Lee's effort to structure the patient visit shows four points important in caring for older patients. First, he recognizes the presence of an acute problem that requires immediate attention and may have been the main reason why the person has come to see the doctor. Ascertaining the chief complaint is an excellent way to begin the assessment of any patient. Second, Dr. Lee also makes an opening to allow the patient to present other concerns that may actually be more important to the person than the chief complaint. Mr. Abboud may be less concerned about his breathing than he is about his urinary incontinence, but he is embarrassed to bring it up at the beginning of the meeting with the doctor. If Dr. Lee becomes obsessed on the differential diagnosis of shortness of breath and not keep an ear out for what the patient is trying to tell him, he could miss key information. As an example, it is not uncommon that some patients will stop or cut back on their diuretics because they despise wetting their pants, finding it impossible to make it to the toilet in time because of the rapid diuresis caused by potent drugs like furosemide. A physician who probes no deeper than the chief complaint of shortness of breath may make things worse by prescribing more diuretics, never realizing that the patient is not even taking the lower dose of prescribed diuretic. Third, Dr. Lee mentions the time available for the appointment and how he plans to structure that time. Being forthright about time constraints and a plan for the office visit serves as one possible approach in striking a balance between a desire to be attentive to the total experience of the patient and the need to be efficient with limited time. Fourth, in making the effort to structure the time of this appointment, Dr. Lee makes it clear that he is concerned about the person, that he does want to know about the patient's concerns, and that the process may well require more than one visit.

The challenges in obtaining a history from a complicated patient like Mr. Abboud can be daunting. Not only is limited time a problem, Dr. Lee has, in welcoming Mr. Abboud and his daughter and eliciting the chief complaint, identified a variety of problems that require further investigation. How does a physician obtain an adequate history when confronted with shortness of breath, urinary incontinence, falls, a possibility of Parkinson's disease, a question of cognitive impairment, and the differential diagnosis of proximal muscle weakness?

Dr. Lee thought for a moment before asking his next few questions. He wanted to hear, as much as possible, what Mr. Abboud was feeling and thinking. At the same time, a number of concerns were running through Dr. Lee's mind. Foremost was the etiology of the shortness of breath: how serious was the problem, would it require immediate hospitalization, or could reasons for the current problem be identified and simple interventions planned?

'Mr. Abboud,' he asked, 'Please tell me how long you have had the problem with your breathing?'

'Oh, Margaret knows the details better than me, but things have been bad since my heart attack this summer.' Mr Abboud turned to his daughter.

Margaret responded, 'Well, Dr. Crane, Dad's old doctor, said he had heart failure. Dad had a big heart attack this summer. It was touch and go for a while. He seemed to gather strength after the hospitalization but in the last few months he has been going downhill.'

Dr. Lee, concerned about the time course of the decline and whether there could be an acute cause of the shortness of breath, asked Mr. Abboud and his daughter: 'How have you been since your last visit with Dr. Crane? Has there been any change?'

Mr. Abboud and his daughter turned to each other, he answered: 'Oh, it has been about the same. I do not do much and if I try to do a lot I get winded. I am spending more time in my chair in front of the TV. It gets very dull.' Margaret added: 'My dad finds it hard. There is not much to do. He is alone a lot of the day. He goes to bed early and stays in bed till about seven in the morning.'

Dr. Lee turned to Mr. Abboud: 'How are your spirits? Some people can get very down when they have a heart attack. What do you think?'

Mr. Abboud look downward and began to sigh. 'I don't sleep well. I fall asleep and then wake up after a couple of hours. Then I toss and turn. Usually I have to go to the toilet several times. It is hard.'

Margaret added, 'I think Dad is depressed.'

Dr. Lee answered: 'I want to explore that more in a minute. For now, let me ask a few more questions about the breathing. Mr. Abboud, what medicines do you take?'

'I take three different pills. One, a yellow one, I take in the morning. Another one I take twice a day. Then there's the diuretic, I am supposed to take that twice a day.'

'Do you know the names of the pills or their strengths?'

'Margaret does.'

'Dad takes his digitalis in the morning. He is supposed to take his furosemide then, it is 40 milligrams. He also is supposed to take a captopril. I do not know the strengths, but Dr. Crane gave me a list of his medicines and a copy of his most recent ECG.'

Dr. Lee responded: 'Mr. Abboud, I have a lot more questions for you. I need to examine you, and I want to review your records. But before we go any further, can you give me a sense of how you understand your health problems? What is your impression of how you are doing?'

Mr. Abboud began, 'Well, Doctor, it's hard. Margaret is wonderful and my boy tries to help, but I am old and don't think I have much time left. Since my wife died a few years back I have been lonely. I helped her. She had cancer.' Mr. Abboud stopped for a moment, his voice cracking. 'Well, now I just seem to be falling apart. I had the heart attack. I never felt right afterwards. I was weak. The furosemide makes me piss so fast it is hard to get to the bathroom on time. Some days I would rather be short of breath than wet. I don't like to take the medicines. I am getting weaker. I feel trapped in my chair.'

Dr. Lee is forced to proceed on multiple fronts at once as he gathers information on Mr. Abboud. This case is complex but it is not particularly unusual. Taking a history from a sick older person is often challenging. There are at least four reasons to consider.

First, illness behavior and presentation in the elderly can differ from younger patients. As previously mentioned, older people are often reluctant to approach physicians. They may minimize their symptoms out of fatalism, embarrassment, or a concern that they are inappropriately 'bothering' the physician. Symptoms may be non-specific, normal clues to the presence or absence of serious illness are missing, and older people frequently minimize their symptoms. Frequently, multiple medical problems will coexist in the same person, making it difficult to sort out the cause of signs and symptoms.

Second, there are certain syndromes that are more common in the elderly. Recognition and questioning regarding the presence of these problems can streamline history taking and focus on common, but often neglected, conditions among older persons. A standard historical approach that focuses on the history of present illness, past medical history, and review of systems can leave the physician frustrated and baffled and the patient unhelped. Physicians may not routinely think about falls, weight loss, incontinence, and memory problems as syndromes that should be asked about with all older patients. Some of these problems are sources of embarrassment that may prevent a patient from asking for help. All have a differential diagnosis and potentially reversible etiologies. Asking about these syndromes can elucidate the major problem and allow the doctor to focus on sources of distress, physical decline, and substantial morbidity.

Third, although one should not assume cognitive impairment as an inevitable accompaniment of aging, delirium and dementia are frequent problems. Although the patient's responses to questions during the history may give clues to the underlying condition, the specific responses are less likely to provide factual information. When faced with vague, conflicting answers or peculiar behavior, the physician has to rapidly differentiate if this is a new problem, a consequence of another

medical problem, or evidence of a long-term cognitive decline.

Fourth, information often must be gathered from family members and friends to provide a full picture. As Dr. Lee has relied on Mr. Abboud's daughter Margaret for key pieces of information, doctors can piece together bits of factual information from those closest to the patient. This is particularly the case with a sick older person, who may not recall all the facets of her or his history. Again, there needs to be caution lest physicians assume that old people are somehow stupid and incapable of giving answers. The point is not a presupposition of prejudice, but a recognition of complexity that may take the help of those closest to the patient to help unravel.

Illness behavior and presentation

Recognizing the danger of sweeping statements about older patients, it is a frequent experience of those who care for the elderly that old people often do not get sick in the same ways as younger persons (Hodkinson 1973; Besdine 1988). How do the elderly differ?

First, common illnesses tend to have uncommon presentations. Signs and symptoms classically associated with certain illnesses may be absent. Myocardial infarction in those over 80 more commonly presents with shortness of breath than with chest pain. Older individuals may have serious infections without an elevation in temperature.

Second, non-specific signs and symptoms often are the only evidence of serious illness. Weight loss, confusion, falls, and urinary incontinence are syndromes that have a broad differential as to etiology. An older person with pneumonia may be brought to the doctor because of sudden confusion and falling.

Third, serious disease in older persons may be mistakenly ascribed to changes associated with old age. This mistaken attribution may be twofold. First, is the often erroneous attribution of an older person's complaints to old age. Fatigue, aches and pains, difficulty walking, and other symptoms can be dismissed in a cavalier manner as 'old age'. That provides no explanation and blocks the potential cure or amelioration of reversible illness. An attempt should be made to obtain a specific diagnosis, or several diagnoses, to explain a problem. The second possible mistake is to assume that a common complaint necessarily is caused by an old problem. Back pain in a woman with osteoporosis may be due to a compression fracture. The fracture and pain could also be due to multiple myeloma or metastatic cancer. The lesson to be gleaned is not that every older woman with back pain requires a bone scan, blood count, and serum protein electrophoresis. Rather, the prudent course is that when a sign or symptom in an older person persists despite appropriate therapy, it is important to reconsider the diagnostic possibilities rather than be bound to a diagnosis.

Recognition of some of the factors that may affect illness presentation and behavior in the elderly provides a valuable resource for physicians in history taking and considering an approach to older persons. First, recalling that common illnesses may have unusual presentations, physicians will have a high degree of suspicion when older persons present. Breathlessness in an older person, for example, requires not just an exam of the lungs and a chest radiograph but may need an electrocardiogram. Likewise, confusion should not be ascribed to dementia simply because the person is old, but requires a thorough investigation with special attention to treatable medical illness. Second, knowing that illness in the elderly may present with non-specific signs and symptoms, an alert physician will be vigilant regarding the possibility of serious reversible illness even in older persons who have complaints that are vague and non-localizable. Third, the presence of more than one disease should be considered. In the example of the older woman with osteoporosis, the possibility was suggested that back pain may be due to another process like multiple myeloma or a metastasis from a solid tumor. In the case of Mr. Abboud, the presence of congestive heart failure does not rule out the possibility of depression, hypothyroidism, or other treatable conditions. Multiple diseases are not uncommon among older patients.

While considering the patient's description of symptoms, the wise physician is one who asks the patient for his or her thoughts as to the cause of the symptoms. Inquiring about the patient's explanatory model serves four important functions (Kleinman *et al*. 1978). First, it may reveal the answer. Mr. Abboud revealed that he felt he was dying. He admitted that non-compliance with his medical regimen and depression were major factors in his current situation. Dr. Lee realized he had a possible explanation for Mr. Abboud's current problem, needed to further evaluate the possibility of depression, and must simultaneously consider a medical regimen that would be followed by Mr. Abboud, while trying to evaluate and ameliorate his annoying urinary incontinence. Second, as dramatically illustrated by Mr. Abboud and Dr. Lee, the patient's explanation gives the physician an insight into how the patient views his or her illness and the experience of aging. A patient may tell you that the symptoms are

what he or she expects from aging. The physician can then have the opportunity to educate the person about the specific disease and what can be done to help. Perhaps more importantly, it allows the physician the chance to understand the person better and consider ways that he or she may be of help. Third, the physician's inquiry about the patient's explanation shows respect for the patient's experience. Fourth, understanding the patient's explanation serves as an important framework to discuss treatment and improve compliance with recommended therapy.

Armed with a patient's explanatory model, the physician has a chance of understanding how his or her patient experiences illness. A more complete picture requires evaluation of functional status. Functional assessment is an especially useful tool to determine the types of help that an older person may need to maintain or improve independence.

That evening, Dr. Lee thought more about how the conversation with Mr. Abboud and his daughter had progressed. He was concerned that Mr. Abboud could be suffering from a major depression complicating and influencing his problems with congestive heart failure. It appeared that non-compliance with the prescribed medical regimen and diminished activity were creating a vicious cycle for Mr. Abboud. Mr. Abboud felt sick, he moved less, was thus getting progressively further out of shape, making him able to do even less, making it difficult to get to the bathroom, leading him to skip doses of diuretic, making him even sicker, et cetera. Dr. Lee had specifically asked Mr. Abboud about thoughts of death, suicide, and plans for suicide. He, and Mr. Abboud's daughter, were relieved that even though Mr. Abboud did think about death, he had rejected suicide as an option, feeling it was against his religious faith and would be a cruel legacy for his children.

Dr. Lee returned to the standard medical questions, inquiring more about Mr. Abboud's shortness of breath, associated symptoms like chest pain, sputum production, fever, and the presence or absence of orthopnea. He considered other causes for the shortness of breath like infection and pulmonary embolism. Realizing that he could not totally exclude these possibilities at present, he moved on to other concerns.

Dr. Lee asked specifically about urinary incontinence, fecal incontinence, weight changes, appetite, difficulties with memory, and falls. He asked some follow-up questions to explore the diagnosis of depression, specifically inquiring about any previous psychiatric history, vegetative signs of depression, and the presence of any signs of delusional or paranoid thinking.

After these questions, Dr. Lee realized that he had sketched in the outlines of the important historical information. But more was needed. He began to ask questions about Mr. Abboud's ability to perform household chores, care for himself, and manage his affairs.

Functional assessment

Caring for older persons requires more than a knowledge of their symptoms and medical history. It requires a sense of how they are doing. Medical diagnoses give little information about the impact of illness on a person's life. They provide no information to the physician and other caregivers as to whether or not the person is struggling to maintain independence at home or if some interventions could make the person safer and more comfortable. Functional assessment is an important part of caring for older patients. It considers more than diagnoses but aims at developing a picture of what the person is capable of doing, how much help he or she needs, and what interventions can be made that will maximize independence and/or safety.

The elements of functional assessment are varied but usually include questions about activities of daily living (ADLs), instrumental activities of daily living (IADLs), ascertaining supports from family members or friends, and observation of the patient doing simple tasks like walking, getting up from a chair, and dressing. Some individuals include questions about finances and other elements of the social history regarding health habits, like tobacco and alcohol use. Other parts of the history and physical exam included under functional assessment include tests of vision, hearing, and cognitive status. Standardized instruments exist for performing comprehensive functional assessments. These are useful in research settings and in obtaining more precise information than a simple screen. For the purposes of a busy general practice, however, the simple screen outlined above is rapid and provides useful information (Lachs et al. 1990).

Questions about ADLs and IADLs give the physician information about the different tasks a person performs during the day. ADLs are basic functions and include transferring (moving from bed to chair, for example), dressing, grooming, bathing, toileting, eating, and maintaining continence. IADLs are those tasks that allow a person to maintain an independent existence, thus the term 'instrumental.' There are a multitude of IADLs, common ones include preparing and shopping for food, managing the finances, using the telephone, driving or arranging for transportation, and simple housekeeping chores like cleaning and laundry. Individuals who are failing for whatever reason tend to lose the ability to perform the instrumental activities before the ADLs. Inquiry about IADLs may reveal a functional decline in a patient who otherwise seems more or less stable when seen in the doctor's office. Early recognition of a functional decline allows the physician to screen for reversible illness and consider community assistance like a homemaker, meal delivery through the Meals-on-Wheels program, and other supports that keep the older person safe and in his or her home while efforts are made, if at all possible, to reverse the decline. Loss of the functions that enable one to perform ADLs happens either later in a chronic illness or as a consequence of a sudden, severe illness. A failure to perform in the ADLs is evidence of a serious condition that may herald an ominous prognosis. Individuals who cannot get to the bathroom, bathe, and eat usually cannot live by themselves. Frequently, they may require some sort of institutional care if their needs are more than family members can assist with or there are no family members or friends to help. Difficulties with ADL's should alert a physician that there is a need for immediate investigation as to the cause and a plan developed as to how the person can be cared for.

Questioning a patient about ADLs and IADLs can be done by proceeding through a list or asking the person to recount his or her day. With the latter, the physician can gain a greater sense of what the person's life is like but it requires further questioning about specific activities. The clinician needs to find out if the person can walk unassisted at home, use the toilet, feed him/herself, maintain continence, dress, and bathe. Likewise, specific questions about who does the shopping, cooking, cleaning, and other errands are needed in the effort to find out how the person performs with IADLs.

Some parts of the functional assessment are easily done during the physical examination. As previously noted, careful observation throughout the patient encounter provides much information about how well the person walks, transfers, dresses, bathes, grooms, and maintains continence. During the exam, hearing can be specifically checked by whispering a question in the patient's ears and awaiting an appropriate answer. Vision can be screened by asking the patient to read an item in a newspaper or book. Dexterity at tasks like dressing is easily accomplished by remaining in the room after the exam and watching the patient dress. Obtaining the information necessary for functional assessment is not particularly difficult. It is a process of informed observation and questioning regarding specific tasks that often are otherwise ignored. Finally, mental status testing provides important information regarding the presence or absence of cognitive deficits or affective disorders (e.g., depression).

Dr. Lee, sitting in the study, reviewed the elements of functional assessment for Mr. Abboud. There were a number of deficits and concerns. Mr. Abboud basically lived a bed-to-chair existence. He walked with difficulty and was frequently incontinent. He could not get in and out of the tub on his own. His hygiene left something to be desired. Mr. Abboud was dependent on his daughter for almost all IADLs. He could use the phone, but rarely made a call himself.

The physical exam was remarkable for evidence of congestive heart failure. Mental status testing suggested Mr. Abboud had a short-term memory loss. He was attentive, able to recite the months of the year backwards. Mr. Abboud was also oriented, knowing the address of the doctor's office, the day of the week, month, year, and season. He missed the date by two days. He could only remember one out of three items when asked to recall them after three minutes.

Dr. Lee still had concerns about how he had managed Mr. Abboud earlier in the day. There were no clear answers. His functional assessment, coupled with the history and physical exam, revealed a man who was very dependent on his daughter, doing poorly, with congestive heart failure, at risk for a fall and further injury at home, and, perhaps depressed. Dr. Lee had some questions about an early dementing illness but decided further assessment of this possibility would have to wait. He felt that the main issue was whether to admit Mr. Abboud to the hospital and sort out these issues or if a trial could be made of therapy outside the hospital. The worsening shortness of breath raised the issue of a new myocardial infarction, although non-compliance with his medical regimen seemed a likely explanation for the current decline.

Dr. Lee obtained an electrocardiogram and drew some blood to check a complete blood count, electrolytes, urea nitrogen, creatinine, digoxin level, and thyroid functions. He asked Mr. Abboud to give him a urine sample. He would have to wait until tomorrow to know the results of the blood tests. The electrocardiogram was unchanged from the one taken last month in Dr. Crane's office, making it less likely that there had been another myocardial infarction. Mr. Abboud's urine, however, was cloudy and smelled foul. A quick glance under the microscope revealed numerous white cells, red cells, and bacteria.

In putting the case together, Dr. Lee felt that the most important parts of the case were treating the congestive heart failure, getting Mr. Abboud some help at home, and closely monitoring his progress. Given the finding of the apparent urinary tract infection, it seemed reasonable that treatment of the infection might alleviate some of the urinary incontinence and make Mr. Abboud more willing to comply with his medical regimen. Probably at least as important as medications, however, is getting the visiting nurse association involved in assessing the situation at Mr. Abboud's apartment, monitoring his medications, and discussing with the daughter the need for further assistance.

Dr. Lee, after making sure there were no allergies, gave Mr. Abboud a prescription for trimethoprim/sulfamethoxazole to be taken twice a day. He explained that he thought a urinary tract infection could be worsening the problems with the urinary incontinence. Dr. Lee also asked Mr. Abboud to give the furosemide another try. He asked him to take two of the 40 mg tablets when he returned home, and suggested he remain near the bathroom. Mr. Abboud was to continue on his digoxin, one pill daily. The captopril would be re-evaluated in the next visit, scheduled for a week's time. Dr. Lee felt it likely was a good medication for him but he was uncertain of compliance and did not want Mr. Abboud taking large doses of diuretic and then having a sudden hypotensive episode after a sporadic dose of captopril.

Finally, Dr. Lee called the Visiting Nurse Association and made a referral for a nurse to visit Mr. Abboud the next morning. He reviewed with Mr. Abboud and his daughter the rationale for the visiting nurse, made sure they had the number of his office to call if there were any problems, and asked them to make an appointment for a return visit in one week.

Before leaving the office that afternoon, Dr. Lee had called Mr. Abboud to see if he was comfortable. The furosemide had worked and the breathing was a touch easier. Dr. Lee realized how tenuous Mr. Abboud was and made a mental note to call him later in the week. He would need to check his labs, reassess the possibility of depression, and reconsider his mental status on subsequent visits. Dr. Lee wondered if Mr. Abboud would improve enough to remain at home or whether he might need, in time, long-term care in a nursing home.

Mr. Abboud, and the way Dr. Lee managed his case, gives an insight into the potential complexity of caring for older persons. The care plan required attention to a number of different concerns: an unexpected urinary tract infection, worsening congestive heart failure, issues of dementia and depression, and serious deficits in function with problems of incontinence, mobility, dressing, and hygiene. Dr. Lee was unable to fix everything on one visit. He had to leave unanswered some of his questions. The key decisions faced by Dr. Lee involved whether or not to admit Mr. Abboud to the hospital and, if not, how to improve his breathing, treat his infection, and provide more assistance at home. Another physician may well have admitted Mr. Abboud to the hospital. As Dr. Lee mused that evening, he was not sure he had made the best decision. An in-hospital admission would allow for more rapid assessment of Mr. Abboud's condition, more vigorous treatment of the congestive heart failure, and an opportunity to further address the questions of depression and dementia. As Dr. Lee realized, however, admitting an older person to the hospital often runs the risk of precipitating a delirium in the unfamiliar environment, creating complications from new and powerful medications, and the possibility of further loss of function caused by bed rest and the hospital routine. The option to care for Mr. Abboud at home requires the physician to call the patient and family, provide for the visiting nurse to check out the situation, and arrange for early follow-up.

Aging well

As he sat in his apartment that evening, Dr. Lee recalled his history and physical with Mr. Haas. He had a few questions about Mr Haas' care but nothing like the concerns he had faced in his care of Mr. Abboud. Mr. Haas had told him he had moved to his own apartment to be near his son and grandchildren, that he felt well and that, although the move was difficult in leaving his old home and associations, he was excited by the opportunity. He enjoyed his grandchildren and had a close relationship with his son. Mr. Haas had found an apartment within walking distance of his son's home, in a building that provided a number of services if needed, and was already getting to know some of his neighbors.

Dr. Lee reviewed with Mr. Haas his medical problems of prostate cancer, hypertension, and arthritis. Mr. Haas brought a summary of his medical conditions from his previous doctor. The prostate cancer had been detected three years ago when a physical exam found a nodule which, on biopsy, proved to be cancerous. Mr. Haas was treated with radiation and has done well. His blood pressure was well controlled by atenolol, an antihypertensive agent taken once a day. The arthritis could be annoying, causing pain in the knees and hips that sometimes limited activity. Currently, Mr. Haas was experiencing only minor discomfort. In response to Dr. Lee's questions, he reported his complete independence with regard to both ADLs and IADLs. His only medications, other than the atenolol, were one aspirin a day and an occasional ibuprofen when his arthritis troubled him.

Physical exam was remarkable for the lack of problems. Blood pressure was 150/80 mmHg with a pulse of 60. Pulmonary, cardiovascular, and abdominal exams were normal. A rectal exam revealed guaiac negative stool with an enlarged and somewhat firm prostate but without nodules. On today's exam, Mr. Haas had no pain in his knees and hips and full range of motion. Neurologic exam included a screen of mental status showing intact cognition and normal affect. The remainder of the exam was non-focal and remarkable only for some decreased sensation distally in the lower extremities.

Dr. Lee reviewed Mr. Haas' laboratory results as summarized by his former physician. All values were within normal limits on an exam three months ago, including a prostate specific antigen. His electrocardiogram showed evidence for left ventricular hypertrophy. The remainder of Mr. Haas' medical record included immunizations for influenza, pneumococcal vaccine, and tetanus.

Dr. Lee asked Mr. Haas if he had any particular questions or concerns before concluding the exam. Mr. Haas responded: 'No, Doctor, I just wanted to check in with you. I feel fine but I felt it important to make sure there was a doctor who knew me in case I had a problem.'

Dr. Lee again reflected how different Mr. Haas was from Mr. Abboud. Clearly, more than age was at play. He wondered why two individuals of the same age could be so different. In speaking with Mr. Haas, he had been interested to hear how he had always taken care about his diet and some exercise, usually walking a mile or two daily and avoiding fatty foods. He also was interested to hear that both of Mr. Haas' parents had lived into their eighties. Dr. Lee never had the opportunity to ask Mr. Abboud much about his lifetime habits or his family history. But the question remained, what happens with age and why the diversity?

Although some people seem to age more successfully than others, aging is a universal human experience. The causes of aging appear to be genetic. Some scientists postulate that aging occurs as a consequence of genes that have a variety of effects over time. The genes that have a favorable effect for an individual early in the life span may, after many years, produce less favorable effects. From an evolutionary perspective, such genes with a variety of effects, called pleiotropic, would have favored survival and reproduction and thus given a reproductive advantage to those with the genes. In later years, after reproduction is completed, the negative influence of a gene may occur and lead to the aging and death of the individual. Thus, genes that are favorable during youth and reproduction may also have a deleterious effect with advanced age (Martin 1992).

The existence of pleiotropic genes does still not explain exactly how or what happens with age. One difficulty is sorting out the influence of disease from aging. Many of the changes that people may commonly associate with aging—frailty, memory loss, dependence—are usually the consequence of some specific disease process rather than the aging process. A genetic basis for aging could occur via genes that directly cause aging or by genes that lead to changes in repair processes or defense mechanisms thus allowing disease.

A variety of theories about aging exist that can be correlated with a genetic basis for the aging process. A number of these theories overlap. It seems likely that none explain the whole picture and that aging represents a multi-factorial process. Among some of these theories are the free radical theory, the protein error catastrophe theory, and somatic DNA mutation theory. The free radical theory postulates that cellular damage occurs as a consequence of accumulation of reactive intermediates that cause oxidation of molecules. The protein error catastrophe theory links aging to inaccurate transcription and translation leading to defective protein production. In time, there is a build up of these defective proteins leading to aging and death. The somatic DNA mutation theory suggests that the repair processes that correct errors in the genetic material of somatic cells become less efficient and error-laden DNA accumulates. The DNA, bearing defective codons, leads to abnormal proteins, termination of transcription, and a variety of other errors with deleterious effects. Obviously, these theories are not mutually exclusive. A lack of efficient scavenging systems for free radicals could be the consequence of DNA mutations or protein errors. DNA mutation provides an obvious explanation of how protein errors could build up. Other theories of aging which ascribe aging to changes in immunologic or neuroendocrine function, could, likewise, explain the neuroendocrine and immunologic changes of aging on the basis of changes that are postulated by the free radical, protein error, and DNA mutation theories. Research efforts are currently directed toward understanding the genetics of aging in the hope of finding evidence for these theories. A problem for researchers is the lack of a clear marker for aging. It is extremely difficult to untangle aging from the diseases that so commonly accompany the aging process (Vijg and Wei 1995).

Other than a vague answer that points, somehow, to the effects of genetics, there is no clear cut answer as to why one individual ages well and another poorly. It seems likely that there is no one answer. Mr. Haas may have done so well because of the genes he inherited from his long-lived parents. It may be, however, that less crucial than the genetic inheritance he obtained were the habits of daily exercise and sensible diet.

Dr. Lee realized it was getting late and time to go to bed. He made a mental note to make sure he gave a call to Mr. Abboud in the morning and see how he was doing. He also would need to check with the Visiting Nurse Association about the results of the home visit. He wondered what the new day would bring. It came as a surprise to him when he reflected that much of caring for older people was meticulous observation and careful follow-up. The scientific basis of practice still needed development. In the interim, taking good care of the elderly meant what it has always meant to be a good doctor: thoughtfulness, care, caution, and a willingness to listen carefully to what the patient says.

Questions for further reflection

1. How would you alter your approach to an introductory history and physical between an 85-year-old man and a 45-year-old man?

2. Describe at least three ways in which disease presentation and illness behavior may differ in an older person compared to a younger individual.

3. How might functional assessment assist you in the care of older patients and other individuals with complex problems?

References

Besdine R.W. (1988). Clinical approach to the elderly patient. In *Geriatric medicine*, (2nd edn), (ed. J.W. Rowe and R.W. Besdine), pp. 23–36. Little Brown, Boston.

Cassel C.K. and Brody, J.A.(1990). Demography, epidemiology, and aging. In *Geriatric medicine*, (2nd edn), (ed. C.K. Cassel, D.E. Reisenberg, L.B. Sorenson, and R.J. Walsh), pp. 16–27. Springer, New York.

Gaylord, S.A. (1991). Demography of aging. In *Geriatrics review syllabus*, (ed. J.C. Beck), pp. 1–3. American Geriatrics Society, New York.

Hodkinson, H.M. (1973). Non-specific presentation of illness. *British Medical Journal*, 4, 94–6.

Kleinman, A., Eisenberg, L., and Good, B. (1978). Culture, illness, and care: Clinical lessons from anthropologic and cross-cultural research. *Annals of Internal Medicine*, 88, 251–8.

Lachs, M.S., Feinstein, A.R., Cooney, L.M., Drickamer, M.A., Marottoli, R.A., Pannill, F.C., *et al.* (1990). A simple procedure for general screening for functional disability in elderly patients. *Annals of Internal Medicine*, 112, 699–706.

Levkoff, S.E., Cleary, P.D., Wetle, T., and Besdine, R.W. (1988). Illness behavior in the aged. *Journal of the American Geriatrics Society*, 36, 622–9.

Martin, G.M. (1992). Biological mechanisms of ageing. In *Oxford textbook of geriatric medicine*, (ed. J.G. Evans and T.F. Williams), pp. 41–8. Oxford University Press.

Soldo, B.J. and Manton, K.G. (1988). Demography: Characteristics and implications of an aging population. In *Geriatric medicine*, (2nd edn), (ed. J.W. Rowe and R.W. Besdine), pp. 12–22. Little Brown, Boston.

Suzman, R., Kinsella, K.G. and Myers, G.C. (1992). Demography of older persons in developed populations. In *Oxford textbook of geriatric medicine*, (ed. J.G. Evans and T.F. Williams), pp. 3–14. Oxford University Press.

Vijg, J. and Wei, J.Y. (1995). Understanding the biology of aging: The key to prevention and therapy. *Journal of the American Geriatrics Society*, 43, 426–34.

2

Drug therapeutics in the elderly

Mark Monane

Olga Abrahamsen had been married to her husband Frank for 63 years. They had met and married during the Great Depression. Life had been tough in the beginning. Both were the oldest in their respective families, and even as they tried to start a life together they had to help out their many younger brothers and sisters. But they had both worked hard and they had been able to scratch out a living. In time, they were able to raise their own family and grow more prosperous together.

Olga cried a bit as she sat at her kitchen table. She thought about Frank and how hard it was to be alone at 85. He had died two weeks after having a large stroke. She was, in some ways, relieved that he had not survived with multiple deficits. After so many years, however, it was hard to believe that she was alone. Their children were good. They called and visited. The grandchildren and now the great-grandchildren brought a smile to her. Oh, but she missed Frank.

Mrs. Abrahamsen walked to her bathroom and realized that it was time to take her pills. She pulled out her eye drops and her thyroid pills. Mrs. Abrahamsen wondered how much of the eye drops got in her eyes and how much she spilled on the floor. She also wasn't sure if she had taken the thyroid pill yesterday. The doctor had told her, a few years before, that she was hypothyroid and that she needed to take a pill every day. 'Oh well,' Mrs. Abrahamsen thought, 'I doubt missing it for one day will kill me.'

After swallowing her dose of thyroid medication, Mrs. Abrahamsen walked back to the kitchen to clear the breakfast dishes. She thought about how anxious and depressed she had been since Frank's funeral. Last week her friend Eleanor had been over and given her a nerve pill. She did not know the name of it, but it had made her feel relaxed, although she had been unsteady on her feet the next day. Mrs. Abrahamsen decided to call Dr. Raj Kumar and ask him for something for her nerves and something that would help her sleep. Dr. Kumar had been her and her husband's doctor for many years. He had been so kind to her during Frank's illness. Mrs. Abrahamsen liked how much he cared for her, and how he always seemed to give her a pill to help her out when she had a problem. She gave Dr. Kumar's office a call.

Dr. Kumar called her back later that morning. Mrs. Abrahamsen told him how hard it was to sleep and that she felt anxious and all wound up since Frank's death. She also confided that she was dreading the talk with the lawyer about their finances, as that topic had always frightened her after being so poor in the Depression. Dr. Kumar listened and felt sad for Mrs. Abrahamsen. Hearing the tension in her voice, he told her he would call the pharmacy with two prescriptions, and that he wanted her to make an appointment with the receptionist to see him in two weeks. After finishing with Mrs. Abrahamsen on the phone, he called her drug store and ordered diazepam 5 mg three times a day as needed for anxiety and flurazepam 30 mg at bedtime as needed for sleep.

Mrs. Abrahamsen took her new flurazepam prescription that evening. She had taken the diazepam late in the afternoon and it had made her very sleepy. But she wanted to make sure she had a good night's sleep and so she decided to take the flurazepam at 9 p.m., just before heading to bed. Next morning, at 7 a.m., she awoke to her alarm clock. She felt sluggish and got up slowly. She stumbled on her way out of bed and landed heavily on her right side, feeling a sharp pain as she hit the floor of her bedroom. Mrs. Abrahamsen panicked as she realized she could not get up. She still felt sleepy, her thigh hurt, and as she tried to maneuver herself to an upright position, she slipped down on the floor. Finally, Mrs. Abrahamsen was able to crawl near her bedside table and pull the phone down to the floor. She called the operator and asked for help.

Dr. Kumar took a call from the hospital emergency room, where the doctor on duty told him that Mrs. Abrahamsen was in the emergency room, had suffered a right intertrochanteric fracture, and would need a hip repair.

Introduction to pharmacotherapy

Mrs. Abrahamsen's fall and hip fracture are a classic example of a bad outcome from drug therapy. Before considering some of the specifics of what happened in her case, it is worthwhile to understand the role of pharmacotherapy in the care of older persons. Pharmacotherapy, the pharmacologic management of a variety of acute and chronic conditions, is one of the most common, and cost-effective, means of treatment for older persons. Clinicians need a good working knowledge of the goals, risks, and benefits of pharmacotherapy in elderly patients. Such an understanding requires consideration of three interrelated topics: (1) the older person; (2) the role of medications; and (3) pharmacology and the older patient.

The elderly patient

The aging of the population is a key fact for physicians. Large numbers of individuals will be reaching 65 in the coming years and, along with this growing number of older persons, there is an increase in life expectancy. In considering pharmacotherapy for these increasing numbers of old and very old individuals, the physiology of aging is an issue of concern. A number of changes commonly occur that have substantial effects on drug handling. For instance, there is frequently a linear decline in hepatic, renal, and musculoskeletal function that lead to marked changes in the amount of drug that is circulating and active (drug availability) as well as to the effects of the drug (drug action). These changes will be discussed in more detail below. Despite this tendency of decreasing function, however, there is tremendous variability in aging individuals. This is a constant theme in geriatrics. Treatment strategies should be tailored toward the individual older person, recognizing the inherent variability from individual to individual.

William Osler once stated that: 'The desire to take medicines is one of the principal features that distinguishes man from animals' (Lasagna 1973). While somewhat flippant, Osler's quotation helps to emphasize that medication use is a human activity that involves human relationships. Medication use is driven both by physicians and patients. While clinicians would like to think that the medications they prescribe will make their older persons feel better and be healthier, it is important to realize that medication use is also associated with the risk of adverse drug reactions such as falling, dry mouth, urinary retention, constipation, and confusional states.

The role of medications

In a 1981 survey of out-patient care in the United States, drug therapy was the only form of treatment provided in 40% of encounters for patients 75 years of age and older. In another 30% of encounters, drug therapy was used in conjunction with non-drug treatment, such as exercise and non-pharmacological

management of hypertension. Thus, for patients 75 and over, almost three-quarters of all office encounters end with the prescription of some medication (NACS 1983).

Writing a prescription does more than dispense a drug or specify a pharmacologic treatment. The prescription has a number of important influences besides those associated with a drug's properties. Handing over a prescription to a patient, for example, can empower a patient with the ability to control symptoms based on the knowledge and skill of the clinician. For the clinician, handing over a prescription to a patient provides an opportunity for appropriate physical contact and can create a situation where the physician can teach the patient about their condition, the importance of the medication, and how it should be used. For many clinicians and patients, handing over a prescription is seen as the way a visit to the doctor should end. It is a tangible product of the visit and a sign that the doctor has 'done something'. Unfortunately, not every prescription is appropriate therapy. Patients, however, may feel cheated if the doctor does not give them a medication. Many physicians find it awkward to take the time to discuss other treatments or to resist a patient's demands for a pill or tablet to relieve a problem. In Mrs. Abrahamsen's case, the prescriptions for diazepam and flurazepam were efforts on the physician's part to respond to her distress in the period immediately after her husband's death. The adverse outcome suggests that other choices would likely have been better.

Prescription drug use in the elderly raises a number of important economic issues. There is increasing national expenditure for pharmaceuticals due to the increasing number of older persons in the population, the disproportionate drug use among older compared to younger persons, and the introduction of new, powerful, and expensive medications for management of many of the chronic conditions that afflict the elderly. In 1965, approximately $10 billion was spent for drug use in the United States. In 1980, this figure grew to $20 billion while by 1990 it had approached $50 billion. Current figures estimate that close to $60 billion are spent yearly for the management of drug therapy. This $60 billion cost for yearly drug therapy is about the same as the cost of long-term care for the elderly in the United States. Despite increasing interest in health care policy and reform of the health care industry in America, little attention has been paid to the combined costs of long-term care and drug therapy (Levit et al. 1990).

Efforts at reducing expenditures for drug payments have had unanticipated outcomes. In a project in the state of New Hampshire, a program was instituted that limited the reimbursement for prescriptions for older persons to up to three drugs. Those persons who were prescribed more than three drugs were responsible for the costs of these medications above the reimbursement cap. As a result of the state program, the use of medications decreased sharply among older patients. While there was a net cost savings for the state in expenditures on drug therapy, there was a concomitant rise in long-term care and hospital admissions. Unfortunately, the cost savings achieved by capping reimbursement to a limit of three drugs were overwhelmed by the costs incurred from the increased hospital and nursing home admissions that had been precipitated by patients who had become ill as they required more than the three drugs for which the state provided reimbursement. In a choice between drug therapy and long-term care, restrictions on reimbursement for medications will lead to increased long-term care utilization (Soumerai *et al.* 1991).

Pharmacology and the older patient

The epidemiology of medication use also represents an important aspect of pharmacotherapy in the elderly. Even though the elderly make up only 12–13% of the population, they are responsible for between 25 and 30% of all drug use in the country. Patterns of medication use vary greatly among individuals and depend on whether they are at home, in the hospital, or in a nursing home. As previously noted, the older individual takes an average of four prescription medications and four over-the-counter medications on a daily basis. Of course, as always when dealing with older persons, diversity is the rule, and there is no such thing as an average geriatric patient. There are approximately 25% of older persons taking no medications on a regular basis, while other persons will consume 15 or more different medications every day. Much of this medication use may be appropriate, as many older persons have a number of concurrent chronic conditions that require drug therapy. For example, an older individual, such as Mrs. Abrahamsen, may have diagnoses of glaucoma, osteoarthritis, and hypothyroidism, all common conditions that require pharmacotherapy. This appropriate setting for multiple drug use in the management of common chronic diseases is referred to as 'polymedicine'.

Polymedicine is contrasted to polypharmacy: a pattern of excessive and inappropriate use of prescription and non-prescription medications. Three factors commonly contribute to polypharmacy. First is a lack of a primary care physician for the older person. The patient may visit many health care providers and specialists but lack a physician who supervises the care. Second is a failure of a physician to recognize an adverse event as possibly drug-related and responding to a new event or complaint not by discontinuing the medication but by prescribing another. A vicious cycle may be the result. Third, older persons may continue to take prescribed medications that were indicated at one time but now are not needed. Health professionals should make a concerted effort to prevent and identify these risk factors for polypharmacy. A practice of regularly asking patients about their medication use is an appropriate first step, along with a periodic 'brown bag' review, when the patient is asked to bring in all the medicines in the house that are in the medicine cabinet or on the bedside table. A number of patterns of medication use may indicate polypharmacy: medications with no apparent indication; medications contraindicated in the elderly, such as long-acting benzodiazepines (e.g., flurazepam); medications that are redundant in action (e.g., two sedative hypnotic agents); inappropriate dosages of medications (e.g., doses of medication for hypertension or diabetes that may be appropriate for a younger person but are too high for many elderly individuals); medications prescribed to treat a complication for a medication that has not been discontinued (e.g., cimetidine for an ulcer in a person who continues on a non-steroidal antiinflammatory agent); or medications that have potentiating or possible conflicting effects (e.g., warfarin and aspirin).

There may be good reasons for an older person to take several medications, including some that initially make an astute clinician suspicious of polypharmacy. *The key to prevention of polypharmacy is increased communication among all the doctors who treat a patient, the pharmacist, other community-based caregivers, and the patient.* Polymedicine may be necessary in the older individual and should lead to a carefully considered multiple drug regimen with a goal of optimizing a person's condition and functional status. Polypharmacy, however, is never desired.

Over-the-counter medications are an important consideration in caring for older persons and avoiding polypharmacy. These medications represent 60% of all medications taken and represent one-half of the total medication cost. Common over-the-counter medications include cold preparations, which may have several active ingredients including diphenhydramine, acetaminophen, and the non-steroidal agent ibuprofen. Many persons do not consider over-the-counter drugs as medications, because they can be obtained without a doctor's visit or a prescription. When taking a drug history, one must ask specifically about prescription as well as any over-the-counter medications that have been purchased by or for the patient.

The cost of medications is a major concern for older persons. Seventy-seven per cent of all out-patient drug costs are paid for by the elderly (Leibowitz 1985). Although most individuals in the United States over 65 years of age are covered by Medicare for hospital coverage and doctor visits, the cost of out-patient prescription medications is borne by the individual. Although geriatrics is often concerned with overmedication of the elderly, another very real concern is that older persons may have difficulty paying for their medications. For many older individuals on a fixed income, a new prescription can be a difficult expense to afford and must be balanced between rent, food, and costs for utilities.

Despite the disproportionate amount of medication consumed by the elderly, little is known about how the commonly used medications affect elderly individuals. In the United States, pre-market testing is a federal requirement by the Food and Drug Administration before a drug can be approved for use by the general population. Older individuals are usually poorly represented in such pre-market tests. The lack of inclusion of the elderly in pre-marketing trials has several reasons. There may be a lack of older volunteers for these types of studies. There may well be a bias, however, on the part of pharmaceutical companies to exclude the elderly and those with chronic illness in favor of young and healthy persons with less likelihood of adverse drug reactions and poor clinical outcomes. Because of the lack of information to guide the clinician on drug therapy in older patients, several post-marketing surveillance trials are currently underway that include elderly patients and those with concomitant illnesses. Post-marketing studies on the sedative-hypnotic agent flurazepam have shown that this agent is associated with increased drug reactions with increased dose, regardless of the patient's age. But there also is an important interaction between age and the medication dosage: 66% of all patients over the age of 70 who were given greater than 30 mg per day of flurazepam experienced an adverse drug reaction.

Drug handling

Pharmacokinetics

Obviously, Mrs. Abrahamsen's broken hip represents an adverse outcome. How can it be explained? Although the factors that ultimately contributed to a broken hip are multiple, the provoking factor that led to her fracture was her new medications. An understanding of the basic principles of pharmacology can help to explain why this is so. Pharmacokinetics, which relates the dose of a drug to its concentration in the blood, describes 'what the body does to the drug' (Avorn and Gurwitz 1990). After administration, by whatever route, a drug reaches a concentration in blood that changes over time. The concentration of the drug ultimately determines, and is in equilibrium with, the concentration of the drug at various sites of action. Pharmacokinetics can be divided into four separate entities: absorption, distribution, metabolism, and excretion.

The absorption of a drug depends on the route which is used to administer it. Routes of administration include oral, transdermal, intravenous, intra-arterial, intramuscular, or rectal. Some means of administration maximize the percentage of the dose that concentrates on the desired target organ while minimizing the amount that enters the systemic circulation. Examples include local administration of timolol eyedrops for glaucoma, using pseudoephedrine in a nasal spray, or applying hydrocortisone in a topical cream. These maneuvers, however, do not totally prevent at least a portion of the dose from entering the systemic circulation and producing effects distant from the target organ. For most drugs that are administered orally, the route of absorption is almost identical in the older person as it is in younger individuals. Despite common age-related changes in the stomach and its physiology, such as decreased gastric acid and less villi in the mucosal surface, absorption of an oral medication is not affected by age. Clinicians can be assured that in the absence of some intrinsic pathology, an elderly patient will absorb an oral medication as well as a younger patient. The same, however, is not the case with age-related changes and the distribution, metabolism, and excretion of many drugs.

The distribution of a drug describes the process that begins when a drug has been absorbed and reaches the systemic circulation. The drug will be mixed with the intravascular space and come to a rapid equilibrium with other tissues in the body. With normal aging, there are substantial changes in body components that affect drug distribution. Body fat as a proportion of total body weight increases by over 35% from age 20 to age 70. Concurrently, lean body mass and total body water decrease by about 20%, with plasma volume decreasing by 8–10%. These changes of increased body fat and decreased lean tissue mass, total body water, and plasma volume have marked consequences on the distribution of drugs in the elderly. Diazepam and flurazepam, the drugs that Mrs. Abrahamsen took before her disaster, are two commonly used benzodiazepines. They are markedly lipophilic and have a wide volume of distribution because they tend to concentrate in fat after moving from the systemic circulation. Because of their redistribution to body fat, these drugs tend to be stored in the fat, rather than being rapidly cleared from the body. Other benzodiazepines, such as lorazepam, oxazepam, and temazepam, are less lipophilic than diazepam or flurazepam.

Normal aging can greatly alter the metabolism of many drugs. Hepatic biotransformation is needed for metabolism of highly lipophilic drugs that are not cleared by the kidneys. Most lipophilic drugs are transformed in the liver to more hydrophilic metabolites that can then be renally excreted. Hepatic biotransformation involves two types of reactions. Phase I reactions include oxidative and hydroxylation reactions mediated by the mixed-function monooxygenase system, also known as the cytochrome p450 system. The classic change in normal aging is a decrease in the efficiency of phase I reactions. The cytochrome p450 system is responsible for metabolism of common drugs such as cimetidine, isoniazid, and theophylline. Diazepam and flurazepam are also metabolized by this system in a phase I process. The metabolism of these agents is markedly slower in the older person compared to the younger individual. The second type reaction, phase II reactions, are generally not affected by aging. Phase II reactions include synthetic reactions, in which the drug molecule is altered by the addition of glucuronic acid, glycine, sulfate, acetyl groups, or glutathione. These phase II metabolites are usually pharmacologically inactive. Drugs metabolized by phase II processes are preferred in older patients over those metabolized by the mixed function microooxygenase system. Among the benzodiazepines, lorazepam, oxazepam, and temazepam are metabolized via phase II pathways.

Excretion is a process involving the kidney, including glomerular filtration, tubular secretion, and, sometimes, tubular reabsorption. The kidneys generally clear the circulation of hydrophilic drugs and metabolites. The kidney can excrete drugs that are present in the protein-free ultrafiltrate and can actively secrete weak organic acids or weak organic bases into the renal tubules. Reabsorption of the drug or its metabolite from the tubular fluid may also occur. With normal aging there is a decrease in the number of functional nephrons, with a resultant decrease in glomerular filtration and tubular secretion and reabsorption. As a result, drugs that depend on renal clearance tend to have a much longer half-life in older persons. Plasma clearance of certain drugs by the kidneys can be decreased by as much as one half in an 80-year-old individual compared to a 30-year-old (Rowe *et al.* 1976). Diazepam and flurazepam are hepatically metabolized into active metabolites which are, in turn, renally cleared. The half-life of a drug is directly related to the volume of distribution for the drug and inversely related to its clearance. The

effective half-life of the parent compounds is increased with decreased renal function, resulting in an increased potential for adverse effects in those persons.

As can be seen, there are a number of reasons why diazepam and flurazepam were less than optimal choices for management of Mrs. Abrahamsen's anxiety and insomnia. Both drugs are lipophilic and thus have a large volume of distribution through the increased body fat. Recall that with aging, body fat is increased. Diazepam and flurazepam tend to stay in the fat well beyond the time desired for the management of insomnia. Drug clearance of these two agents is also decreased in the older person because of decreased phase I reactions as well as decreased functional nephrons to clear the active metabolites. These metabolites exert effects that are similar to the parent compounds and prolong the effective half-life of diazepam and flurazepam. For example, in a younger person, the half-life of diazepam is about 36 hours. Due to the summing of the processes described above, however, the half-life can extend to as high as 72–96 hours in an older person such as Mrs. Abrahamsen. Flurazepam is even worse. It has a half-life of 74 hours in the younger individual and as long as 120 hours in an older woman! The pharmacokinetics of other benzodiazepines are not as adversely affected by aging. Oxazepam has a half-life of 7 hours in the younger individual and about 8 hours in an older individual. The paucity of change with aging is a consequence of the hydrophilic properties of oxazepam and relatively small volume of distribution, phase II conjugation reaction for metabolism, and the absence of active metabolites after conjugation. Lorazepam and temazepam have similarly attractive profiles which make them and oxazepam better choices for sedative-hypnotics in older persons than agents such as diazepam and flurazepam. Although clinicians should not reach for a drug as a first response to patient complaints of anxiety or insomnia, when a choice is made among benzodiazepines for the management of these problems, knowledge of pharmacokinetics and age-related changes suggests prescribing a short-acting medication.

Pharmacodynamics

In contrast to pharmacokinetics, pharmacodynamics attempts to describe 'what the drug does to the body' (Avorn and Gurwitz 1990). Unfortunately, pharmacodynamics in the older patient is not well understood. Pharmacodynamics requires an understanding of the pharmacokinetics of the agent being considered. This information is lacking in many cases. In addition, pharmacodynamic outcomes require the testing of effects on target organs, which can be difficult to measure. Measuring the amount of sedation produced by a benzodiazepine is an example of a pharmacodynamic effect but it is technically more difficult to do than drawing a blood level. Nevertheless, some important information has been gathered in attempts to describe the relationship of aging to pharmacodynamics.

There is evidence that older people may have altered sensitivity to benzodiazepines and other agents when compared with younger individuals. In one study, individuals about to undergo elective cardioversion were given diazepam for sedation. The researchers measured the diazepam concentration necessary to create a level of sedation appropriate for the procedure with response to vocal stimuli but not to painful stimuli. For those patients 70 years of age and older, the concentration of diazepam required to reach the same level of sedation of patients aged 30 to 50 was much lower. Even when controlling for pharmacokinetic changes, this finding suggests that there is an increased intrinsic sensitivity to benzodiazepines in the older person. The mechanism of the increased sensitivity is not understood; it may be due to either a post-receptor event or a second messenger phenomenon, perhaps affecting the gamma-amino-butyric acid (GABA) system (Reidenberg *et al.* 1978). Along with benzodiazepines, an increased sensitivity to agents and altered pharmacodynamics has been demonstrated with opiates, anticholinergics, dopamine antagonists, and antihypertensives. Conversely, decreased pharmacologic effects due to decreased sensitivity have been found with beta blockers, beta agonists, and insulin. Further work is needed to help to decipher the changes in pharmacodynamics that occur with normal aging.

Non-pharmacological approaches

In hindsight, it is clear that Dr. Kumar should not have prescribed Mrs. Abrahamsen diazepam or flurazepam, and certainly not the combination. Although he could have chosen a shorter-acting benzodiazepine, such as oxazepam, a number of non-drug approaches could have been tried first. Even when pharmacotherapy is prescribed, these non-pharmacological maneuvers may increase the desired effect of the medication and lead to a lower dose with a good result and less chance of side-effects. Although Mrs. Abrahamsen's management was complicated by her grieving, good non-pharmacological approaches to the management of insomnia include limiting of caffeinated beverages to the morning, keeping regular times for bedtime and rising, and having a period of 'winding down' from the events of the day before going to bed. Avoidance of daytime napping is crucial, as napping lessens the need for nocturnal sleep. A sensible program of regular exercise can also appropriately

increase the tiredness of the person and promote a more normal sleep–wake cycle. Exercise should not be done within two hours of the desired bedtime. These non-pharmacological approaches are often successful but may be resisted by the patient who desires a pill to manage their sleep disorder. Patients may be annoyed at a physician's reluctance to prescribe medication and complain: 'I didn't need you to tell me this . . . I could have done this without seeing you' or 'My sister has trouble sleeping and her doctor gave her a pill.' Taking the time to emphasize the risks and benefits of pharmacotherapy as well as describing the role of non-pharmacological therapies may improve patient understanding. It may be worthwhile for the physician to pick up the prescription pad and write down some specific non-pharmacological suggestions; for example: 'Caffeinated beverages, as needed, only before noon daily.' As mentioned, the prescription has important influences beyond the drug prescribed. It serves to bestow knowledge from the clinician to the patient. The handing over of the prescription may be a therapeutic act in and of itself. Although clinicians may think that oral communication is the most effective way to transmit information to a patient, writing down a non-pharmacological approach for the patient to refer to after the office visit is also a useful strategy. Unfortunately, doctors may sometimes find themselves in the situation of refusing to prescribe a medication and running the risk of having the patient go to another doctor who is more willing to give them a medication.

Drug utilization

Mrs. Abrahamsen recovered well from her hip fracture and, after a brief time in rehabilitation, returned to her own home. She adjusted to the loss of her husband and, with the support of her family, was able to maintain her independence. Mrs. Abrahamsen continued to see Dr. Kumar on a regular basis. He noted, several months after her hospitalization, that she had consistently elevated blood pressure readings. He reviewed the notes from her last three visits and saw that her pressures were in the range of 180/100 mmHg each time. Dr. Kumar, on the last visit, spoke to Mrs. Abrahamsen and told her that she had hypertension and gave her a prescription for hydrochlorothiazide 25 mg per day. He asked her to review her medications with him, to make sure they both agreed on what she was taking. Mrs. Abrahamsen went through her list: 'I take the Synthroid for my thyroid every morning, and then there are the timolol eye drops for the glaucoma twice a day, and I take acetaminophen if I have aches and pains. If the aches and pains are bad, then I take the ibuprofen you gave me. Oh yes, there is that stool softener, the docusate. Now tell me about this hydrochloro ... whatever you call it.' 'Your blood pressure is up and I want you to take this to get it back in the normal range,' answered Dr. Kumar.

Mrs. Abrahamsen returned for follow-up visits in two weeks and then six weeks after receiving Dr. Kumar's prescription. He wanted to see how her blood pressure had responded as well as check her electrolytes on the diuretic. Each time Mrs. Abrahamsen's pressure remained elevated. Her electrolytes showed little change. Dr. Kumar was surprised to see her potassium had actually gone up a bit, from 4.1 to 4.7 mEq/dl while on the hydrochlorothiazide. He wondered if she was actually taking her diuretic.

Compliance

The most likely cause of treatment failure in Mrs. Abrahamsen's case is non-compliance with her medication regimen. Compliance relates the extent to which a patient's behavior, such as taking medications, following diet, or making lifestyle changes, actually coincides with the advice given by the doctor. Some find the term 'compliance' objectionable as it implies obedience or a servile response to the authority of the physician. Alternative choices to describe the correspondence between a patient's medication use and what the doctor prescribed include words like adherence, fidelity, or maintenance. For the purpose of this discussion, we will continue using the word 'compliance', but being aware of its shortcomings.

Compliance with medication therapy can be measured by direct and indirect methods. Direct methods include measuring blood levels and checking urine for the drug and its metabolites. These direct measures, however, are not available for many common medications. Indirect methods of determining compliance are relatively inexpensive and worthwhile in detecting compliance in the clinical setting. Indirect methods include patient interview, doctor estimate, pill counting, or checking pharmacy records for refill information. Evidence from the clinical literature indicates that compliance with medications for short-term therapy is approximately 80–90%, while compliance for chronic therapy, as is the case with treatment of hypertension, reaches only 50–60%. Factors that determine patient compliance include patient knowledge, the patient–physician relationship, and the complexity of the therapeutic regimen.

Patient knowledge is a key factor in determining compliance. Inadequate patient education is a major obstacle in achieving appropriate medication use. Educating patients about their medication, including the dose, intended effects, side-effects, and scheduling, is a time-consuming process. In a busy office practice, it can be very difficult for physicians to take the time to explain and teach patients about their drug regimen. Despite the problems, it is important for the clinician to educate the patient, especially with asymptomatic conditions, such as hypertension, and to emphasize the importance of taking the prescribed drug even in the absence of symptoms. When the major deficit in reaching compliance is a lack of knowledge, educational strategies and an opportunity for patients to ask questions are reasonable interventions.

The physician–patient relationship can be a powerful determinant of compliance. It is important for the physician to stress the use of medication before the patient leaves and ask the patient about how he or she has progressed (or not) with the medication at the time of the next visit. As part of an ongoing relationship, the clinician can show concern about the patient, increase the patient's knowledge of the prescribed medications, and monitor compliance. Simply assuming that writing a prescription means that the patient will take what is prescribed or take it as directed is naive.

The single most important strategy for improving compliance is simplification of the therapeutic regimen. This can be done by either decreasing the total number of medications or decreasing the frequency with which medications are to be taken each day. Compliance is not adversely affected by age. As Helen Hayes said: 'Age is not important unless you're a cheese.' Compliance is actually highest for people over 75 years of age; still, the average is only about 50–60% (Monane et al. 1993). Although there is no relationship between compliance and age, compliance is clearly related to the number of medications. Compliance decreases linearly with an increase in medications. Since older persons often have multiple medications for the management of multiple chronic diseases, they may appear to be less compliant because of the increased drug burden, not as a consequence of age *per se*.

The key issue regarding compliance is the relationship between compliance with prescribed medications and clinical outcomes. Although this issue has not received as much attention as attempts to measure medication compliance, there is evidence suggesting poorer outcomes for patients who are non-compliant. Studies have found that hospitalization rates (Col et al. 1990) and emergency room visits (Schneritski et al. 1980) are higher for those who do not adhere to their medical regimen. In an investigation of the relationship between compliance with beta blocker therapy and coronary heart disease, the investigators found that the risk for myocardial infarction was four times greater in patients with compliance less than 80% compared to those patients who had 100% compliance (Psaty et al. 1990). Not all research, however, has indicated that compliance with medical regimen is related to the outcome. In a study of patients with streptococcal throat infections and compliance with antibiotic therapy, many patients who resolved their infections did so despite inadequate doses of medication. Conversely, many patients who took the medication exactly as prescribed did not clear their infections. An ideal model in which improved patient outcome is solely mediated by compliance with the prescribed regimen allows a clear relationship between medical care and improved outcomes. The relationship is, in reality, not so clear. Although improved outcome is affected by compliance, other factors also play a role, including comorbid diseases, other medication use, and lifestyle issues, such as smoking and non-pharmacological interventions.

In considering compliance issues, one should acknowledge what can best be called intelligent non-compliance. It may sometimes be in the person's best interest not to take what the doctor has prescribed. When taking a medicine is associated with an adverse drug effect, many patients will no longer take the prescribed drug. In the case of Mrs. Abrahamsen, she would have done well to have not taken the dose of flurazepam that night when she already felt sleepy from the diazepam. Ironically, her compliance likely contributed to her hip fracture.

After reviewing her laboratory results and her apparent lack of response to the hydrochlorothiazide, Dr. Kumar asked Mrs. Abrahamsen to come in for a repeat visit to discuss her blood pressure. She returned to the office a month later and appeared ill. Mrs. Abrahamsen complained of difficulty breathing, swollen ankles, weakness, and a feeling that she was not urinating as often as was normal for her. Dr. Kumar was concerned as he noted her increased respiratory rate, heard the râles in Mrs. Abrahamsen's lungs, and found pitting edema in her shins. He sent blood for electrolytes, which revealed a sodium 128 mEq/l, a potassium of 6.3 mEq/l, a blood urea nitrogen of 60 mg/dl, and a serum creatinine of 3.0 mg/dl.

Her values a month earlier had shown a sodium of 136 mEq/l, potassium of 4.7, blood urea nitrogen 28 mg/dl, and creatinine of 1.4 mg/dl.

Dr. Kumar reviewed Mrs. Abrahamsen's medications. He suspected that she was not taking her hydrochlorothiazide. Although hyponatremia and dehydration might be adverse reactions to hydrochlorothiazide, the current set of electrolyte results would be unusual. The clinical signs of fluid overload, coupled with these laboratory results were suggestive of acute renal failure. Dr. Kumar wondered what could be causing the problem. He studied his notes on Mrs. Abrahamsen and saw that she was prescribed ibuprofen for pain not relieved by acetaminophen. 'Mrs. Abrahamsen,' he asked, 'have you been using the ibuprofen much lately?' Mrs. Abrahamsen seemed a bit surprised by the question, as she had not mentioned that her arthritis had been acting up. 'Well yes, Doctor, I have had a lot

of pain in my knees lately and the acetaminophen did not touch it. I have been taking the ibuprofen three times a day every day for the last two weeks.'

Adverse drug reactions

As Dr. Kumar suspected, acute renal dysfunction caused by a non-steroidal anti-inflammatory agent is the likely diagnosis for Mrs. Abrahamsen. The mechanism of renal failure involves inhibition by the non-steroidal antiinflammatory agents of prostaglandin biosynthesis. Non-steroidal antiinflammatory drugs (NSAIDs), like ibuprofen, inhibit cyclooxygenase and decrease the synthesis of prostaglandins from arachidonic acid. Decreased prostaglandin synthesis has numerous physiological effects. These include decreased renal blood flow, decreased glomerular filtration, decreased renin secretion, and decreased excretion of sodium and water. The inhibition of prostaglandin synthesis occurs with the use of NSAIDs in both young and old individuals. The phenomenon of NSAID-induced renal dysfunction, however, is observed much more commonly in older patients. As is a common theme with typical aging, the older person has less functional and physiological reserve. In the elderly, there is a lower baseline rate of synthesis of prostaglandins. With the use of NSAIDs, there is increased vulnerability for prostaglandin synthesis to drop to a level where renal dysfunction and clinical signs and symptoms appear. While younger persons also have diminished prostaglandin synthesis after taking NSAIDs, renal failure is rarely seen because their levels of prostaglandins and the amount of prostaglandin biosynthesis usually remain above the threshold for clinical impairment.

Side-effects from medications are common in the elderly. The high incidence of drug-related adverse effects in older persons is due to the lack of physiological reserve with aging, the presence of multiple chronic illnesses, and the large number of medications commonly taken by many older patients. It is important to differentiate adverse reactions to medications from what may mistakenly be considered normal changes of aging. Although xerostomia, or dry mouth, may occur with aging, it may be caused by many commonly used medications, including diphenhydramine. Constipation, another common problem in the elderly, may be caused or aggravated by the use of anticholinergic medications or calcium channel blockers, such as verapamil (Wei 1989). Confusion should never be considered a normal part of aging. Many medications, including anticholinergic agents and several over-the-counter preparations can commonly cause confusion in older persons. In an older man, urinary retention is frequently ascribed to prostatic enlargement. Although the prostate may be enlarged, the urinary retention may well be caused by a medication. The appropriate clinical course is to discontinue the medication, allow a period of bladder drainage with an indwelling catheter, and see if the man can void. All too often, urinary retention leads to a prostate resection. Medications which can cause urinary retention include those that increase sympathetic tone such as pseudoephedrine (commonly used in cold preparations) as well as medications that block bladder contractility, including calcium channel blockers and anticholinergics.

The risk of drug interactions, not surprisingly, increases with an increased number of drugs (prescription and non-prescription) being taken. Interactions may occur that either augment the action of one or more medications (such as the interaction between ciprofloxacin and theophylline); diminish the effect (as occurs when a patient takes phenobarbital and phenytoin); or negate the effect of one or more medications (using an inhaled beta-adrenergic preparation while on a beta blocker). These interactions can have adverse or beneficial drug effects. In caring for the elderly, certain drugs are considered high risk because of their narrow therapeutic index: there is only a slight difference between the effective therapeutic levels and the toxic range. Examples of drugs which are high risk include warfarin, digoxin, cimetidine, phenytoin, and theophylline. The risk is increased in the presence of coexisting conditions which may affect the distribution, metabolism, and excretion of these agents. Despite the risk, these drugs are often useful therapeutic agents in the elderly. Their use, however, requires scrupulously careful monitoring by the physician.

As mentioned previously, individuals who see multiple physicians are at increased risk of adverse drug reactions. There may well be little communication among the various physicians and little knowledge of what other doctors have prescribed. Likewise, older persons may patronize multiple pharmacies so that no one pharmacist may know all the medications that the person is prescribed. The solution to these potential dangers is relatively simple: one physician should co-ordinate all prescriptions and one pharmacist should keep track of a person's medication use.

Mrs. Abrahamsen was briefly hospitalized as her ibuprofen was stopped. Her signs of fluid overload resolved with gentle diuresis and her renal function improved over the next several days. Before she left the hospital, Dr. Kumar spoke with her. 'Mrs. Abrahamsen, you have had a lot of trouble because of some of the pills that you have taken. Let's take some time when you get out of the hospital to go over your medications, stop what you don't need, and make sure that you are taking what I think you need.

I think that you have not been taking the blood pressure medicine I asked you to take and I need to do a better job in explaining why I think that's important.' Mrs. Abrahamsen responded: 'I think that's a fine idea. I know I'm an old lady, but I think I do pretty well and I don't like being sick from the medicines. And we will talk about that hydrochloro . . . whatever you call it.'

Guidelines for proper prescribing

The first question to ask before writing a prescription for a medication is: 'Is a drug really necessary here?' Many symptoms do not require a medication and reassurance may be all that is needed. As was considered with sleep disorders, a variety of non-pharmacological approaches are possible for many complaints.

A second question to be considered in evaluating a person's complaints is: 'Are the signs and symptoms a consequence of an adverse drug reaction?' Asking a question like this can prevent the vicious cycle of treating unrecognized drug reactions with another drug. Appropriate therapy consists of reviewing a person's medications and removing potentially offending agents, if at all possible. It does take substantial time and effort to review a patient's medications, but the reward of improved quality of life for the patient makes the time spent a wise investment.

Being aware of drug interactions and the possibility of adverse effects, however, should not stop a clinician from prescribing medications for older patients when there are clear indications and the potential benefits outweigh the risks. Every drug has some risk but most have significant benefits! Three rules of thumb can help in writing a prescription.

First, consider if there are any characteristics of the drug that are likely to cause problems for an older person, as compared to a younger person. This is especially true when considering the pharmacokinetics and pharmacodynamics of a medication. One might avoid a drug that undergoes phase I hepatic metabolism if there is another choice with an alternate route available. Obviously, some of this has to be personalized depending on the diversity of the older population. An anticholinergic drug may be a reasonable choice for one older person but may not be for the older man with bladder outlet obstruction. Questions of compliance and how the drug will interact with a person's lifestyle are also important. A drug that is taken three times a day may be impossible for the 70-year-old with early Alzheimer's disease (as well as the absent-minded 40-year-old!), but may be no problem for the precise and careful retired bookkeeper.

A second rule of thumb is to begin with the lowest possible starting dose that makes sense for the person. Older persons may well have a beneficial therapeutic response at a much lower dose because of age-related pharmacokinetic and pharmacodynamic changes. The physician can begin at a low dose, monitor for side-effects, and gradually increase the dose as needed to maximize the beneficial effect while minimizing the risk.

The third rule of thumb is to always follow-up and monitor patients after prescribing a new medication. Scheduling a return visit at the time of giving a prescription may help with compliance. It also gives the physician the opportunity to assess response, inquire about adverse drug reactions, and obtain laboratory tests, as needed, for drug levels or screening for measuring untoward metabolic effects.

Questions for further reflection

1. Distinguishing between pharmacokinetics and pharmacodynamics, consider how the same drug might affect an older person differently than a younger individual.
2. What are some of the pitfalls of medication use in older persons?
3. Using other reference materials as needed, comment on the following: In your office, an elderly woman presents with confusion and tremulousness, and is accompanied by her son. He states that her regular physician had been caring for her and gave her the following medications: cimetidine, a theophylline preparation, and, recently, erythromycin in the setting of an apparent bronchitis.

References

Avorn, J. and Gurwitz, J. (1990). Principles of pharmacology. In *Geriatric medicine*, (ed. C.K. Cassell, D.E. Reisenberg, L.B. Sorensen, and J. Walsh), pp. 66–77. Springer, New York.

Col, N., Fanale, J.E., and Kronholm, P. (1990). The role of medication noncompliance and adverse drug reactions in hospitalizations of the elderly. *Archives of Internal Medicine*, 150, 841–5.

Lasagna, L. (1973). Fault and default. *New England Journal of Medicine*, 289, 267–8.

Leibowitz, A., Manning, W.G., and Newhouse, J.P. (1985). The demand for prescription drugs as a function of cost sharing. *Social Science and Medicine*, 21, 1063–9.

Levit, K.R., Freeland, M.S., and Waldo, D.R. (1990). National health care spending trends: 1988. *Health Affairs*, 9, 171–84.

Monane, M., Gurwitz, J.H., Monane, S., and Avorn, J. (1993). Compliance issues in medical practice. *Hospital Physician*, 29, 35–9.

NACS (National Ambulatory Care Survey) (1983). *1981 summary. Advance data. No. 88.* Department of Health and Human Services, Washington, D.C.

Psaty, B.M., Koepsell T.D., Wagner, E.H., LoGerfo, J.P., and Inui, T.S. (1990). The relative risk of incident coronary heart disease associated with recently stopping the use of beta-blockers. *Journal of the American Medical Association*, 263, 1653–7.fi(#fi

Reidenberg, M.M., Levy, M., Warner, H., *et al.* (1978). Relationship between diazepam dose, plasma level, age, and central nervous system depression. *Clinical Pharmacology and Therapeutics*, 23, 371–4.

Rowe, J.W., Andres, R., Tobin, J.D., Norris, A.H., and Shock, N.W. (1976). The effect of age on creatinine clearance in man. *Journal of Gerontology*, 31, 155–63.

Schneritski, P., Bootman, J.L., Byers, J., *et al.* (1980). Demographic characteristics of elderly drug overdose patients admitted to a hospital emergency department. *Journal of American Geriatrics Society*, 28, 544–6.

Soumerai, S.B., Ross-Degnan, D., Avorn, J., McLaughlin, T.J., and Choodnovskiy, I. (1991). Effects of Medicaid drug-payment limits on admission to hospitals and nursing homes. *New England Journal of Medicine*, 325, 1072–7.

Wei, J.Y. (1989). Use of calcium entry blockers in elderly patients. *Circulation*, 80, IV-171–7.

3

Cognitive decline in the elderly

Germaine L. Odenheimer

Dr. Colin O'Rourke picked up Mr. Pruitt's chart from the nurses' station and brought it over to a desk so that he and Ben Stevens, a senior medical student doing a rotation in geriatrics, could review it together. Dr. O'Rourke had been consulted for evaluation of Mr. Pruitt's mental status and for assistance with discharge planning. There was considerable concern over whether it was safe to discharge Mr. Pruitt to home, as he appeared to be insisting. Mr. Pruitt's primary physician had followed him closely over the years. He had left an extensive admission note, documenting in detail Mr. Pruitt's history.

Dr. O'Rourke and Ben began to review the history:

Mr. Pruitt is a 76-year-old right-handed, college-educated man who, on the day of admission to the hospital, was found by his son to be confused and acting inappropriately. He was asking for his deceased wife and did not appear to recognize his son.

'OK Ben, what do you think so far?' Dr. O'Rourke asked the student. 'I know that you don't know too much about Mr. Pruitt yet, but what is of concern?'

'Well, Dr. O'Rourke, I know that just because Mr. Pruitt is an older person one cannot brush off sudden changes in behavior. Any acute change in behavior in the elderly has to be considered a medical emergency. There are a huge number of possibilities as to why an older person might become confused. It might be something directly affecting the brain, such as a stroke, or it could be something more systemic, such as a heart attack, dehydration, pulmonary embolus, pneumonia, or reaction to a medication.'

'Excellent. Confusion is not normal in older people. It is very important to have a sense of a person's baseline mental status. Too often a physician will see a patient who is confused and assume that, because the patient is old, the problem is Alzheimer's disease or some other dementing illness and will therefore miss an acute process that could be causing the confusion. Dementia does not come on suddenly. Delirium does. Let's read on.'

Together, they continued to review the chart:

Mr. Pruitt's past medical history is remarkable for a colon cancer treated with surgery some 30 years ago. The tumor was found during a barium enema and at surgery had not penetrated the bowel wall. There has been no evidence of metastases. Mr. Pruitt also has a history of coronary artery disease with an episode of presumed angina one year ago. He has systolic hypertension with pressures between 180 and 200 mmHg. Migraine headaches have been a frequent problem with four to six bouts a month for the last 40 years.

Mr. Pruitt's medications are amitryptiline (Elavil) 50 mg at bedtime, carisoprodol (Soma) 350 mg prn migraine, and sumatriptan (Imitrex) injection for migraine not relieved by carisoprodol.

Mr. Pruitt has a 20 year history of smoking. He had quit for 10 years but restarted one year ago, at the time of his wife's death, and currently he smokes one pack per day. He drinks alcohol occasionally. Mr. Pruitt will also occasionally use marijuana if he has difficulty sleeping. In the past there was a history of oral narcotic overuse for his migraines. There is no history of intravenous drug use. He drinks six to ten cups of coffee per day. Since his wife's death, his diet has been haphazard with junk food, cookies, chips, and ice-cream comprising the major part of his calories. Mr. Pruitt was sexually active with his wife until her death. She died from metastatic breast cancer last year. Since her death, he has had a relationship during which he had unprotected sex.

Mr. Pruitt's family history is remarkable for migraine in an aunt and brother. There also is a family history of suicide with his mother killing herself at age 27 and a brother dying of a self-inflicted gunshot wound at 63. Another brother, now in his eighties, had a stroke last year and suffers from osteoarthritis.

The social history includes an occupation of furniture building for 30 years as well as being a writer, with a second novel published two years ago. Mr. Pruitt had been an avid fisherman, baseball fan, and ballroom dancer. Over the last year, his son, Brian, has noted a decline. Brian rented an apartment near his father's home when he realized the difficult time his father was having after the death of his wife. Brian stops by for lunch daily and checks in with his father.

Physical examination is limited because of lack of co-operation and reveals a thin, disoriented, obstreperous man in mild distress, coughing thick yellow sputum. He is unsteady on his feet and smells of urine. Vital signs include a regular pulse of 100, a blood pressure of 180/70, respirations of 23, with lungs congested to auscultation.

Dr. O'Rourke skimmed through the rest of the chart, summarizing it for Ben: 'Mr. Pruitt refused further examination, saying he did not trust doctors or hospitals. His son Brian calmed him. A chest X-ray showed a right lower lobe pneumonia. He was admitted and they began intravenous antibiotics. OK Ben, let's talk a bit about delirium and consider what the history tells us about Mr. Pruitt.'

Mr. Pruitt is in an acute confusional state, or delirium. The history reveals that he has been highly functional, although suffering a decline since his wife's death, until the day of admission when his son noticed a sudden change. The sudden clouding of consciousness is characteristic of delirium. In Mr. Pruitt's case, the change was quite obvious. He was unable to co-operate with instructions from the physicians and other hospital staff. Mr. Pruitt was also suspicious and his complaints against the hospital staff displayed a paranoid tinge. Delirium is a common and morbid condition among hospitalized elderly which may go unrecognized. Insidious metabolic derangements may lead to subtle cognitive changes and inattention, rather than the florid symptoms displayed by Mr. Pruitt. About one-quarter of elderly patients who are admitted to the general medical or surgical services may be delirious (Rockwood 1989). Even when the diagnosis is excluded at the time of initial hospitalization, nearly one-third of those over 65 will develop a delirium during the course of their hospital stay (Schor et al. 1992). Delirium presents major management challenges. Patients with delirium are at high risk for iatrogenic complications from medications and physical restraints. Delirium is associated with increased mortality and lengthened hospital stay. Predisposing factors for delirium include advanced age, vision and hearing impairments, polypharmacy, psychosocial stress, and intracranial disease, such as stroke or dementia (Lipowski 1989). The differential diagnosis of delirium includes infections (especially pneumonia), cardiac conditions (especially congestive heart failure or myocardial infarction), medications, and metabolic derangements (Lipowski 1989; Rockwood 1989). Many of the conditions that can commonly afflict older persons, such as urinary tract infection, fecal impaction, and hip fracture, can adversely affect mental status. In the frail elderly person, almost any illness and a vast array of drugs can cause a delirious state (Schor et al. 1992). People with dementing illnesses may be particularly susceptible to delirium. The appearance of a delirium in the face of an otherwise relatively trivial illness, such as an upper respiratory tract infection or cataract surgery, may indicate that a person has an underlying dementia.

There are several key features that identify a person who is suffering from a delirium. Persons have a sudden change, usually over hours to days, in the ability to pay attention to conversation or do tasks that require mental focus. They may suffer from visual hallucinations. The condition tends to fluctuate during the course of the day, with persons sometimes being relatively lucid and co-operative while at other times being obstreperous and agitated. A classic example of delirium is delirium tremens, which can develop in alcohol withdrawal. Not all persons with a delirium, however, are agitated, as is commonly the case with delirium tremens. Some people may be quiet, withdrawn, and seem to be doing relatively well according to hospital staff, only to be found to be totally confused and unable to care for themselves.

'Dr. O'Rourke, can you tell me why Mr. Pruitt became delirious?'

'Well, Ben, there are a number of possible reasons. Let's go through some of the points of his history. First, he has a history of cardiovascular disease with hypertension and angina. A history of cardiovascular disease should alert you to the possibility of cerebrovascular disease as well, even in the absence of definitive history of stroke. In some people who have had a stroke, they can become quite confused when they are also systemically ill. I am not saying that this is the case for Mr. Pruitt, but we need to keep it in mind.'

'Next, let's think a bit about his medications. All of his medications, amitryptiline, carisoprodol, and sumatriptan can cause confusion. Amitryptiline is a tricyclic antidepressant that is sometimes used in migraine. It has very potent anticholinergic properties and is associated with increased rates of confusion, especially in the elderly. Carisoprodol is a muscle relaxant. It can contribute to a confusional state. Sumatriptan can worsen cardiovascular disease and has also been reported to cause mental confusion. Again, I do not know if any of these are the culprit with Mr. Pruitt.'

'Some of his social habits raise a few questions about whether they might have contributed to Mr. Pruitt's acute confusion. Alcohol is often a contributor to cognitive impairment. Although Mr. Pruitt is supposedly a social drinker, that can mean different things to different persons. It is important to obtain corroborative information regarding alcohol consumption from a reliable source other than the patient. He is also a smoker, which increases his risks for cerebrovascular disease. Mr. Pruitt's nighttime marijuana habit is intriguing. Marijuana may persist in the body for days to weeks and could contribute to confusion. He uses the marijuana for poor sleep. The insomnia could have a variety of causes, including all the coffee he drinks, or it may be a sign of depression. Mr. Pruitt has had a sexual relationship in the past year since his wife died and did not use a condom. Although I doubt that Mr. Pruitt has an acquired immune deficiency syndrome (AIDS) associated dementia, the history of unprotected sex puts him at risk for sexually transmitted diseases, including human immunodeficiency virus (HIV). This is a good example of the fact that older persons are still sexual persons and that one cannot assume that older men and women are celibate, monogamous, or exclusively heterosexual.'

'Mr. Pruitt's family history is positive for migraine, suicide, arthritis, and stroke. Migraine tends to run in families. Complicated migraines occur with focal neurologic deficits which usually resolve. Occasionally, persons with complicated migraines will have a stroke. Mr. Pruitt, however, appears to have 'simple' migraine without any associated neurologic symptoms. The history of suicide in the family is worrisome, not so much for Mr. Pruitt's current presentation as it is something to be mindful of, for his future. An older white man, with recent loss of a spouse, living alone, is at particular risk for suicide.'

'Mr. Pruitt's social history gives us some important information. He has a variety of interests and although he seems to have had a tough time since his wife has died, I definitely get the sense of an active, involved man who has suddenly changed. It may turn out that there is an underlying cognitive problem, some sort of dementia, going on with Mr. Pruitt. Clearly, though, he has had a major change from a high level of function. I think that it is most reasonable to conclude that he has a delirium.'

'Given what we know so far about Mr. Pruitt from reading his chart, I would bet that his delirium is due to the pneumonia. We will need to review his labs and keep in mind some of the other points of his history. But my first plan would be to treat the pneumonia and see what happens.'

'Do you have any questions, Ben?'

'Well, Dr. O'Rourke, I was wondering if there are any clinical observations or maneuvers that can help with the diagnosis of delirium?'

'Sometimes delirium is fairly florid, Ben, and the diagnosis is not very hard to make. But many times the diagnosis depends on careful observation and a high index of suspicion. One trick that I have learned is to be particularly careful with patients from whom I have difficulty getting a history. By this I do not mean talking about difficult subjects, but when a person will not answer a simple question or seems unable to give a coherent explanation. Cognitive and behavioral states are sensitive markers of recovery. Is the patient attentive and appropriate with the staff? Does the patient remember and more or less comply with the treatment plan? Does the patient's behavior fluctuate throughout the day and evening? A simple bedside test that can be useful for following the severity of delirium is a forward digit span task. Ask the patient to repeat a series of numbers after you. Begin with three digits and increase the length of the series by one digit after each successful repetition.

When a patient is unable to correctly repeat five digits, it is indicative of inattention. Inattention is a primary characteristic of delirium. Even patients with moderately advanced dementia retain the ability to repeat at least five digits. A reduced digit span is a red flag that something bad is going on, usually a metabolic or toxic condition, and there is a need to look carefully at the patient.'

'Let's see what's happened to Mr. Pruitt since admission and figure out what we can do to help.'

Dr. O'Rourke and Ben Stevens reviewed the chart. Over the next five days, Mr. Pruitt got better. His agitated behavior calmed down and his overall condition clearly improved with intravenous antibiotics. In the evenings, however, the nurses' notes commented that Mr. Pruitt seemed disoriented and sometimes wandered in the hall. Mr. Pruitt's internist felt that the pneumonia had improved to the point that it would be reasonable to switch to oral antibiotics and plan for discharge—except for concerns over the confusion. Mr. Pruitt was becoming increasingly insistent on returning home. His son, Brian, expressed concerns that his father was not ready to return home, and worried that his dad would not be able to drive safely or live independently.

Dr. O'Rourke and Ben read the note in the chart from Mr. Pruitt's internist this morning:

'Mr. Pruitt is adamant on returning home but I am concerned about his episodes of confusion and about his ability to care for himself. I will obtain a geriatric consultation to help evaluate his mental status, determine if it is safe for Mr. Pruitt to go home, and sort out some of the issues for his ongoing care. I am also concerned that Mr. Pruitt may be depressed, considering his complaints of difficulty sleeping and the recent loss of his spouse.'

What is the point of a geriatrics consultation? Medical issues are important. Functional and psychosocial issues, however, take on greater prominence in geriatric medicine than is commonly the case in the other medical models. Ideally, a multi-disciplinary team of specialists in geriatric medicine, functional assessment, and psychosocial assessment and management work together to co-ordinate an optimal care plan. This team can consist of a physician and a nurse trained in geriatrics, a physical therapist, an occupational therapist, and a social worker. The lack of such a team, however, does not mean that an individual physician cannot do a thorough assessment or rely on the assistance of other professionals.

'Tell you what, Ben,' Dr. O'Rourke said, getting up from the nurse's station where he and the medical student had been reviewing the chart, 'I have a busy schedule of out-patients this afternoon. I would like you to do this consult and speak with me later in the day. We won't get it all done today, but I think that you can get a good handle on the issues.'

'OK Doctor, I'll get going.' Ben watched while Dr.

O'Rourke headed down the hallway. He began to feel a bit nervous, wondering if he could figure out what was going on with Mr. Pruitt. 'Hmmm . . . I wonder just what are the issues with Mr. Pruitt,' Ben thought as he looked at the open chart before him. 'I guess there are four big questions that I need to address. First, can Mr. Pruitt live independently? Second, what about Mr. Pruitt's wishes? He wants to leave now. Can we stop him from leaving the hospital? Should we try to stop him? Third, what has really been going on with Mr. Pruitt at home? What has his function been like and what sort of functional ability does he have now? Fourth is a big question. What is Mr. Pruitt's diagnosis? Does he still have some element of a delirium? Is depression playing a role? Could he have a dementing illness underlying some of his problems?'

Ben continued to review Mr. Pruitt's chart and make some notes. Then he prepared to examine Mr. Pruitt and hoped to be able to clarify the issues that he had identified.

Physicians are frequently asked whether a patient can live independently. No one can guarantee the safety of a patient on discharge from the hospital. There are, however, some considerations that can help to clarify issues of independence and safety. Some typical concerns are staying at home without supervision, driving an automobile, and using heavy equipment (as Mr. Pruitt would be in his furniture shop). Such complex skills are referred to as instrumental activities of daily living (IADLs) and require learning, experience, and the ability and motivation to plan. Other IADLs include cooking, cleaning, shopping, washing clothes, maintaining the house, managing money, using the telephone, taking medications, and using public transportation (Lawton and Brody 1969). Many individuals will be able to perform the more basic activities of daily living (ADLs), which are transferring from bed to chair, toileting, grooming, bathing, dressing, and feeding, but will have difficulties with IADLs. A person who is unable to perform ADLs is not safe to return home alone and will require some type of assistance and, perhaps, placement in an assisted-living situation. Significant deficits in IADLs may likewise require further assistance or other discharge plans. A careful functional assessment and the use of formal assessment tools can serve to identify service needs including criteria for admission and discharge to a variety of care settings, and to monitor clinical improvement or decline.

Direct observation of the patient in his or her own setting, such as at home, at work, in the car, etc., is the best way to assess a person's functional capabilities. Clinicians often rely on reports from the patient or family about the patient's ability to perform designated skills (Lawton and Brody 1969). Such data collection may be efficient, but the information may not be accurate. In general, families tend to underestimate and patients tend to overestimate their capabilities when compared with direct observation by trained staff of functional skills (Rubenstein et al. 1984). Direct observation of a patient allows for evaluation of the interaction of medical and psychosocial factors that may have an impact on functional status and can lead to appropriate management interventions (Ramsdell et al. 1989).

Functional impairments have a variety of causes. Common age-associated conditions can dramatically impair function, such as Alzheimer's disease, Parkinson's disease, stroke, depression, arthritis, coronary artery disease, and visual as well as hearing impairments. Hearing loss is also independently associated with decreased mobility, although the reasons for this relationship remain undetermined (Bess et al. 1989). Slowness in task performance is a useful predictor of long-term care needs (Williams 1987). Functional decline may help to identify dementing disorders much earlier and more accurately than by traditional clinical assessments (Morris et al. 1991).

Cultural standards also play important roles in functional ability. IADLs are influenced by educational, cultural, and gender differences. For example, cooking and laundering have typically been the responsibility of women. Home maintenance and driving, however, are performed more often by men, especially in the current cohort of elderly. The loss of a spouse can leave the survivor vulnerable and in difficult straits. Regardless of the functional abilities of the individual, cultural standards and social support systems play critical roles in the determination of the need for home services or nursing home placement.

Treatment teams, with skills in functional assessment and knowledge about the available social services, are invaluable in caring for the functionally declining elderly patient. In the event that such a team is not available, the practitioner should observe at least those functions that are of concern to the patient or family. If the services are available, the patient may also be referred for home and driving assessments. In conducting evaluations about the ability of a person to return home, reasonable and acceptable alternatives must be sought when the results of the assessment suggest that a person would not do well living alone or without further assistance. Sometimes this may mean nursing home placement. Semi-independent living arrangements may also be pursued. Visiting nurses, home health aides, and food delivery programs may provide enough support to maintain a frail person at home for extended periods.

Mr. Pruitt's insistence on returning home when

there are significant questions about his ability to manage safely raises a variety of ethical and legal concerns. Patients can refuse treatment and need not agree with the recommendations of physicians. Such refusals, however, should be informed refusals and should provide evidence that the patient understands the physician's recommendations, the alternatives, and the risks. Individuals are allowed to make their own choices, even if they appear to be bad choices from the viewpoint of physicians and other health care professionals. An older person who understands the risks and does not pose a significant danger to others may choose to go home despite the clinical impression that the patient will do poorly at home. If the treatment team feels that the risk is great, the physician may ask to have the patient sign out of the hospital against medical advice (AMA) to indicate that the team disagrees strongly with the patient's decision. The situation is different, however, when a person appears to be incapable of appropriate decision making. For example, it is inappropriate to allow a delirious person to leave the hospital in the middle of the night against medical advice. When a patient appears unable to make appropriate decisions regarding medical care, the physician can act without the patient's permission, if the situation requires action in order to save the life of the patient. This decision is made easier when a family member is willing to take responsibility for approving such action.

Decisions involving the safety of discharge and the capability of patients to decide for themselves are often ambiguous. In the case of a life-threatening illness or when a person is a risk to others, physicians normally act to preserve life. This may require a court decision that the person is incompetent to decide his or her own affairs, and that appointment of a legal guardian to make decisions on behalf of the patient is warranted. More commonly, however, persons with cognitive deficits are capable of decision making in some areas, but not in all. Because of an inability to understand the ramifications of a decision, a patient may be incapable of handling finances or making a decision to return home. That same patient, however, may be capable of refusing an invasive procedure. In common practice, concerns about decision making capability and competence are usually raised in the context of a patient who chooses a course which counters medical judgement. When there is a supportive family, clinicians and family members can attempt to work out solutions that keep the person safe and provide him or her with maximum independence. This is not always possible. A confused person who leaves the gas on in an oven cannot be allowed to blow up an apartment building. Creative solutions, such as disabling the stove and providing home meals, can provide alternatives to forced nursing home placement and legal guardianship.

Ben Stevens went to Mr. Pruitt's room. Mr. Pruitt was initially pleasant and interactive, but he soon became quite irritable. Ben proceeded as quickly as he could, figuring that if he missed some portion of the exam he could return at a later time. General medical examination was remarkable for bilateral cataracts and râles and rhonchi at the right lung base. Neurologic examination was notable for bilaterally decreased hearing. Ben tried to do a Mini-Mental State Examination (MMSE) (Fig. 3.1) but Mr. Pruitt became angry. It was clear that he was inattentive, had a poor memory, and had mild word-finding problems. His digit span was reduced, being able to do only four digits forward.

Ben called Mr. Pruitt's son, Brian, and asked him about how he thought his father had been doing. Brian revealed that his father had left the gas stove on in his house twice in the past six months. While driving, he had also become confused at intersections but had not had any accidents. Once, Mr. Pruitt had gone the wrong way down a one-way street. Brian said that his father maintained his own finances but he did not know how well that was going. He also revealed that his father had been very upset recently when a customer refused to accept or pay for a piece of custom furniture that Mr. Pruitt had built. The son was not sure of the basis for the conflict. Brian expressed concern over his father's eating habits. He was not eating well and rarely attempted to cook for himself, despite having been the primary cook of the family.

Ben returned an hour later and tried again with the MMSE. Mr. Pruitt was more co-operative this time. He was oriented to all questions. He lost three points for errors on serial 7s (a task where the patient is asked to subtract the number 7 from 100 and continues to subtract 7 from each answer, up to five subtractions, i.e., 100 . . . 93 . . . 86). He also lost one point when being tested for recent memory. (The patient is asked to repeat three unrelated words, like 'cup', 'pencil', and 'airplane', and then to recall them a few minutes later.)

Observing that Mr. Pruitt seemed to be in a fairly good mood, Ben decided to try some other cognitive testing that he had learned about by spending time on this rotation with the neuropsychologist. He found that Mr. Pruitt had difficulty learning new information from both verbal and visual sources. There was a mild problem in naming items. In copying complex drawings, Mr. Pruitt revealed deficits in visuospatial skills with sloppy, imprecise figures that showed little attention to detail. On the Boston Naming Test, Ben showed Mr. Pruitt drawings of common items. Mr. Pruitt had moderate naming problems, even for items of his trade, calling a hammer 'that thing you hit a nail with'.

Ben asked Mr. Pruitt some questions about his mood and how he was feeling. He reviewed the criteria for depression, easily remembered by the mnemonic

SIG E CAPS, for sleep, interest, guilt, energy, concentration, appetite, psychomotor changes, and suicidality [Jenike 1989]. Mr. Pruitt admitted to difficulties with sleep, lack of interest, lack of energy, and difficulty concentrating. He denied feelings of guilt but did seem to ruminate about his wife's death and how difficult he found it to care for her when she was ill. He denied suicidality, saying how angry it had made him that his mother had taken her life, and that he would not do that to his son.

Ben met Dr. O'Rourke at 5 p.m. and presented his findings. He felt that Mr. Pruitt still had some evidence for a delirium but he was also concerned about an underlying dementing illness. He also thought Mr. Pruitt was depressed. Ben asked Dr. O'Rourke about his impression and how to address the questions of further diagnosis, testing, and discharge planning.

'Well, you did a good and thorough job,' Dr. O'Rourke began, 'geriatric consultations take a long time since there is so much information to gather and the questions being asked are usually quite broad. I am particularly pleased that you followed up on the MMSE. As you know, the MMSE is an easy screening device that investigates multiple areas of cognitive function: orientation, registration, recent memory, attention, naming, reading, following commands, and visuospatial ability. It is scored on a scale of 30. It is a common belief that a MMSE score of 25 or greater is normal. That's not correct. You can have individuals who will have a 30 who still have a problem. The score itself, however, is much less telling than the items missed. Missing the exact date may have little significance. But forgetting one of the memory items may be very significant. In addition, in a man with a college education who has been in business for himself, the mistakes on the serial 7's are also likely to be meaningful. In Mr. Pruitt's case, the errors may be secondary to inattention, poor motivation, depression, language impairment, or confusion. The further testing you did, and the information you got from the son, makes me concerned about a dementing illness.'

'The differential diagnosis for cognitive decline in an adult includes depression, delirium, dementia, and the amnestic syndrome. I agree with you that Mr. Pruitt shows evidence of a resolving delirium and I also think he might be depressed. Let's talk about the other possibilities.'

Cognitive function remains relatively stable with normal aging. Reduced neuronal reserve in aging brains, with the loss of cortical neurons, may increase the sensitivity to cognitive side-effects of medications or metabolic aberrations, which may cause acute cognitive dysfunction or delirium. Persistent cognitive decline characterizes the syndrome of dementia. Alzheimer's disease is the most common etiology of dementia. Significant dementia occurs in approximately 11% of the population over 65 years of age and in nearly 60% of nursing home residents. Of those diagnosed with dementia, up to 30% have been reported to have reversible conditions, 45–70% have dementia of

Alzheimer type (Cummings and Benson 1992) and 10–25% are related to cerebrovascular disease. Nearly 25–50% of community-dwelling, over 85-year-olds meet the clinical criteria for Alzheimer's disease (Evans *et al.* 1989). Some physicians consider dementia to be over-diagnosed, claiming that cognitive decline is psychiatric 20% of the time (Garcia *et al* 1981). In follow-up, however, many of the so-called psychiatric cases evolve into classic dementia. Multiple causes of cognitive dysfunction may coexist and a number of dementias are potentially reversible. Table 3.1 lists some reversible or contributing factors to cognitive dysfunction.

The *Diagnostic and statistical manual of mental disorders* (4th edition) (DSM-IV, APA 1994), was developed to clarify and standardize behavioral and diagnostic terminology by providing explicit inclusionary and exclusionary criteria. The DSM definitions are widely but not universally accepted because they may be overly restrictive (Cummings and Benson 1992). In the interest of uniformity, our discussion will be based on DSM-IV definitions. As Dr. O'Rourke noted, the differential diagnosis for cognitive decline in an adult includes the amnestic syndrome, dementia, depression, and delirium. Depression and delirium have already been discussed.

The amnestic syndrome is characterized by memory impairment that does not occur in the context of dementia or delirium. When alcohol is the likely etiology, the syndrome is called alcohol amnestic disorder (APA 1994). In Mr. Pruitt's case, he missed one out of three memory items. In the context of his inattention, it is likely that he is unable to attend to the task closely enough to learn the new information. If, however, an individual is attentive and misses one item it should prompt further testing to see if there is subtle recent memory loss.

The diagnosis of dementia, by DSM IV standards, requires impairment of memory plus at least one of the following: impaired abstract thinking, judgement, higher cortical functions, or a personality change. In addition, the memory and/or other changes must be substantial enough to interfere with function either at work or socially. The dementia syndromes characterized in DSM-IV are dementia due to Alzheimer's disease, dementia due to cardiovascular disease, and dementia due to other medical conditions (APA 1994).

Dementia due to vascular disease, sometime referred to as multi-infarct dementia (MID), is defined clinically as dementia with a stepwise deteriorating course and a patchy distribution of deficits. In addition, there is evidence of stroke(s) by focal neurologic signs and symptoms, supported by history, physical examination, and relevant laboratory and radiologic tests. Multi-infarct dementia is part of a broader category

MINI-MENTAL STATE EXAMINATION (MMSE)

Add points for each correct response.

			Score	Points

Orientation

			Score	Points
1.	What is the:	Year?	_____	1
		Season?	_____	1
		Date?	_____	1
		Day?	_____	1
		Month?	_____	1
2.	Where are we?	State?	_____	1
		County?	_____	1
		Town or city?	_____	1
		Hospital?	_____	1
		Floor?	_____	1

Registration

3. Name three objects, taking one second to say each. Then ask the patient to repeat all three after you have said them.   _____   3

   Give one point for each correct answer. Repeat the answers until patient learns all three.

Attention and Calculation

4. Serial sevens. Give one point for each correct answer. Stop after five answers.   _____   5
   Alternate: Spell WORLD backwards.

Recall

5. Ask for names of three objects learned in question 3. Give one point for each correct answer.   _____   3

Language

6. Point to a pencil and a watch. Have the patient name them as you point.   _____   2

7. Have the patient repeat "No ifs, ands, or buts."   _____   1

8. Have the patient follow a three-stage command: "Take a paper in your right hand. Fold the paper in half. Put the paper on the floor."   _____   3

9. Have the patient read and obey the following: "CLOSE YOUR EYES." (Write it in large letters).   _____   1

10. Have the patient write a sentence of his or her choice. (The sentence should contain a subject and an object and should make sense. Ignore spelling errors when scoring).   _____   1

11. Have the patient copy the design. (Give one point if all sides and angles are preserved and if the intersecting sides form a guadrangle).   _____   1

_____ Total 30

In validation studies using a cut-off score of 23 or below, the MMSE has a sensitivity of 87%, a specificity of 82%, a false positive ratio of 39.4%, and a false negative ratio of 4.7%. These ratios refer to the MMSE's capacity to accurately distinguish patients with clinically diagnosed dementia or delirium from patients without these syndromes.

SOURCE: Courtesy of Marshall Folstein, MD. Reprinted with permission.

Fig. 3.1

Table 3.1 Reversible or contributing factors to cognitive dysfunction

Drugs	Psychotropic agents
	Neuroleptics
	Antidepressants
	Anxiolytics
	Sedative/hypnotics
	Anti-Parkinson agents
	Antihistamines
	Most over-the-counter cold medications
	Antihypertensives
	Cardiovascular agents
	Hypoglycemics
	Anticonvulsants
	Analgesics
	Other
	Alcohol
	Cimetidine
	Steroids
Toxins and heavy metals	Alcohol
	Lead
	Mercury
	Thallium
	Manganese
Depression	'Pseudodementia'
Organ dysfunction	Hepatic dysfunction
	Renal/Fluid imbalance
	Hyponatremia
	Dehydration
	Thyroid (hypo/hyper)
	Pituitary (hypo/hyper)
	Cardiovascular/low flow states
	Congestive heart failure
	Hypotension
	Arrhythmias
	Pulmonary (hypoxemia)
	Hematologic
	Polycythemia vera
	Anemia
	Pernicious anemia
Metabolic disorders	Hypo/hypernatremia
	Hypo/hyperglycemia
	Hypercalcemia
	Hyperlipidemia
	Porphyria
	Wilson's disease
Intracranial disease	Infectious
	General paresis (syphilis)
	Chronic meningitis (cryptococcal, tuberculous)
	AIDS and related infections
	Lyme disease
	Whipple's disease
	Structural
	Brain abscess
	Hydrocephalus
	Subdural hematoma
	Intracranial neoplasm
	Immunologic
	AIDS dementia
	Systemic lupus erythematosis
	Granulomatous angiitis
	Remote effects of carcinoma (limbic encephalitis)
	Multiple sclerosis
	Other
	Seizures
Miscellaneous	Sensory deprivation (e.g., intensive care unit psychosis)
	Fecal impaction
	Sepsis

of vascular dementias which includes arteriosclerotic disease of the small vessels in the deep white matter of the brain, as well as multiple embolic strokes from the large vessels of the heart. The etiology of small vessel disease is typically chronic hypertension. The description of the clinical syndrome of multi-infarct dementia is in evolution. In general, it is proposed that patients with multiple subcortical infarcts may exhibit frontal lobe and subcortical features of bradykinesia (slowed movement), bradyphrenia (slowed thought), and depression.

The diagnosis of dementia due to Alzheimer's disease is based on clinical evidence of dementia together with an insidious onset and a progressive downhill course. In general, this description suggests the diagnosis of Alzheimer's disease. Uncommon diagnoses, such as Pick's disease, also fall into this category. This description may also characterize the 'subcortical' dementia of Parkinson's disease (Cummings and Benson 1992). When significant subcortical features are seen, especially early in the course of the illness, the diagnosis of Alzheimer's disease may be less likely.

Alzheimer's disease is the most common cause of primary degenerative dementia and was clinically and pathologically described in 1907. The pathologic lesions of Alzheimer's disease are senile plaques and neurofibrillary tangles. These changes are also found in brains of older individuals with normal aging, but usually to a substantially lesser extent than is found in Alzheimer's disease. This observation has led some to suggest that Alzheimer's disease may be a manifestation of accelerated aging.

Alzheimer's disease is characterized by an insidious onset and progressive decline of cognitive, behavioral, and functional status. Recent memory loss and word-finding difficulties are the most common presenting signs. Visuospatial skills also decline, as do judgement and sequencing skills. The physical and neurological examinations are typically normal in the early stages except for subtle changes in muscle tone and gait. Laboratory studies are usually unremarkable, including computerized tomography, which may or may not show excessive cortical atrophy. As the disease progresses, aphasia becomes more prominent. Patients may get lost in previously familiar settings. At times, they may not recognize their loved ones and eventually may develop incontinence. In the end stages of the disease, they are bed-bound and lose the ability to co-ordinate swallowing. Death is usually secondary to infections associated with aspiration pneumonia, urinary tract infections, or skin breakdown.

The clinical course of Alzheimer's disease and the associated pathologic lesions are similar regardless of age of onset. The past distinction between 'presenile' (onset younger than age 65) and 'senile' (onset over

age 65) appears to be arbitrary. The National Institute for Neurological and Communicative Disorders and Stroke and the Alzheimer's Disease and Related Disorders Association (NINCDS-ADRDA) established clinical parameters for the diagnosis of Alzheimer's disease (see Table 3.2) that consider the probability of the accuracy of the diagnosis, based upon the typicality of history and other findings and the coincidence of confounding disorders (McKhann et al. 1984). There is no absolute clinical gold standard for making a definitive diagnosis of Alzheimer's disease.

The etiology of Alzheimer's disease is unknown although there are a number of theories. The trigger for Alzheimer's disease may be multi-factorial. Some major theories regarding the cause of the disease include genetically determined accelerated aging, toxin exposure, atypical slow virus infection, and head trauma. The genetic theory rests on several observations. Virtually all patients with Down's syndrome (a genetic disorder associated with abnormality of chromosome 21) develop Alzheimer's pathologic changes by the age of 35. This association between Down's syndrome and Alzheimer's pathology led to the discovery of a genetic marker on chromosome 21 for the amyloid protein prevalent in the brains of Alzheimer's patients. Another piece of evidence for a genetic predisposition to Alzheimer's disease comes from a subgroup of Alzheimer's patients with an autosomal dominant pattern of inheritance. In addition, identical twins have strong concordance in the development and age of onset of the disorder. Concordance, however, does not always occur. This suggests that some cases of Alzheimer's disease are not genetically determined or that exogenous factors trigger the development of the disease in genetically predisposed individuals. Identifying such triggers could lead to appropriate preventative measures. The triggers may be toxic, infectious, and/or traumatic.

Some investigators, noting that significantly elevated aluminum levels have been found within the brains of persons with Alzheimer's disease, postulated that aluminum acted as a toxic agent in causing the disease. Aluminum binds to degenerating neurons, however, regardless of the cause of degeneration and the high levels of aluminum appear to be an epiphenomenon. There is no increase in the prevalence of Alzheimer's disease in regions of high aluminum environmental exposure. Chelation therapy is an unproven and expensive approach offered by some clinics with a claim of reducing levels of heavy metals and thus treating a multitude of ailments including Alzheimer's disease. No controlled studies of this treatment modality have been done and there is little theoretical support for this potentially dangerous treatment.

The discovery of a transmissible agent responsible

Table 3.2   NINCDS-ADRDA criteria for diagnosis of Alzheimer's disease

A. The clinical picture is compatible with the DSM definition of dementia
B. The certainty of diagnosis is characterized by one of the following:

Probable:
1. Clinical impression of dementia is supported by neuropsychological tests.
2. Deficits in two or more areas of cognition.
3. Progressive worsening of memory and other cognitive functions.
4. No disturbance of consciousness.
5. Onset between 40 and 90 years of age.
6. Absence of systemic disorders or other brain diseases that could account for the progressive deficits in memory and cognition.

Supported by:
1. Progressive aphasia, apraxia, and agnosia.
2. Impaired ADLs and change in behavior.
3. Family history of similar disorder especially if pathologically confirmed.
4. Normal CSF, normal or non-specific slowing on EEG, atrophy on CT with evidence of progression by serial studies.

Other features consistent with the diagnosis:
1. Plateaus in the course.
2. Depression, insomnia, incontinence, delusions, illusions, hallucinations, catastrophic outbursts, sexual disorders, and weight loss.
3. Increased muscle tone, myoclonus, or gait disorder in advanced disease.
4. Seizures in advanced disease.
5. Normal CT.

Features that make the diagnosis unlikely:
1. Sudden onset.
2. Focal neurologic signs early in the course.
3. Seizures or gait problems early in the course.

Possible:
1. Diagnosis of dementia in absence of other neurologic, psychiatric, or systemic disorders that could account for the clinical picture with atypical onset, presentation, or clinical course.
2. The systemic or brain disorder present is not felt to be the cause of dementia.
3. A single progressive cognitive deficit occurring in the absence of another identifiable cause.

Definite:
Clinical evidence of probable Alzheimer's disease and pathologic confirmation by biopsy or autopsy.

Subtypes:
1. Familial
2. Onset before age 65
3. Presence of trisomy-21
4. Coexistence of other relevant conditions such as Parkinson's disease or amyotrophic lateral sclerosis

Adapted from McKhann et al. (1984)

for Jakob-Creutzfeldt dementia inspired the search for an infectious etiology to Alzheimer's disease. Another tantalizing finding is that neuritic amyloid plaques characteristic of Alzheimer's disease are seen in slow viral spongiform encephalopathies (Deutsch et al. 1982). Although Alzheimer's disease has not been shown to be infectious, proponents argue that a suitable experimental host has not been found or that the incubation period may be longer than the life span of the experimental animal.

It has been observed that boxers may develop progressive dementia long after they have discontinued fighting. The pathologic changes in the brains of some of these individuals are characteristic of Alzheimer's disease. In addition, retrospective surveys have suggested that victims of Alzheimer's disease are more likely than controls to have had a history of significant head injury (Mortimer et al. 1985).

The evaluation of a person with cognitive decline begins with a careful history from both the patient and a significant other. The history should focus on systemic disorders, medication and alcohol use, nutritional status, psychiatric or behavioral changes, functional decline, family history, and social support systems.

The physical examination and laboratory studies

should support or dispel the initial diagnostic impression. The physical exam focuses on the cardiovascular and neurologic systems. The mental status examination provides information regarding the patterns of strengths and deficits that will suggest diagnostic and therapeutic strategies. The cognitive domains of critical interest include attention, memory, executive function, and higher cortical functions, such as language and visuospatial skills (Odenheimer 1989).

The laboratory studies recommended by a National Institutes of Health consensus conference (NIH 1987) include: measures of electrolytes, glucose, calcium, phosphorus, triglycerides, cholesterol, renal, hepatic and thyroid function, vitamin B$_{12}$ and folate levels, serology for syphilis, urinalysis, chest X-ray, and electrocardiogram (see Table 3.3). A computerized tomogram (CT) scan of the head is adequate for detecting most significant structural lesions. Magnetic resonance imaging (MRI), however, may be more sensitive than CT for identifying small vascular lesions.

Dr. O'Rourke suggested that Mr. Pruitt's physicians obtain a CT scan as well as thyroid function tests, vitamin B$_{12}$ and folate levels, serology, and liver function tests. He

Table 3.3 Recommended evaluation of dementia

A. Laboratory studies
CBC with differential
Erythrocyte sedimentation rate
Electrolytes, blood urea nitrogen, creatinine, glucose
Calcium, phosphorus
Liver function tests
Triglyceride, cholesterol
Thyroid functions (including TSH)
Vitamin B$_{12}$, folate
Serology (VDRL or RPR and FTA-abs)
Urinalysis

If indicated:
Serum protein electropheresis
Arterial blood gases
Cortisol
Drug levels, toxic screen
Heavy metal testing
HIV testing

B. Radiological studies
Chest roentgenogram
Computerized tomogram of brain (CT scan)

C. Special studies
Lumbar Puncture
Electroencephalogram (EEG)
Brain biopsy
Magnetic resonance imaging of brain (MRI)
Single photon emission computerized tomogram of the brain (SPECT scan)

Adapted from NIH (1987)

reviewed the rest of Mr. Pruitt's laboratory studies and found them to be within normal limits, including electrolytes and blood gases. Dr. O'Rourke also noted that the ongoing presence of delirium made it impossible to diagnose a dementing illness with certainty, but that it was an issue of concern. Also, he felt that it was unsafe for Mr. Pruitt return home alone while he was delirious. Dr. O'Rourke further noted that even after the delirium cleared, Mr. Pruitt's history suggested that a return to home would require further supports and might be difficult to maintain for the long term.

In conversation with Ben, Dr. O'Rourke asked him to do a few more things. 'I would like you to check on Mr. Pruitt over the next several days and see what happens with his cognitive state. I think his delirium will basically clear. It is important to try and do a functional assessment. Go for a walk with Mr. Pruitt around the hallway, check his endurance, range of motion, and co-ordination. If you can, go see him at meal time and see how he does with food. You should also ask the nurses how he is doing with using the toilet, whether he is continent or not, and how he does with bathing and dressing. I will speak with Mr. Pruitt's internist and see if we can arrange a meeting with his son and find out what he is able to do to help his dad.'

Ben, with the assistance of the physical therapist, found that Mr. Pruitt did well walking and had fair endurance. Upper extremity range of motion, strength, and co-ordination were all normal. Mr. Pruitt was independent in feeding himself but had a poor appetite. He was independent in dressing but needed assistance with bathing. The nurses noted that Mr. Pruitt had been incontinent of bowel and bladder in the first days of his admission but that the incontinence had cleared with the improvement of his mental status.

Repeat mental status testing showed improved attention with a digit span forward of 6, although deficits persisted in executive tasks, memory, naming, and visuospatial functions. The CT scan showed mild bifrontal atrophy as well as a small lacunar infarct in the left subinsular cortex.

At the family meeting two days after the initial consultation, Dr. O'Rourke spoke with Mr. Pruitt and his son. He asked Mr. Pruitt how he felt his memory was and was not surprised that Mr. Pruitt was aware of some deficits. Dr. O'Rourke listed his concerns about some areas of function: medication administration, finances, shopping, food preparation, homemaking, and transportation. Because of the problems with driving as noted by his son, Dr. O'Rourke felt that Mr. Pruitt was an unsafe driver and strongly recommended that he stop driving. Mr. Pruitt was adamant that he wanted to return home and his son Brian expressed a willingness to help him with that goal if it was not unrealistic. Dr. O'Rourke outlined the need for assistance with medications, shopping, food preparation, homemaking, and transportation. He said that a social worker would assist in making the appropriate referrals to obtain a homemaker and visiting nurse. The visiting nurse would make a home visit and see if there were other safety concerns in the house.

In addition, the gas stove should be disabled or removed and some other less dangerous cooking device, such as a microwave oven, could be installed. Dr. O'Rourke suggested that Mr. Pruitt consider allowing his son Brian to be designated his power of attorney, to handle his finances and assist with legal needs. It would also be worthwhile for Mr. Pruitt to discuss his wishes regarding health care in the event of a life-threatening illness and name someone as a proxy in the event that he was unable to participate in the decision making. Brian said he was willing to help out his father with these tasks and would make a point of continuing to check on him daily.

Mr. Pruitt appreciated the assistance that was being offered and agreed that he needed help with his finances. He expressed his thanks to his son for all the help that he was offering. He did not, however, agree with the suggestion that he stop driving. Dr. O'Rourke reiterated his recommendation that Mr. Pruitt not drive and suggested two alternatives: he could be tested by the state department of motor vehicles or he could be referred to a local rehabilitation hospital that had a program for checking the skills of older drivers. Brian told his father that he would take the keys away from him until he went to the rehabilitation facility for testing.

In further discussion with Mr. Pruitt's primary care physician, Dr. O'Rourke suggested that Mr. Pruitt might have a depression and that a psychiatric consultation could be helpful. It would be important to follow Mr. Pruitt closely.

After the family meeting, Ben asked Dr. O'Rourke a few questions.

'Why did you have the meeting with Mr. Pruitt present?'

'Well, I don't always do that, but I try to because I think that it is important that the patient not be left out. After all, it is his life we are talking about. I gave him the opportunity to talk about his memory difficulties but he did not ask me about the possible diagnosis. I will make a point of speaking by phone with his son and let him know that I am concerned that his father has a dementing illness which may be Alzheimer's disease.'

'Dr. O'Rourke, things seem to be working out fairly well in that Mr. Pruitt's son is willing and able to help his father and his father seems to be willing to allow him to help. What would you have done if there was not a supportive family, or if Mr. Pruitt insisted on going home without any supports?'

'Both those possibilities would be very problematic. First, if you have a patient with a cognitive decline without family, or with a family that will not help out for whatever reason, it is very hard to allow that person to go back into the community. A person without support structures can end up in a nursing home much earlier than another person who might be much more seriously impaired but who has people to help him or her. Second, if Mr. Pruitt had insisted on leaving the hospital and going home without any help,

we would have had to decide if the situation could be so dangerous that we would need to go to court, ask that he be declared incompetent, and appoint a legal guardian. Luckily, Mr. Pruitt seems to recognize the need for help and is willing to allow his son to have legal control over his affairs. Some provision for control of finances and legal business and for setting up an advance directive for health care are very important items to discuss with patients and families.'

'One last question about Mr. Pruitt. What should be done to optimize his care?'

'Good question, Ben, that requires a careful plan. Let me explain.'

Optimizing the care of demented or delirious individuals includes attention to medical, functional, and psychosocial aspects of an illness. Medical issues include: (1) the prevention or slowing of progression of the disorder; (2) treatment of 'reversible' dementias; 3) clarification of treatment goals; (4) treatment of intercurrent illness; (5) symptomatic treatment; (6) experimental approaches; and (7) ongoing care (Table 3.4) (Odenheimer 1989).

1. *Prevention or slowing of progression*: general good health practices may improve or protect cognitive function. These practices include adequate nutrition, exercise and rest, minimal use of medications, alcohol, tobacco, or other drugs, as well as treatment of conditions known to cause or contribute to cognitive decline, such as vitamin B_{12} deficiency, hypothyroidism, hypertension, hypercholesterolemia, and cardiac arrhythmias.

2. *Treatment of 'reversible' dementias*: 10–30% of all dementias may have reversible etiologies. In a review of 200 out-patients referred for a dementia evaluation, 30% improved with appropriate intervention during the first year of follow-up (Larson *et al.* 1985). The most common treatable causes were drug toxicity, depression, hypothyroidism, and other metabolic diseases. In delirium, the underlying metabolic derangement is identified in 80–95% of the cases (Lipowski 1983). The treatment of potentially reversible factors is superficially straightforward: correct the metabolic or deficiency states, stop the offending drug, drain the subdural hematoma, treat the depression, etc. Unfortunately, these interventions may only partially restore function or prevent further cognitive decline. Although some cognitive deficits reverse quickly after appropriate therapy is instituted, many older patients may not achieve their optimal responses to treatment for many months.

3. *Clarification of treatment goals*: as dementia progresses, treatment goals often change. The patient

should be involved in the discussion about treatment goals. Plans should be made to establish an advance directive for the time when the patient will not be able to participate in decision making. Plans for treatment of intercurrent illnesses or symptoms are best made in the context of the dementing illness. Thus, more aggressive care may be appropriate early in the course of the dementia and a palliative approach more appropriate at later stages.

4. *Treatment of intercurrent illness*: treating intercurrent illnesses or coexisting problems can optimize the residual cognitive potential of the patient and stabilize quality of life for the patient and caregiver. A number of common, relatively easily treated, problems in demented patients are often overlooked: visual or hearing impairments, urinary tract infection, fecal impaction, pneumonia, seizures, depression, cardiac, and pulmonary disease.

5. *Symptomatic treatment*: the symptoms associated with dementia that typically come to medical attention include anxiety and depression, delusions and hallucinations, agitation and aggressive behavior, insomnia and nocturnal wandering, as well as lethargy. Medications that are used for behavioral control are often associated with increased confusion and the risk of falls. Response to these medications is highly variable and may be related to the sedative effects of psychoactive agents. Caution must be exercised in the use of any sedative or anticholinergic drug because of the risk of further impairing cognition (Cummings and Benson 1992). The side-effects must be weighed against the possible benefits of the drugs. This is often determined through trial and error.

6. *Experimental approaches*: a short-acting benzodiazepine such as oxazepam (Serax) is a reasonable choice for treating anxiety in the elderly. Buspirone (Buspar) is not a benzodiazepine and may also be effective in treating anxiety but needs further investigation (Shader *et al.* 1987). Depression can be successfully treated with a variety of antidepressants or, in severe cases, with electroconvulsive therapy. The choice of an antidepressant agent is based on its side-effect profile. A patient who is agitated may benefit from a sedating antidepressant, such as doxepin (Sinequan, Adapin) or trazodone (Desyrel). A lethargic or hypokinetic depressive patient may benefit from an activating antidepressant such as desipramine (Norpramin) or fluoxetine (Prozac). Thioridazine (Mellaril) or haloperidol (Haldol) at bedtime are both popular choices for the management of delusions, hallucinations, and agitation. Agitated and aggressive behaviors are particularly difficult to treat. Besides

major tranquilizers, short-acting benzodiazepines may sometimes provide benefit. Lithium and valproic acid have been reported to be useful in treating aggression. Data in demented patients, however, are not compelling. Propranolol has been used in the management of aggression in head injury patients with promising anecdotal results and its use in degenerative dementia is under investigation. Sleeplessness may respond to short courses of diphenhydramine (Benadryl), chloral hydrate, or a short-acting benzodiazepine like temazepam (Restoril). Wandering is best managed by environmental cues such as a 'STOP' sign on a door, a complicated lock, or an alarm system. Stimulants, such as methylphenidate and amphetamines, are occasionally used to treat lethargy in advanced stages of dementia.

The management of Alzheimer's disease remains principally supportive. No cure is known. The use of drugs to alter the natural history or the clinical picture has been largely unsuccessful. The first drug approved for the treatment of Alzheimer's disease is tacrine (Cognex). It has been shown to produce modest improvement in a small group of patients. Long-term benefits have not been demonstrated.

7. *Ongoing care*: clinicians often minimize the follow-up of patients with incurable diseases. Close follow-up, however, is warranted in order to provide appropriate medical interventions for patients and therapeutic support for caregivers. In addition, the types and extent of interventions will often be modified as the disease progresses.

Functional issues underlie the concerns of patients and families. Can the patient drive safely or continue to work at technical or dangerous occupations? Can the person live alone or even be left alone for any length of time? Will the person remember to take medications or even remember to eat? Issues like these must be specifically addressed by the clinician or critical aspects of the impact of the disease will be missed. Specialists in rehabilitation, such as occupational therapists or geriatric nurse practitioners, can offer strategies that maximize preserved function in the effort to improve and maintain independence for prolonged periods. A home visit can be invaluable for determining appropriate environmental adaptations for enhancing independence. Vocational evaluations may help patients find work with reduced demands. Driving assessments can help determine situations that patients should avoid or when driving should be discontinued altogether.

The psychosocial aspect of dementia is enormously important. Tremendous stress befalls the family of patients with dementing illnesses. Support groups and counseling may help caregivers cope or find assistance in caring for themselves in addition to the patient. There are numerous legal considerations that must be addressed as the patient deteriorates, such as preparing a will, designating power of attorney, or assigning legal guardianship. A social worker may be better prepared to deal with these issues than a physician, and is invaluable in identifying alternative living arrangements or home assistance. If patients retain the capability of understanding the issues, they should be included in these discussions. Individuals with dementia may also benefit from the opportunity to discuss their feelings regarding their illness.

Over the next year, Mr. Pruitt showed more signs of an underlying dementing illness. He had increasing difficulty with word-finding and showed worsening of his short-term memory. Because of his history of hypertension and the presence of a lacunar infarct on his CT scan, it was difficult to be dogmatic that he was suffering from Alzheimer's disease, as there is likely a component of vascular dementia. The progressive course, in the absence of recognized vascular events, argued for the possibility of Alzheimer's dementia, although the two could coexist (mixed dementia).

Despite worsening of his cognitive function, there were some improvements in Mr. Pruitt's living situation. Home supports were in place with a homemaker, visiting nurse, ongoing visits from his son, and home delivery of meals. Mr. Pruitt's primary care physician, with the consultation of a psychiatrist, discontinued his amitryptiline and began sertraline with good results. His spirits improved and he was less cranky. Decaffeinated coffee was substituted and Mr. Pruitt showed improved sleep habits. Mr Pruitt cut back on cigarettes and discontinued marijuana use. Mr. Pruitt's physician began low dose daily aspirin and started antihypertensive therapy. The visiting nurse helped to monitor the medications.

Mr. Pruitt, after failing his driving test at the rehabilitation hospital, had initially been very upset about the loss of his car. With the passage of time, however, he seemed to forget the loss. He would, however, complain about the limitation on occasion. Mr. Pruitt's son Brian had made a number of arrangements with his father and the family lawyer, including establishing durable power of attorney and a health care proxy. He had also begun to consider the need in time for nursing home care for his father given his slow but progressive deterioration. Brian had also consulted a therapist himself, as a way to deal with the stress of caring for his father.

Summary

Decline in cognitive function is common in the elderly. Although available therapeutic agents have not lived up to our expectations, there are many approaches that

Table 3.4 Principles of management of patients with cognitive dysfunction

I. Medical principles
  A. Prevention
    1. *General health considerations*
       Maintain physical and mental activity
       Nutrition
       Avoid excess alcohol and/or tobacco
    2. *Treat systemic medical conditions*
       Replace deficient vitamin B_{12}
       Treat thyroid excess or deficiency
       Control hypertension
       Aspirin in cardiac or cerebrovascular disease
       Control arrhythmias
       Anticoagulation in atrial fibrillation
       Minimize psychoactive drugs
  B. Treat reversible etiologies
  C. Clarify treatment goals
  D. Treat intercurrent illness
  E. Symptomatic treatment
     Anxiety/depression
     Delusions/hallucinations
     Agitation/aggression
     Insomnia/wandering
     Lethargy
  F. Experimental treatment
  G. Ongoing care
     Medical/pharmacologic changes
     Cognitive/physical changes
     Functional/psychosocial changes

II. Functional principles
  A. Emphasize strengths
  B. Assess the environment and suggest adaptations
  C. Avoid stressful situations (crowds, travel)
  D. Use memory aids
  E. Prepare the patient and family for change
  F. Vocational and driving assessments when indicated

III. Psychosocial principles
  A. Family counseling
     Identification and resolution of family conflicts, anger, and guilt
     Placement or home care options
     Legal/Ethical concerns (guardianship, power of attorney, wills, driving)
  B. Social service resources
     Community health care resources
     Legal and financial counseling
     Support groups
     Reference list of readings

can greatly enhance the quality of the lives of victims of dementing disorders and their families. Optimal management of the demented individual is based on addressing medical, functional, and psychosocial factors (see Table 3.4). Strength to cope with this devastating process is bolstered by ongoing clinical care and family participation in support groups. Medical treatment of dementia is aimed at identifying and reversing those conditions that could be causing or contributing to intellectual decline. The use of medications in managing behavioral disturbances is difficult and often leads to serious side-effects. Ongoing care of individuals with dementing illness requires careful planning for the patient's future and awareness of the stress on family members.

Questions for further reflection

1. What clinical features may help to distinguish delirium from dementia?
2. How would one care for an acutely confused older person in the hospital setting?
3. An 84-year-old man presents with his wife for evaluation of a progressive memory loss over the last two years. He recently became lost driving back from church to his home of the last forty years. Consider what further history would be important to obtain, points of emphasis on the physical and neurological examinations, and appropriate laboratory and ancillary studies to order.

References

APA (American Psychiatric Association) (1994). *Diagnostic and statistical manual of mental disorders*, (4th edn), pp.123–63. American Psychiatric Association, Washington, D.C.

Bess, F., Lichtenstein, M.J., Logan, S.A., Burger, M.C., and Nelson E. (1989). Hearing impairment as a determinant of function in the elderly. *Journal of the American Geriatrics Society*, 37, 123–8.

Cummings, J.L. and Benson, D.F. (1992). *Dementia: a clinical approach* (2nd edn). Butterworth–Heinemann, Boston.

Deutsch, S.I., Mohs, R.C., and Davis, K.L. (1982). Theoretical note—a rationale for studying the transmissibility of Alzheimer's disease. *Age and Ageing*, 3, 145–7.

Evans, D., Funkenstein H., Albert M., Scerr. P., Cook, N., Chown, M., *et al.* (1989). Prevalence of Alzheimer's disease in a community population of older persons. *Journal of the American Medical Association*, 262, 2551–6.

Garcia, C.A., Reding, M.J., and Blass, J.P. (1981). Over diagnosis of dementia. *Journal of the American Geriatrics Society*, 9, 407–10.

Jenike, M.A. (1989). *Geriatric psychiatry and psychopharmacology*, pp. 34–6. Year Book Medical Publishers, Chicago.

Larson, E.B., Reifler, B.V., Sumi, S.M., Canfield, C.G., and Chinn, N.M. (1985). Diagnostic evaluation of 200 elderly outpatients with suspected dementia. *Journal of Gerontology*, 40, 536–43.

Lawton, M.P. and Brody, E.M. (1969). Assessment of older people: self-maintaining and instrumental activities of daily living. *The Gerontologist*, 9, 179–86.

Lipowski, Z. (1983). Transient cognitive disorders (delirium, acute confusional states) in the elderly. *American Journal of Psychiatry*, 140, 1426–36.

Lipowski, Z. (1989). Delirium in the elderly patient. *New England Journal of Medicine*, 320, 578–82.

McKhann, G., Drachman, D., Folstein, M., Katzman, R., Price, D., and Stadlan, E.M. (1984). Clinical diagnosis of Alzheimer's disease: report of the NINCDS-ADRDA work group under the auspices of Department of Health and Human Services Task Force on Alzheimer's disease. *Neurology*, 34, 939–44.

Morris, J.C., McKeel, D.W., Storandt, M., Rubin, E.H., Price, J.L., Grant, E.A., et al. (1991). Very mild Alzheimer's disease: informant based clinical, psychometric, and pathologic distinction from normal aging. *Neurology*, 41, 469–78.

Mortimer, J.A., French, L.R., Hutton, J.T., and Schuman, L.M. (1985). Head injury as a risk factor for Alzheimer's disease. *Neurology*, 35, 264–7.

NIH (National Institutes of Health) (1987). *National Institutes of Health consensus development conference statement: differential diagnosis of dementing diseases*. U.S. Department of Health and Human Services, Bethesda, MD.

Odenheimer, G. (1989) Acquired cognitive disorders of the elderly. *Medical Clinics of North America*, 73, 1383–1411.

Ramsdell, J.W., Swart, J.A., Jackson, J. E., and Renvall, M. (1989). The yield of a home visit in the assessment of geriatric patients. *Journal of the American Geriatrics Society*, 37, 17–24.

Rockwood, K. (1989). Acute confusion in elderly medical patients. *Journal of the American Geriatrics Society*, 37, 150–4.

Rubenstein, L.Z., Schairer, C., Wieland, G.D., and Kane, R. (1984). Systematic biases in functional status assessment of elderly adults: effects of different data sources. *Journal of Gerontology*, 39, 686–91.

Schor, J.D., Levkoff, S.E., Lipsitz, L.A., Reilly, C.H., Cleary, P.D., Rowe, J.W., *et al.* (1992). Risk factors for delirium in hospitalized elderly. *Journal of the American Medical Association*, 267, 827–31.

Shader, R., Kennedy, J., and Greenblatt, D. (1987). Treatment of anxiety in the elderly. In *Psychopharmacology: the third generation of progress*, (ed. H. Meltzer). Raven, New York.

Williams, M.E. (1987). Identifying the older person likely to require long-term care services. *Journal of the American Geriatrics Society*, 35: 761–6.

4

Stroke

J. Grimley Evans

Dr. Geert van Gossen received the telephone call from Mr. William Gardner. His wife, Susan, 82 years old, had fallen about an hour ago and was now unable to move her right arm or leg. He had seen her fall and ran over to help her. She did not strike her head. She was a heavy woman and he could not get her up off the floor by himself. Mrs. Gardner's speech was unintelligible. Dr. van Gossen asked Mr. Gardner a few questions. Had there been any loss of consciousness? Any jerking movements? Any evidence of tongue biting or urinary incontinence? Mr. Gardner responded that his wife was awake and there were no jerking movements, tongue biting, or incontinence. Dr. van Gossen continued with his questions. Did Mrs. Gardner have a headache? How had she been feeling over the last few days? As near as Mr. Gardner could tell, Mrs. Gardner had felt well. She did complain, two days ago, of having some difficulty while writing a letter to their daughter. Her right hand had been clumsy and numb for about five minutes and she thought that it was just a 'cramp.' There had been no mention of a headache.

Dr. van Gossen instructed Mr. Gardner to stay with his wife. He advised him to make his wife as comfortable as possible and to keep her warm with a blanket. He would arrange for the ambulance to bring her to the emergency room of the hospital.

Mrs. Gardner's most likely diagnosis is a stroke. The physician's first concern, however, must be to exclude other possibilities that could explain her focal neurological signs (see Table 4.1). The history of onset is particularly important to allow Dr. van Gossen to establish that Mrs. Gardner's fall was a consequence of the neurologic deficit and not the other way around. Having witnessed his wife's fall, Mr. Gardner provided assurance that her deficit was not due to striking her head. Hypoglycemia can produce focal neurologic signs in an older person, particularly if there has been a pre-existing stroke. In the vast majority of cases, however, hypoglycemia would occur as a consequence of insulin use or use of an oral hypoglycemic agent. Mrs. Gardner is not a diabetic and has not been taking any medication. Another possibility to

be excluded is that of an epileptic seizure followed by Todd's paresis, where the area of the brain that has been the focus of an epileptic discharge may take up to 24 hours to regain full function. Mrs. Gardner, however, did not lose consciousness and showed none of the signs associated with a seizure. An additional concern is that the apparent stroke could be due to hemorrhage into a cerebral tumor. At Mrs. Gardner's age, a metastatic tumor would be more likely than a primary glioma. A review of Mrs. Gardner's record made metastatic cancer an unlikely possibility. She has had no symptoms suggestive of cerebral disease, elevated intracranial pressure, or cancer. The most likely source of a cerebral metastasis would be a breast or bronchial carcinoma. Mrs. Gardner has been perfectly well with no complaint of headache. She has been a lifelong non-smoker, making bronchial carcinoma less likely, nor had she noticed any breast lumps. Recent mammograms have been negative. Colorectal carcinoma is also common but is more likely to present with metastases to the liver rather than to the brain.

In the absence of features suggestive of another diagnosis, Mrs. Gardner's clinical picture is most consistent with the diagnosis of a stroke with a premonitory transient ischaemic attack (TIA) two days before. About 10% of strokes are preceded by TIAs, and although transient strokes are suggestive of thromboembolic pathology, they can sometimes be produced by small hemorrhages.

Although some patients with stroke can be managed at home, there are two main indications for admission

Table 4.1 Some common possibilities other than stroke that may cause the sudden onset of focal neurologic signs in an older person

Head trauma
Hypoglycaemia
Seizure with Todd's paresis
Mass lesion (e.g., brain tumor, either primary or metastatic, brain abscess)

to the hospital. The first is the severity of the stroke and the lack of adequate care in the home. The second is the need for further investigations and the possibility of acute interventions.

Dr. van Gossen examined Mrs. Gardner in the emergency room of the hospital. He checked that she had not deteriorated since her husband phoned and that there were no immediate airway problems. Mrs. Gardner was alert and breathing normally. There was no pooling of saliva in her mouth, although she did cough when turned or when sat up during the physical examination. Dr. van Gossen moved quickly to assess Mrs. Gardner's cardiovascular state with particular regard to blood pressure, cardiac rhythm, and the possibility that her stroke might have been associated with a myocardial infarction. Mrs. Gardner's blood pressure was 170/90, her pulse was irregularly irregular and the electrocardiogram showed atrial fibrillation with some non-specific ST-T wave changes. There were no acute changes of myocardial infarction.

Although Mrs. Gardner had not complained of chest discomfort, the frequency of painless myocardial infarction increases with age. An association of stroke with myocardial infarction can occur as a result of an embolus from a mural thrombus in the left ventricular cavity, usually one or two days after the heart attack. Simultaneous myocardial and cerebral infarcts, however, do occur. The mechanism is not always clear but can include emboli from the left atrium or (usually in younger patients) paradoxical embolism from the venous system through a patent interatrial foramen.

What would Dr. van Gossen have done if there were acute changes of a myocardial infarct? There is clear evidence of the benefit from giving thrombolytic therapy to patients within a few hours of a myocardial infarct. This benefit has also been clearly demonstrated for patients aged over 70 years. Indeed, in terms of lives saved per thousand patients treated, thrombolysis is more effective in older patients than in younger. In the ISIS-2 trial of streptokinase and aspirin for suspected myocardial infarction (1988), the number of lives saved per thousand patients treated were 25 for patients aged under 60, 70 for those aged 60–69, and 80 for those aged 70 and over. This is because the fatality of myocardial infarction increases with age and, for any given percentage reduction in fatality, the gain in number of lives saved will increase with the background level of fatality (ISIS-2 1988).

There is a theoretical possibility that thrombolysis might also be of benefit in acute stroke if occluded arteries could be reopened before irreversible neuronal damage has occurred. Trials of thrombolysis in acute stroke are currently under way but the results have not yet been published. It is well established, however, that one of the more serious complications of thrombolysis

in myocardial infarction is cerebral hemorrhage. Even if Mrs. Gardner's stroke were thromboembolic, there would still be a risk of thrombolytic agents inducing haemorrhage in the infarcted area of her brain. Present practice, therefore, would preclude the use of thrombolysis in her case except perhaps as part of a formally established controlled trial. Even then, Mrs. Gardner's difficulty with speech would make it problematic to obtain informed consent. Mr. Gardner might be willing to enter his wife into a clinical trial but studies have shown that proxy decisions made for older people, even by close family members, may not correspond with what the older person would wish.

Dr. van Gossen continues his physical examination, focusing on items of the exam that are important for the immediate management of Mrs. Gardner's care. He is well aware that the examination should be as full as possible but can be very tiring for a patient with a stroke. Dr. van Gossen notes that Mrs. Gardner has obvious problems with speech. He is concerned about the possibility of a receptive dysphasia that would hinder her ability to co-operate with care. Mrs. Gardner does not respond well to requests to carry out particular movements. Mr. Gardner, however, reminds Dr. van Gossen that his wife is deaf and that her hearing aid has been misplaced. The electronic communicator, which every well-equipped medical ward has available, is brought into action. It then becomes clear that although Mrs. Gardner can understand and carry out simple requests like 'stick out your tongue,' more complex requests like 'touch your lips with the three fingers of your left hand' lead to no response at all or a repeated response to the previous request. She is able, however, to imitate gestures. Mrs. Gardner can say 'yes' and 'no' but is not consistent in their use.

Dr. van Gossen notes that Mrs. Gardner has some spontaneous movements of her right foot but has no movements of her right arm. The lower part of her face also droops on the right side. Because of her dysphasia, it is difficult to evaluate sensory modalities. Dr. van Gossen observes that Mrs. Gardner never looks spontaneously to the right and if someone speaks to her from her right side she initially searches for the speaker by looking to the left.

Dr. van Gossen assesses Mrs. Gardner's swallowing by helping her up to the sitting position and offering a small amount of water in a cup. She swallows the water with some difficulty and coughing.

Although Mr. Gardner's history suggested that his wife had fallen without injury, Dr. van Gossen checks for any signs of visible bruising, head injury, or signs of bony injury, especially a hip fracture.

Dr. van Gossen continues with the cardiovascular examination, the blood pressure and pulse having been taken on Mrs. Gardner's arrival. Mrs. Gardner's fundi are examined for features of retinopathy or arterial occlusions. He notes the presence of peripheral pulses and listens carefully for carotid bruits. No bruits are heard. The heart sounds are good, without a murmur or an S3 gallop. The remainder of the exam is normal. Lungs are clear to auscultation. Abdomen is soft, without masses, and there is no evidence of a distended bladder. Mrs. Gardner's legs are equal in size, without edema, or any prominence of the superficial leg veins.

Dr. van Gossen's examination focuses on four key areas in the person with a new stroke: communication, swallowing, remainder of the neurologic evaluation, and essential portions of the general physical examination (see Table 4.2). In assessing communication, the first concerns are to establish what the patient can understand and what the patient can say. As in Mrs. Gardner's case, not all difficulties with communication are related to a stroke but may be due to a problem with hearing. Mrs. Gardner's ability to perform simple actions and imitate gestures, but difficulty in doing more complex tasks, is important information for the nursing staff who care for her. Mrs. Gardner's difficulties with speech will mean she may easily become frustrated and have problems communicating her needs. Special care will be required to ensure her comfort. It is wise to assume that Mrs. Gardner understands part of what is said to her and that her comprehension is likely to improve. Caregivers and visitors must, therefore, be careful with what they say in front of her, make repeated attempts to explain what is happening, and be encouraging and supportive. The more people communicate with her the faster will be the recovery of her speech.

Assessment of swallowing is important in order to protect the stroke patient from inhalation of food and water. It is conventional practice to assess the gag reflex in evaluating persons with strokes but the presence or absence of this reflex may not be very relevant in the protection of the airway from aspiration. As Dr. van Gossen does with Mrs. Gardner, placing the patient in a sitting position and encouraging a small drink of water gives valuable information. Mrs. Gardner's swallow and cough reveal that she has a problem with swallowing but also shows that her airway is protected by the sensory and motor limbs of the cough reflex. Swallowing difficulties are common in early stroke and when they are of this mild degree, are likely to recover. Mrs. Gardner, like many stroke patients, must be regarded as at risk of an inhalation (or aspiration) pneumonia and for the time being must have a 'nil per os' (npo, or nothing by mouth) order written. Maintenance of fluid balance by intravenous line will be required.

The neurologic evaluation of a stroke patient, in addition to determining communication and swallowing ability, requires attention to localizing deficits in motor ability, sensation, and visual fields. In Mrs. Gardner's case she has a dense right hemiparesis, sensory findings are unclear because of her speech problems, and there is a right hemianopsia or visual inattention. These findings, together with the difficulties with language, suggest a localization of the stroke to the distribution of the left middle cerebral artery. Perhaps of more importance than anatomical localization, knowledge of the deficits make it clear that Mrs. Gardner will not initially be able to stand

---

Table 4.2 Four key areas of emphasis in examining the stroke patient

1.	Neurologic evaluation	Cranial nerve abnormalities (visual field cut, gaze palsies, facial weakness)
		Changes in muscle tone
		Areas of weakness or paralysis
		Sensory deficits
		Co-ordination
		Ability to walk
		Balance
		Reflexes
2.	Communication	Determine if the patient can speak and follow simple or complex directions
3.	Swallowing	Assess carefully with small sips of water
4.	Physical Examination	Special attention to:
		Cardiac rate and rhythm
		Fundi for retinopathy
		Carotids for bruits
		Lungs: listen for congestion or consolidation, evidence of heart failure
		Cardiac exam: gallops or murmurs, valvular disease
		Abdomen: check for distended bladder
		Extremities: evidence of venous thrombosis

or, probably, sit unsupported. Individuals who wish to visit or speak with her should stand on her left side.

The physical examination should be tailored to identify any conditions that may be associated with stroke (e.g., valvular heart disease, peripheral vascular disease), any evidence of a complication of the stroke (e.g., aspiration pneumonia), any conditions that may predispose to complications (e.g., deep venous thrombosis), and any evidence of injury from the initial event. Practically speaking, this means a thorough exam with emphasis on the features noted by Dr. van Gossen in his examination of Mrs. Gardner: checking of fundi, auscultating for carotid bruits, palpating peripheral pulses, listening for rales or rhonchi or any sign of pneumonic consolidation, a careful exam of the heart for murmurs or gallops, checking the abdomen for a distended bladder, and examining the extremities for evidence of deep venous thrombosis or any signs of venous disease that might predispose to thrombosis. The skin should be examined, especially over areas of pressure, for evidence of pressure sores. Mrs. Gardner was only down on the floor for about an hour. Other individuals with stroke may be down for hours or even days, raising the possibility of complications of a 'long lie' that include hypothermia, pressure sores, and rhabdomyolysis. With a fall occurring at the time of the stroke, the physician should look for any evidence of trauma or fracture. Head injury, a broken hip, or severe soft tissue injury can all occur with the fall and may be missed if examiners are narrowly focused only on the neurologic evaluation of the stroke.

Dr. van Gossen prepares to order some initial laboratory studies. Mrs. Gardner has already had an electrocardiogram (ECG), revealing atrial fibrillation, as part of her immediate evaluation. He asks for blood to be drawn for cardiac enzymes (creatine phosphokinase with isoenzymes and lactate dehydrogenase), a complete blood count, erythrocyte sedimentation rate (ESR), blood glucose, and measurement of blood urea nitrogen, creatinine, and electrolytes. Because of the atrial fibrillation, he requests that thyroid function tests be performed. Dr. van Gossen also asks that a computerized tomogram (CT) of the head be done as soon as possible.

Laboratory testing in the evaluation of patients with stroke aims at identifying possible contributing causes to the stroke and establishing a baseline to guide the appropriate use of intravenous fluids and other medications. An ECG and cardiac enzymes will help to establish whether the stroke is associated with myocardial infarction. In Mrs. Gardner's case, the presence of new onset atrial fibrillation provides further concern about the possibility of infarction although there are no other acute changes in the ECG. A full blood count with ESR is appropriate in order to search for the possibility of the stoke being due to a cryptic giant-cell arteritis or to a hyperviscosity syndrome such as polycythemia, myelomatosis, or Waldenstrom's macroglobulinemia. If anemia is present, it has implications for further investigation but mild anemia should not be corrected by blood transfusion in the acute phase of stroke. Although there might be some advantage in increasing the oxygen-carrying capacity of the blood by raising the hemoglobin, the benefit is outweighed by the benefits of the lower blood viscosity associated with mild anemia. Measurement of urea, creatinine, and electrolytes will provide a baseline for later assessment of fluid and electrolyte balance as well as providing some index of renal function. This is relevant to the possible prescription of renally cleared drugs as well as the use of intravenous fluids. Determination of blood sugar is useful partly to exclude the possibility of hypoglycemia (already discounted as improbable in Mrs. Gardner's case) as well as to look for hyperglycemia. The stress of a cerebrovascular accident can precipitate hyperglycemic hyperosmolar decompensation in someone who is a diabetic (known or unknown). There is also evidence that the presence of hyperglycemia can be detrimental in an acute stroke by extending the amount of cerebral damage. The maintenance of normal or only mildly raised blood glucose levels is, therefore, part of the early management of stroke and may require a 'sliding scale' dosage of insulin with titration of the amount of insulin depending on frequent checks of blood sugar.

Dr. van Gossen reviews the CT scan of Mrs. Gardner. There is mild cerebral atrophy in proportion to her age but no identifiable lesion compatible with her present neurologic signs. There is, however, evidence of a small old infarct in the right basal ganglia. Dr. van Gossen ponders several therapeutic questions in his mind: What is the significance of this report? Given the presence of atrial fibrillation, should anticoagulation be begun? Would it be appropriate to carry out Doppler scanning of the carotid arteries at this stage? What is Mrs. Gardner's prognosis?

The CT scan provides three important pieces of information. First, there is no evidence of hemorrhage. At this early stage of a stroke, a CT scan will normally exclude hemorrhage but may show no features of thromboembolic stroke. A repeat CT scan in several days would reveal the presence of the infarct. Second, the old infarct in the right basal ganglia shows that Mrs. Gardner is at risk of repeated cerebrovascular events and in more than one vascular field. Studies have shown that there is a higher prevalence of 'silent' cerebral infarcts in patients who are in atrial fibrillation than in individuals in sinus rhythm

(Petersen 1987). The finding increases the likelihood that anticoagulation will be appropriate if she remains in atrial fibrillation. Third, the scan shows no evidence of a cerebral tumor underlying the clinical picture of stroke. Although Mrs. Gardner's clinical history makes a tumor metastatic to the brain or a primary brain tumor unlikely, the CT scan provides added evidence. The great majority of tumors causing symptoms will be detectable on an enhanced CT scan. As in Mrs. Gardner's case, urgent CT scans usually are done without contrast, an enhanced scan is performed only if there is evidence of a lesion. Although a few tumors will not be detectable on an unenhanced scan, it is reasonable in managing Mrs. Gardner to consider the possibility of tumor to be extremely unlikely, and only reconsider that possibility should her clinical course prove to be atypical for stroke.

Dr. van Gossen knows that Mrs. Gardner's stroke is not hemorrhagic and that she is in atrial fibrillation. He wonders at what stage anticoagulation should be introduced. Unfortunately, there is no one definite answer to this common but complex clinical problem. There is a fear that anticoagulation that is begun soon after a thromboembolic infarct in the brain may bring about a hemorrhagic transformation. On the other hand, there is evidence that the risk of stroke recurrence is highest soon after a stroke, making anticoagulation a reasonable option (Barnett *et al.* 1995). An echocardiogram may be helpful in some cases. If it reveals a large left atrium with thrombus, then it would be reasonable to infer that the risk of cerebral embolism is sufficiently high for the benefits of anticoagulants to outweigh the risk of hemorrhagic transformation of the cerebral infarct. Unfortunately, a conventional echocardiogram cannot definitively exclude the presence of thrombus (Moreyra *et al.* 1995). In the absence of clear evidence, such as the demonstration of thrombus on an echocardiogram, most clinicians tend to compromise by waiting two to five days before introducing anticoagulants. Some feel it safe to start sooner if not immediately with aspirin but there is insufficient evidence of safety and effectiveness for all patients (Sivenius *et al.* 1991; SALT Collaborative Group 1991; Dutch TIA Trial Study Group 1991). In caring for stroke patients, another concern is the prevention of deep venous thrombosis. A few relatively small trials have suggested that subcutaneous heparin may be beneficial in reducing the incidence of radiologically detectable deep vein thrombosis. However, the trials were too small to evaluate the benefits regarding long-term consequences of deep vein thrombosis or to quantitate the effects on stroke progression.

Doppler scanning of the carotid arteries is usually performed to evaluate the possibility of a surgically correctable stenosis. There are now several trials showing benefit in some circumstances from carotid endarterectomy (Moore *et al.* 1995; ECSTCG 1991; NASCAETC 1991). In Mrs. Gardner's case, her stroke was preceded by a TIA and is 'incomplete' in the sense that she still has some function on the affected side. Although this would be explicable as a consequence of multiple cerebral emboli due to atrial fibrillation, the clinical features also raise the possibility of a source of emboli in her carotid artery. The potential embolic source would be an atheromatous narrowing of the carotid artery, usually at the origin of the internal carotid, causing tight stenosis and leading to thromboemboli traveling into the cerebral circulation. Given that Mrs. Gardner still has some function on her right side, further thromboemboli from a stenotic left internal carotid could cause further damage.

The absence of a carotid bruit does not exclude the possibility of a significant carotid stenosis. Doppler evaluation of the carotid arteries is widely available and is a painless and non-invasive procedure. It should not be undertaken, however, if the patient is not a candidate for endarterectomy, if a surgical team with a good record for endarterectomy is not available, or if the patient is unwilling or unable to give informed consent. Most clinicians would probably wait for Mrs. Gardner's condition to stabilize and then consider Doppler scanning if she makes a good recovery from her stroke. At that point, the possibility of an elective endarterectomy, depending on the result of the Doppler studies, might be relevant for consideration in an effort to prevent further strokes. These issues should be discussed with Mr. Gardner, who may have read in the popular press about studies of carotid endarterectomy.

Dr. van Gossen's concern about Mrs. Gardner's prognosis has two aspects: (1) the chances for survival from the acute stroke; and (2) the degree of recovery and residual disability that can be expected. Features associated with early fatality from stroke are those that indicate a large amount of brain tissue affected and those which reveal damage in critical areas. Thus, the issue is not only how much of the brain is involved in the stroke but what parts of the brain are affected. Pupillary abnormalities, paralysis affecting extraocular muscles, or divergent strabismus indicate brainstem involvement and an ominous prognosis. Conjugate deviation of the eyes is often associated with a large stroke but is not in itself a particularly dangerous sign. The long-term prognosis post stroke is more problematical. Dense hemiplegias or dense hemisensory deficits are less likely to recover completely compared to minor degrees of impairment. Strokes involving the left side of the body, particularly those associated with flaccid paralysis and sensory inattention, can prove disproportionately disabling.

Urinary incontinence early in the post-stroke period has been associated with poorer functional outcome. The reason for this is not clear but it may reflect frontal lobe damage and associated difficulty in learning the tasks of rehabilitation. There is a tradition in stroke rehabilitation that patients with a non-dominant hemiplegia may develop an easily distracted personality that interferes with co-operation in rehabilitation. It is not clear, however, that this is separate from the problems of sensory inattention, neglect, and, in extreme cases, anosagnosia. In Mrs. Gardner's case, therefore, the prospect for immediate survival seems good, assuming that there is no further stroke or other intervening problem such as myocardial infarction (for which stroke victims are at higher than average risk). Her prognosis for longer-term recovery, however, has to be guarded. In speaking with Mr. and Mrs. Gardner, caregivers should focus on the positive actions that will be taken to facilitate recovery and the assumption that she will return to her own home in due course.

Dr. van Gossen, now at the end of the day, returns briefly to the hospital to check on Mrs. Gardner and to speak further with her husband. There has been little change since admission earlier that morning. She remains stable. Dr. van Gossen asks Mr. Gardner about his wife's previous level of functioning. Mrs. Gardner, prior to her stroke, was fully independent, did the shopping, drove the family car, did the gardening, and had no difficulty going up and down the stairs in her house. Dr. van Gossen recalls Mrs. Gardner as a cheerful and intelligent woman, always pleasant on office visits, who kept up to date with the doings of her children and grandchildren and played bridge once a week with her friends. Mr. Gardner and she have had a strong relationship. He is now retired but remains active in the community and is out of the house most of the day, being involved in various fundraising activities for local charities. Mr. Gardner is very concerned about his wife and asks the doctor what the plan will be to help her.

In developing a plan of action in the care of an older person, there are four aspects of the plan: (1) assessment, including full diagnostic and social and psychological evaluation; (2) the setting of objectives for care; (3) based on these objectives, the development of a management plan; and (4) a program of regular review of the progress of the patient and the management plan, making adjustments as needed (Table 4.3). The objectives of care must be realistically based on the nature and prognosis of the illness and the availability and suitability of treatments. They must also center on the wishes of the patient. When the patient cannot communicate his or her wishes, and there is no relevant advance directive, caregivers may have to reach decisions based on the descriptions of what the patient had said in the past and the type of individual that he or she has been before the illness. It is important to recall that proxy decision making even by close relatives may not be what the patient wanted, and that the responsibility for the decision must rest with the doctor and not with the relatives. The management plan for some severely impaired stroke patients might involve palliative care only because of their poor prognosis. Clearly, however, in Mrs. Gardner's case an active program of care and rehabilitation will be aimed at returning her to her own home. The management plan must deal first with any distressing symptoms that the patient may have, and any diseases that may produce problems in the future. In considering rehabilitation goals, the management plan is usually structured around the gap between what a patient's preferred environment will demand and what the patient will be capable of doing. The plan aims to close this gap primarily by improving the patient's level of function by therapeutic interventions and, secondarily, by reducing the environment's demands through prosthetic interventions.

Developing an adequate management plan requires more than just a knowledge of the diseases and medications. A good sense of an individual's previous level of functioning and the level of social support available

Table 4.3 Four steps in developing a plan of action for stroke patients

1.	**Assessment**	Full diagnostic, social, and psychological evaluation
2.	**Setting of objectives for care**	Realistically based on nature and prognosis of illness Center on patient wishes Depend on availability and suitability of treatments
3.	**Development of a management plan**	Be based on objectives for care Attend to distressing symptoms (e.g., pain) Consider any intercurrent illnesses Improving patient's level of function by therapeutic interventions Reducing environmental demands by use of prosthetic devices
4.	**Regular review of progress**	

after discharge are crucial pieces of information. In Mrs. Gardner's case, as with all stroke patients, it is unlikely that she will do better after her stroke than she did before, but her high level of functioning prior to the stroke, her involvement with a number of social activities, and her independence all suggest a better outcome than would be the case if she were an impaired, isolated individual prior to the stroke. It appears likely that Mr. Gardner will be supportive and encouraging in his wife's rehabilitation. He may be willing to give up much of the community work he does in retirement in order to care for his wife. Such a change in his life, however, would greatly reduce the quality of his life. This would make his wife feel guilty and regret at being a burden to him. Indeed, he may, in due course, need counseling on the dangers of giving up too much to provide care for his wife. The management plan must aim at keeping the care to be provided by Mr. Gardner as low as possible. This will mean that rehabilitative efforts must be directed at improving Mrs. Gardner's condition to the point that she can take some care of herself, and not simply aim at the bare minimum of rehabilitation.

An important part of developing a management plan for stroke patients is careful consideration of any neuropsychologic difficulties that may be a consequence of the stroke. These include depression, and the 'higher order' neurological dysfunctions of agnosia and apraxia. The term 'agnosia' refers to an inability of an individual to appropriately interpret sensory stimuli. Apraxia is the loss of the ability to perform previously well-known tasks and maneuvers. Sensory inattention is more common in non-dominant than in dominant hemisphere stroke. The patient can detect the presence of a visual or tactile stimulus on both sides of the body when these are tested separately, but when symmetrical parts of the body or the visual field are tested simultaneously only the stimulus on the one side, usually the right, is perceived. Unless specifically tested for this, it can be overlooked by an examining doctor. The impairment can be surprisingly disabling.

In many aspects of life, and not just in a neurologic exam, stimuli do appear in symmetrical form. In going through a doorway, for example, a patient may give such a wide berth to the perceived doorpost on her right that she walks into the unperceived one on the left. As another example, she may be conscious of the sensory feedback from her right hand but fail to notice that the left one has fallen into a position where it is about to be shut in a car door. For obvious reasons, caregivers and family members of stroke patients with this type of inattention need to be aware of the problem. Patients with unilateral neglect may fail to maintain insight into their difficulties and foster an unrealistic

expectation of their ability to cope at home. More extreme forms of unilateral distortion of perception can occur, in which the patient actively disowns one side of the body, denying that her arm is hers. Such examples are rare, however.

Visual and spatial agnosias are common forms of agnosia in stroke patients. In visual agnosia, the person can see objects but not recognize them. The person can recognize them however, by manually handling the objects. Relatives and caregivers are often baffled by the person's difficulties and may assume that he or she is displaying a behavioral disorder, being awkward, or demanding unnecessary assistance. Prosopagnosia is a rare and particularly socially disabling syndrome in which the person loses the ability to recognize faces and has to rely on the recognition of voices to identify even family members and close friends. In spatial agnosia, the patient is unable to form or act upon a conceptual map of the environment. Again, a patient, who appears otherwise normal psychologically, but is unable to find his or her way through the family home, can perplex and annoy relatives and friends. Communication difficulties resulting from receptive or expressive dysphasia are obvious enough but dyslexia also needs to be recognized.

Apraxia may be the explanation when a stroke patient has a degree of disability disproportionate to her motor or sensory impairment. So-called 'dressing apraxia' in which the patient cannot put on her clothes appropriately although apparently having the necessary range of movements may be, however, more a problem of perception than a true apraxia.

Personality change may be a behavioral feature after a stroke. Patients who suffer multiple strokes and develop the syndrome of multi-infarct dementia may show profound personality changes as part of the disorder. Dementia and a profound personality change following a single stroke are rare, but increased distractibility may be a feature, particularly of anterior non-dominant hemisphere damage, and can interfere with the learning of skills that are necessary for the process of rehabilitation. Emotional lability may be a distressing feature of strokes in which the upper motor pathways to the brainstem are interrupted. It usually takes the form of tearfulness which may be embarrassing to the patient and disturbing to relatives and friends.

Within a few hours of Mrs. Gardner's admission, the nurse calls Dr. van Gossen to inform him that Mrs. Gardner's blood pressure is rising and has now reached 180/110. Her conscious state has not altered and there has been no change in her pulse. She asks for instructions and wonders if a medication should be given for the blood pressure elevation.

It is common for there to be a rise in blood pressure associated with stroke. This appears to be due, at least in part, to an increase in catecholamine secretion and may be particularly prominent in hemorrhagic strokes. The pressure usually returns to pre-stroke levels within 24 hours. As far as possible, one should avoid treating moderate elevations in blood pressure since the brain's autoregulatory capacity is damaged, probably because of the release of vasoactive substances from damaged neurons. This will particularly affect the regions of ischemic damage. Dropping the blood pressure excessively may lead to a loss of perfusion to the marginal areas around the infarct and result in further damage to brain tissue. In otherwise fit older people who have suffered ischemic stroke, it may be acceptable to aim at keeping the diastolic pressure below 110 mmHg in the absence of the appearance of a third heart sound or other features of compromised left ventricular function. In cases where the blood pressure is rising to levels where it is impairing left ventricular function, or when bleeding from a cerebral aneurysm is suspected, then careful decreases in blood pressure are reasonable. Every effort should be made to avoid a large or sudden drop in blood pressure. The goal is to bring about a small, gradual reduction in the pressure, not necessarily normalization. Small doses of a short-acting calcium channel blocker given on an 'as needed' basis is a common drug regimen in this situation. If the patient did not have a prior history of hypertension, any long-term hypertensive medication probably should not be implemented until two weeks after the stroke, at which time the normal autoregulation of the cerebral circulation will likely have been restored.

Dr. van Gossen reassures the nurse regarding Mrs. Gardner's blood pressure. By morning her pressure has returned to normal levels. Dr. van Gossen meets with the nurse caring for her and discusses some of the aspects of Mrs. Gardner's case. He hopes that after the first 24 hours she will be sat up in a chair, her blood pressure watched carefully for postural instability, and that her swallowing will be tested.

The initial plan for Mrs. Gardner includes scrupulous mouth care while she is npo. Regular and frequent changes of position are done by the nursing staff as the mainstay of pressure sore prevention. Physiotherapy begins with gentle movement of the joints on the affected side, three times a day, to prevent contractures and avoid to a frozen shoulder or the shoulder–hand syndrome. Movement of the legs will also help prevent deep vein thrombosis (DVT), which is a common problem in paretic legs. Given the uncertainty surrounding the value of subcutaneous heparin, graduated pressure stockings are also used as a reasonable, if not well-evaluated, means of reducing the risk of DVT and pulmonary embolus.

Over the next day, Mrs. Gardner makes slow but definite progress. Once sitting up, it is possible to test her swallowing again with a small amount of water from a cup which she holds herself. Careful observation shows that she is slower than normal in making the swallowing movement but she swallows in one movement rather than having to make several attempts and there is no coughing or regurgitation. This is promising. Dr. van Gossen suggests that she be given water in sips only, however, over the next 24 hours and that she be supervised. Intravenous fluids are to be maintained over that time. Then, if all is well, the intravenous line may be discontinued and Mrs. Gardner's oral intake liberalized. Early physiotherapy including standing is begun with two supporters. As soon as possible, she should spend the day dressed in her normal clothes rather than in hospital dress.

After two or three days, Mrs. Gardner's swallowing has improved; she can cough, but does not do so when swallowing, and it is possible to maintain fluid intake by mouth. She is still slow in swallowing solids and a diet of thickened fluids and purée is prescribed. Her speech is recovering and she now only makes occasional mistakes in carrying out requests. She still has difficulty in word finding but is able to make herself understood. The staff has suggested to Mr. Gardner that it may be better to let his wife find a word for herself than for him to supply it when she hesitates in her speech.

Dr. van Gossen continues to evaluate Mrs. Gardner daily and finds that she has a partial right-sided visual field defect with sparing of central vision. He does not find any evidence of any higher order cognitive deficit.

The physiotherapists are now training Mrs. Gardner to walk with a tripod walking aid in her left hand. Mrs. Gardner prefers to use a walking cane, but the therapists stress the advantage that the tripod care remains standing should she need to release it to use her hand for some other task. Her right arm remains paralyzed and is now spastic, making it crucial that she can reach with her left arm. The occupational therapist teaches Mrs. Gardner how to dress herself. She has adapted some of the patient's clothing to make this easier, substituting Velcro fastenings for buttons, but taking care to make the changes discretely so that Mrs. Gardner does not feel stigmatized by unusual clothing that advertises her disability.

Although Mrs. Gardner has been making good progress initially, at the weekly review of her progress it is reported by the nursing staff and the therapists that her progress has slowed. She is complaining of tiredness and exhaustion, and is less willing to co-operate in her therapy sessions. What is the problem?

There are a range of possibilities to be considered when a stroke patient begins to slow in improvement or to worsen. Has some new illness intervened? An extension to the initial stroke may have occurred. In Mrs. Gardner's case, there is no clinical evidence for this. Could there be a cryptic nosocomial infection?

The most likely candidates are chest and urinary tract infections. Physical examination, chest radiograph, and urinalysis show no evidence of this. Is the program of therapy appropriate for the patient? The energy and muscle power requirements of a patient trying to walk with a disability such as a hemiparesis are much greater than those of normal walking. It may be that the current schedule of therapy is too demanding. Apart for the potential to induce physical exhaustion, physiotherapy, like most learning tasks, is more effective if given 'little and often' rather than in long, concentrated sessions. A redistribution of therapy into shorter sessions throughout the day may be necessary. Is the patient suffering from depression? This is very common post stroke (see also Chapter 6 on depression). Although in Mrs. Gardner's case there are no classical features of sleep disturbance and she denies depression, her mood seems flat and she shows little interest in the plans for her return home. It is not unusual for older people to deny depression, as many of them consider it as a sign of a weak personality or a 'lack of moral fibre' to be depressed.

Depression following stroke responds well to antidepressant medication but it needs to be prescribed with care and in the context of a broader approach to the patient's problems. The disruption to the blood–brain barrier which may follow stroke can mean that drugs will enter the brain in higher concentrations than would normally be the case. There is a danger of precipitating central complications, particularly delirium. Mrs. Gardner, however, has shown no sign of cognitive impairment and her stroke, although disabling, is not large. Her risk of delirium is low. With her, and all older persons, initial doses of an antidepressant should be small. Older persons may consider a depressive illness a sign of moral weakness and it is thus necessary to explain to the patient and spouse that the depression is due to organic damage to the brain.

Prescribing medication and giving an explanation as to the cause of depression are only part of the approach to the patient. The physician should also inquire about any possible precipitants to the depression. It is not uncommon for patients to become depressed because they have perceived depression or anxiety in those close to them. Dr. van Gossen would do well to ask Mr. Gardner how he is facing the future. Stroke patients may be depressed over a planned vacation or some other hope for the future dashed by the disability. Other patients may be made depressed by intrusive memories of stroke patients that they have known, perhaps their own parents or a friend. Occasionally, friends and visitors can be tactless, displaying their sympathy by recounting miserable tales of other persons who had suffered a stroke. Caregivers should learn about the patient's knowledge about stroke and the patient's hopes for the future.

Dr. van Gossen prescribes a suitable antidepressant for Mrs. Gardner's depression. He estimates that the beneficial effects of the antidepressant medication can be anticipated in about two weeks time. He plans for her transfer to a rehabilitation ward for a short stay, after which she will return home.

Ideally, Dr. van Gossen would like to arrange a predischarge visit by Mrs. Gardner to her home in the company of the occupational therapist and social worker to identify what is needed in the way of equipment and modifications to the household. After discussion with the other caregivers, however, this might be counterproductive in Mrs. Gardner's present depressed and mildly negative state. A good compromise would be a full assessment of her abilities in the activities of daily living in the ward and occupational therapy department. A visit by the social worker and occupational therapist to Mr. Gardner prior to his wife's return home would also be helpful.

A review of Mrs. Gardner's prospects at transfer to the rehabilitation floor looks quite promising. She can manage to go up and down stairs using the banister rail for support. She can get in and out of a chair of suitable height. Mrs. Gardner can manage getting on and off the toilet if it is of adequate height and there are appropriately placed grab handles. She has some difficulty in managing her clothing, especially a skirt or a dress, but finds slacks comfortable and prefers to wear them.

The occupational therapist provides a detailed prescription for home care. There is a downstairs bathroom in the Gardner's home, so it would be possible for her to live entirely on the ground floor. This would require that the ground floor living room be converted into a bedroom. Unfortunately, this would destroy the Gardner's ability to entertain friends and visitors. If Mr. and Mrs. Gardner prefer to maintain their bedroom upstairs, there will need to be some adaptations. An additional handrail is needed on the stairs so Mrs. Gardner's left hand will have a rail to hold on to when climbing and going down stairs. A walking aid, such as her three legged cane, will need to be provided for use on the first floor as well as one on the second so Mrs. Gardner does not have to carry the cane up and down the stairs. There will need to be a chair of suitable height in the sitting room and the bedroom. The bed will need to be raised on blocks to the same height and the toilets will need raised seats. The shower will need a seat together with a non-slip mat for its floor and a thermostat control that can be reached from the sitting position. In the kitchen, a perching stool will enable Mrs. Gardner to take part in the preparation of food. The occupational therapist also considers how Mrs. Gardner can keep up her correspondence with friends and family. She has difficulty writing with her left hand and is embarrassed by her clumsy efforts. An option would be to purchase a home computer with a word processing package and a printer. This has the added advantage that Mrs. Gardner's grandson can come, set up the equipment, and help his grandmother learn how to use the computer. The challenge of learning a new skill from her favorite grandchild might give Mrs. Gardner particular pleasure. An important part of the preparations for Mrs. Gardner's return home is the purchase of a playing-card holder so that Mrs. Gardner can once again enjoy her bridge games with friends.

In consultation with the occupational therapist and social worker, Mr. Gardner and his family begin the necessary alterations in order to be ready for Mrs. Gardner's anticipated discharge in two weeks time.

In the week before Mrs. Gardner's expected discharge, she begins to complain of a painful right shoulder. The pain is present most of the time, is not particularly related to posture or movement, and has a deep aching quality. It radiates up the side of the neck and down to the elbow.

A painful shoulder on the hemiparetic side is a common problem after stroke. There is a tendency for physicians to overdiagnose an entity of 'thalamic pain'. It is not always clear what is meant by this term. Many seem to believe that it implies that the pain is being generated centrally either as an irritative or a 'release' phenomenon, and that treatment consists of centrally acting medications such as anti-epileptics. These are given with the vague idea that they might inhibit the spontaneous discharge of neurons that is causing the pain sensation. Purely centrally generated pain in this sense is probably rare.

There are three components in the genesis of pain on the side of the body affected by a stroke. First is the peripheral stimulus. The second is the depressive effect. Third is a change in the central threshold or gate for pain. It is the second and third components that give post-stroke pain such a persistent and unpleasant quality, but it is the first component that offers the most potential for effective treatment.

There are a variety of possible stimuli to result in joint pain in stroke patients. Pain can arise from simple spasticity of the muscles around the joint. The joint may become 'frozen,' particularly if there has been inadequate physiotherapy to prevent this. Subluxation of a hemiplegic shoulder is another source of discomfort. Subluxation can occur in a flaccid hemiplegia in which the arm seems to drop out of the glenoid fossa under its own weight. Some form of sling to support the elbow by suspending it from the shoulder or neck may be required in this situation. In a spastic hemiplegia, subluxation can occur when a well-meaning but ill-instructed helper uses a patient's hemiplegic arm to help him or her out of a chair. This produces abduction of the arm but, as it lacks the co-ordinated muscular action around the shoulder, the humerus can prolapse inferiorly. Another potential hazard of using a stroke patient's hemiplegic arm as a handle or lever, and another cause of pain, is fracture of the surgical neck or shaft of the humerus. This may occur as a late complication due to the disuse osteoporosis that may affect a hemiplegic arm, particularly if it is the site of one of the ill-defined sympathetic atrophy syndromes that are the occasional consequence of stroke. Finally, one must remember that shoulders often have reasons to be painful in a person in later life! Osteoarthritis and rotator cuff injuries or calcification which give rise to noticeable but tolerable pain before a stroke may be magnified by neurologic damage and depressive affect into a miserable affliction that dominates a patient's conscious hours.

Management of the peripheral causes of pain follows conventional lines with physiotherapy, analgesia, and specific surgical or injection treatments where indicated. The effect of analgesia may be improved by the use of small doses of tricyclic antidepressants.

Dr. van Gossen prescribes local heat and gentle exercise to her shoulder with mild analgesia for the pain. He also gives her firm reassurance that if the pain persists he will continue to treat her as an out-patient after discharge. The depressive element in Mrs. Gardner's illness has already been noted, and in the week before discharge, a pain in her arm may well become a focus for her growing anxiety about her ability to cope at home. Dr. van Gossen's reassurance of concern and ongoing care is an effort to prevent the pain from becoming a device for delaying discharge. He also tells Mrs. Gardner that the team of caregivers will be keeping in touch with her and her husband after her discharge to make sure that all is going well and to ensure that further help is available should problems arise.

During Mrs. Gardner's last ten days in hospital Dr. van Gossen begins anticoagulant therapy. Careful attention is given to instructing her and her husband about the need for rigorous control of therapy and, in particular, about the dangers of interactions with other over-the-counter medications.

Thanks to the careful planning of the team, the alterations and equipment necessary for Mrs. Gardner's return are ready in time for her planned discharge. Dr. van Gossen gives some thought to the longer-term needs of the Gardners. Mrs. Gardner's son and daughter each agree to spend ten days living with their parents to assist in the transition back home. They will spend no more than ten days so as to prevent being drawn into an overly dependent relationship. The aim of their involvement is to help in the necessary reorganization of the household, assist their father, and oversee the other portions of the care plan. A home health aide will come each day to help Mrs. Gardner with dressing and bathing and make certain that she is adequately fed once her children have left.

Stroke patients and their families need to be made aware of some possible complications that can occur after discharge. Family members may have a tendency to try and do too much. They need to be made aware of the benefits of helping her to be independent and to return to a real and meaningful role in the home. The patient must understand that the stroke has put the patient's body at a mechanical disadvantage but by steady effort her strength and staying power can be increased. If, despite her efforts, Mrs. Gardner's abilities show decline later, she may need short periods of active rehabilitation in a day hospital or as an out-patient in order to boost her skills.

Another possible complication of stroke that may call for attention in the long term is the onset of epilepsy. A stroke may cause a seizure at the time of onset, and, indeed, a few strokes will present as status epilepticus. Usually seizures disappear in the first week or so and, after six months to a year without seizures, consideration can be given to carefully supervised withdrawal of anti-seizure medication. Fits may appear, however, 6–24 months after a stroke in approximately 10% of survivors. In many instances they require drug treatment. The possibility of a seizure months after a stroke needs to be considered lest a seizure, because of the Todd's paresis that may be induced, be misdiagnosed for a further stroke or TIA. Seizures can present as unexplained falls or nocturnal incontinence.

Mrs. Gardner leaves the hospital as planned, quite nervous about returning home and still somewhat down. Dr. van Gossen notes in his appointment book to call her at home in a day or two to check on how the adjustment is going. He also schedules an office appointment for one week to review her progress, and provide an outing for Mrs. Gardner. Mr. Gardner thanks the staff for their care of his wife. He, like his wife, is a bit nervous about the future but is reassured that help is available should it be needed.

Questions for further reflection

1. An older man presents with right hemiparesis and difficulty with speech. What are some of the possible diagnoses? How would you determine this differential diagnosis?
2. One of your patients has been admitted with a cerebral infarct in the distribution of the right middle cerebral artery. He is right-handed. What sort of difficulties might this lesion pose for rehabilitation?
3. You are asked to consult on a patient who suffered a stroke two weeks before but has now stopped making progress in therapy. What sort of concerns would you have?

References

Barnett, H.J., Eliasziw, M., and Meldrum, H.E. (1995). Drugs and surgery in the prevention of ischemic stroke. *New England Journal of Medicine*, **332**, 238–48.

The Dutch TIA Trial Study Group (1991). A comparison of low doses of aspirin (30 mg vs. 283 mg a day) in patients after a transient ischemic attack or minor ischemic stroke. *New England Journal of Medicine*, **325**, 1261–6.

ECSTCG (European Carotid Surgery Trialists' Collaborative Group) (1991). MRC European carotid surgery trial: interim results for symptomatic patients with severe (70–99%) or with mild (0–29%) carotid stenosis. *Lancet*, **337**, 1235–43.

(ISIS-2) Second International Study Group of Infarct Survival Collaborative Group (1988). *Randomised trial of intravenous streptokinase, oral aspirin, both, or neither among 17,187 cases of suspected acute myocardial infarction: ISIS-2. Lancet*, **2**, 349–60.

Moore, W.S., Barnett, H.J., Beebe, H.G., Bernstein, E.F., Brener, B.J., Brott, T., *et al.* (1995). Guidelines for carotid endarterectomy. A multidisciplinary consensus statement from the Ad Hoc Committee, American Heart Association. *Circulation*, **91**, 566–79.

Moreyra, E., Finkelhor, R.S., and Cebul, R.D. (1995). Limitations of transesophageal echocardiography in the risk assessment of patients before nonanticoagulated cardioversion from atrial fibrillation and flutter: an analysis of pooled trials. *American Heart Journal*, **129**, 71–5.

NASCETC (North American Symptomatic Carotid Endarterectomy Trial Collaborators) (1991). Beneficial effect of carotid endarterectomy in symptomatic patients with high-grade carotid stenosis. *New England Journal of Medicine*, **325**, 445–53.

Petersen, P. (1987). Silent cerebral infarction in chronic atrial fibrillation. *Stroke*, **18**, 1098–1100.

Sivenius, J., Laasko, M., Penttila, I.M., Smets, P., Lowenthal, A., and Reikkin, P.J. (1991). The European Stroke Prevention Study: Results according to sex. *Neurology*, **41**, 1189–92.

SALT (The Swedish Aspirin Low-dose Trial) Collaborative Group (1991). The use of low-dose (75 mg a day) aspirin as secondary prophylaxis after cerebrovascular ischemic events. *Lancet*, **338**, 1345–9.

5

Parkinson's disease

J. Grimley Evans

Mr. Robert Newman is a 76-year-old retired accountant who presents to Dr. Ruth Campbell's office with a history of three recent falls and a tremor. His wife had noticed the tremor and was alarmed at the falls. She insisted that her unwilling husband consult with a physician. Mr. Newman, a reserved and private person, has always believed in people taking responsibility for their own health by adopting healthy habits of diet and lifestyle. He has been a lifelong non-smoker and teetotaler.

Dr. Campbell discusses with Mr. Newman the circumstances of the three falls. His first fall occurred on a return from vacation when the shuttle bus taking him from the airplane to the terminal building had to brake rather sharply. The second fall occurred when he was burning some leaves in the garden and his wife called him to the telephone. Mr. Newman turned abruptly to answer her and inexplicably toppled over sideways. Although it was a very cold day, there was no patch of ice to explain the fall. Mr. Newman's third fall occurred inside his home when he turned in the front hall on his way to answer the door bell. He felt that his 'feet were sticking to the ground' and blamed his footwear.

Dr. Campbell notes, while Mr. Newman talks, a tremor of his left hand with a frequency of 3 to 6 times per second. The tremor disappears with intentional movement of the left hand.

Is it significant that Mr. Newman has fallen three times in the last six months? Surveys suggest that between a quarter and a third of people in later life fall once or more during a twelve month period. It would be unrealistic for every fall by an elderly person to lead to a medical consultation. Several attempts have been made to try to distinguish among people whose falls should lead to further assessment and those for whom no action is needed. One method is based on a statistical analysis of the frequency of falls. This suggests that the majority of old people who fall once or twice in a twelve-month period are likely to be simply the victims of bad luck. But individuals who fall more than twice in a year likely have a non-random cause to their falls that may be of medical significance. This type of analysis and the conclusion that one or two falls need not warrant investigation has the disadvantage that some people might suffer potentially preventable injuries in their second fall if the first fall had been ignored. Another approach to exploring falls, therefore, concentrates on the explicability of the fall. Falls that are clearly due to an external hazard, such as a loose carpet, a steep step, or a trailing electrical cord, are not investigated further. Falls that are without an identifiable cause are regarded as requiring further investigation. This approach also has its problems. Falls embarrass and frighten older people who may see them as a sign of failing powers and possibly as a threat of impending institutionalization. An older person may be overly ready to identify some plausible cause for a fall that may obscure the medically relevant condition. An individual, for example, may tell the physician about a fall and place the blame on an irregular paving stone, neglecting to mention a foot drop.

Mr. Newman's three falls in six months, two of which could not be attributed to an external hazard, suggest that there is probably an underlying cause and that further inquiry is appropriate. A detailed history of each fall should be sought in the effort to identify common features. In Mr. Newman's case, all his falls are compatible with the gait problems of Parkinsonism. Typically, the falls of Parkinsonian patients are due to inability to shift their centers of gravity rapidly or appropriately. They are, therefore, more likely to fall over if they receive an unexpected push, or when turning or hurrying.

Mr. Newman's tremor suggests the resting tremor of Parkinsonism. In early Parkinson's disease, the tremor is typically unilateral or more prominent on one side. It later becomes bilateral. The tremor rarely affects the lower limbs. The most common cause of tremor in later life is the benign or senile tremor, which is probably related to familial tremor. The benign tremor is usually more rapid than the Parkinsonian tremor. It is also worse with movement and when the person is nervous or conscious of being under observation. Typically, the tremor of Parkinson's disease disappears on intentional movement.

The history of falls and a resting tremor raise the questions of whether Mr. Newman might have Parkinsonism disease and, if so, what is its cause. Parkinsonism is a syndrome comprising, in various proportions, muscular rigidity, bradykinesia, resting tremor, and postural abnormalities. Several conditions in later life can be mistaken for Parkinsonism, especially those that produce other forms of tremor and those associated with motor defects due to upper motor neuron and corticobulbar dysfunction. The main causes of Parkinsonism include Parkinson's disease itself and drug-induced Parkinsonism, both of which reflect dysfunction of the basal ganglia, and diseases causing more widespread damage to the nervous system (Cummings 1992; Jenner *et al.* 1992). These include multi-system degeneration, particularly progressive supranuclear palsy, the Shy-Drager syndrome, Parkinsonism associated with Alzheimer's disease, or cortical Lewy body disease. Very rarely, some forms of late onset Huntington's disease can also be mistaken at first for Parkinsonism (Young 1992). There may be no family history in such cases. The implications for succeeding generations are obviously serious.

Dr. Campbell asks Mr. Newman if he has noticed any tingling in his hands, muscle stiffness, or aching. Mr. Newman reports that he has noticed 'funny feelings' in the muscles of his forearms and calves and recalls first noticing these feelings soon after his retirement at the age of 69. It was also at that time that he became aware of a tendency to develop 'writer's cramp', noting that it took him more trouble to write clearly and that he was forced to take frequent rests to relieve discomfort in his forearm. He takes no drugs of any sort, neither prescribed nor over-the-counter. His symptoms have been mild and slowly progressive.

Dr. Campbell also questions Mr. Newman about any changes in gait, facial appearance, voice, sleep habits, and swallowing. In further conversation she asks about Mr. Newman's habits and hobbies. Mr. Newman relates that he is a keen member of his church and has acted as treasurer for some years. Mrs. Newman volunteers, with a wry smile, that her husband's work as church treasurer is a mild obsession that is appreciated both by their minister and the fellow parishioners.

Although Parkinsonism is predominantly a motor affliction, some of its symptoms, even in the early stages, may be sensory. Typically, these early symptoms will be parasthesias in the hands and generalized muscle stiffness and aching. These symptoms are assumed to arise from an increase in muscle tone and fatigue. Problems with writing or other activities requiring fine and accurate movements of the hands are typical of Parkinsonism. The 'textbook' sign of micrographia is less common than a general deterioration in the ease of writing and in its legibility.

Mr. Newman's history is compatible with a diagnosis of Parkinson's disease. In the course of a general history and physical examination the doctor will be looking specifically for other features of the disease. There are a number of different findings. The shortening of step in walking and the loss of associated movements, such as arm swinging, are well recognized and often noted by family and friends, if not by the patient. Loss of facial expression is so gradual that it may pass unnoticed by close companions. The daughter who lives far away, for example, may notice this on a Christmas holiday visit whereas those who live close by might notice the difference only on viewing family pictures of some years before. Another feature that may be apparent to a stranger is the staring appearance produced by a reduced rate of blinking. Close companions of an individual with Parkinson's disease may also fail to notice a change in voice towards a quieter and what is often described as a more 'monotonous' tone. The earliest change is a lack of prosody, the absence of the normal grouping of words in the phrasing of sentences that make for the normal expression of speech.

The spouse of an individual with Parkinson's disease may be the first to notice a characteristic early feature of Parkinsonism: a reduced frequency and/or increased difficulty with turning over in bed. The patient may notice this as an increased stiffness and discomfort on waking. In the case of Mr. Newman, Dr. Campbell detects this feature indirectly. Dr. Campbell discovers in speaking with Mr. and Mrs. Newman that he has long had a tendency to snore when lying on his left side. Mrs. Newman always used to give him a gentle nudge and he would turn over without waking up. In the last two years, however, she has noticed that she can only get him to change position if she actually wakes him.

Swallowing difficulties are a common feature of Parkinsonism that should be explored. Occasionally, patients will present with dysphagia before other features have appeared. The difficulty typically takes the form of a problem in co-ordinating the early swallowing action so that food and fluids pool in the pharynx. One manifestation of this may be that the patient has to make two swallowing movements to clear the pharynx. This can be tested for clinically with a glass of water and auscultation over the larynx. The clinical concern is the possibility of aspiration and consequent chest infection. Inefficient swallowing and loss of the regular pattern of unconscious swallowing leads to the troublesome Parkinsonian feature of excess saliva in the mouth. There is no true sialorrhea.

The issue of cognitive decline needs to be pursued.

This always requires tact since memory loss or 'senility' is a fear of many older persons. In many cases, more can be learned from a careful history than from an insensitive and brusque use of one of the mental status tests available in clinical practice. Asking a close relative about a patient's hobbies and interests are particularly significant. Loss of interest in previous activities is a non-specific phenomenon that is a common early sign in both dementia and depression. Mrs. Newman's information that her husband continues to function at a high intellectual level as church treasurer suggests that cognitive loss is not a current problem.

Dr. Campbell performs a thorough physical examination of Mr. Newman, with special emphasis on important aspects of function and differential diagnosis. The vital signs are recorded and the blood pressure and pulse checked for postural changes. She takes extra time in testing Mr. Newman's eye movements. Dr. Campbell checks muscle tone and rigidity. She asks Mr. Newman to take a ten-meter walk, timing how long he takes. Dr. Campbell tests for the presence of primitive reflexes.

The physical examination provides a general appraisal of a patient's health and fitness as well as giving information about function and differential diagnosis. In assessing gait and movement, for example, it is useful to have a standard maneuver that can be observed and timed. This may be a ten-meter walk. Alternatively, one may ask the patient to get up from a chair, walk across the room, turn and return to sit in the chair again. The results can be a useful basis for future assessment of therapy. In Mr. Newman's case there is no evidence from the history of a dementing process, but it is a good practice to note for possible future reference the presence or absence of 'primitive' reflexes such as the pout, root, or palmomental. Vital signs are particularly important to review. A characteristic feature of the Shy–Drager syndrome is dysautonomia and postural hypotension. Postural hypotension, or its absence, needs to be documented with a view to possible therapy for Parkinson's disease with levodopa. Differentiating Parkinson's disease from progressive supranuclear palsy requires examination of extraocular muscle control. A loss of eye movements, particularly the loss of voluntary movements in the vertical direction that can still be elicited by reflex means, is a diagnostic feature of supranuclear palsy. For example, a patient who can no longer voluntarily move his eyes to the right or downward, but can do so as part of the doll's eye maneuver, may have supranuclear palsy. The pattern of muscle rigidity is another key finding. The rigidity of Parkinsonism needs to be distinguished from the spasticity of upper motor neuron denervation and the Gegenhalten of frontal lobe defects. Parkinsonian

rigidity is characterized by steady resistance throughout the range of movement (lead-pipe rigidity) and may show a superadded exaggerated intention tremor resulting in a cogwheel rigidity. In upper motor neuron spasticity, the resistance to movement at first increases under pressure but then may suddenly collapse, like a clasp-knife. In frontal lobe Gegenhalten, the patient seems to resist attempts at passive movement in all directions.

Mr. Newman's exam shows only a minor change in pulse and blood pressure on changing from lying to standing position. There is a very slight impairment of conjugate upward gaze, but this is a common feature of less significance in a 76-year-old individual. Downward and lateral gazes are full. Motor exam reveals a Parkinsonian resting tremor and mild cogwheel rigidity in the left arm but no abnormality in the right. On the walking test he has slowness in getting out of the chair and poor swinging of both arms when walking, more noticeable on the left. When turning he is slow and takes multiple little steps and appears to have some difficulty in initiating his return to the chair. There are no primitive reflexes and the plantar reflexes are bilaterally downgoing. Dr. Campbell finds the history and physical examination strongly suggestive of Parkinson's disease. She considers the possibility of levodopa therapy and arranges for a repeat appointment later in the week for a trial of medication. Mr. Newman undergoes a tapping test with a positive result after receiving levodopa/carboxydopa.

Responsiveness to levodopa is both a therapeutic and a diagnostic issue. Responsiveness is usually less in conditions other than Parkinson's disease that produce the Parkinsonian syndrome. Various forms of tests are in use but all take the basic form of timing or measuring some aspect of function before and after a single dose of a dopaminergic drug. In younger patients, parenterally administered apomorphine, given with domperidone to prevent nausea and vomiting, can be used. With older patients, however, it is more usual to administer an oral dose of one of the levodopa/dopa-decarboxylase inhibitor combinations used for the treatment of Parkinson's disease. The measured test, for example, can be a timed walk over a standard distance. With older patients this may prove insensitive to the effects of therapy if walking speed is limited by another condition such as poor eyesight or arthritis of the hip. Some form of a tapping test may be more sensitive: the patient sits at a table and is required to tap successively between two points 25 cm apart with one hand. The number of taps completed in one minute is counted by observation or electronically. The response of the other hand is then tested. After adequate baseline measurements, the patient is given a single oral dose of 100 mg of levodopa combined with 25 mg of carbidopa (Sinemet). The performance on the tapping test is then monitored at half-hourly intervals over three hours. A

20% improvement in the score is arbitrarily considered to be significant.

Mr. Newman gives a positive result on the tapping test but finds that the Sinemet makes him feel 'uncomfortable'. He says he is unwilling at present to consider taking Sinemet as he does not feel ill enough and would rather keep it in reserve. Dr. Campbell prepares to discuss the future with Mr. and Mrs. Newman. From their earlier meeting, she knows that both of them suspect the diagnosis is Parkinson's disease. Having faced this situation with other patients, Dr. Campbell is aware that she must find out what the Newmans know about the disease, especially as its older name of *paralysis agitans* conveys a frightening image. There may be a need to probe Mr. and Mrs. Newman's ideas about the disease in separate interviews. Some older people, particularly those who have definite views about health and disease, may harbor serious misapprehensions about the origins of neurological illnesses. Even if the specific notion of general paralysis of the insane as a consequence of sexual misdemeanors in youth is not recognized at a conscious level, there may well be a vague idea of an aspect of 'uncleanness' associated with a chronic, crippling disease.

In discussion with Dr. Campbell, the Newmans ask if the disease is likely to be inherited by their children. They are reassured by the information that familial Parkinson's disease, if it occurs at all, is exceedingly rare. Mr. Newman inquires if his lifestyle could have any bearing on the diagnosis. He also wants to know about his prognosis.

Mr. Newman is a somewhat rigid individual with self-reliant attitudes to health and an aversion to medication. He has not taken any prescribed medications that can induce Parkinsonism nor has he been an enthusiast for over-the-counter or health shop medications that might have relevant side-effects. It is possible that his lifelong abstinence from tobacco might be associated with an increased prevalence of his illness. Some studies have shown that non-smokers may have a higher incidence of Parkinson's disease than smokers (Godwin-Austen *et al.* 1982). The reason for this association is unknown but several hypotheses have been advanced. First, smoking is known to induce liver enzymes that increase the first-pass metabolism of certain drugs and chemicals. It is suggested that this enzyme induction might provide a protective effect for smokers by enhanced inactivation of unidentified toxins in food, thereby possibly preventing their absorption into the systemic circulation and distribution to the brain. A second proposed theory, for which there is no evidence, is that the pharmacological effects of nicotine in cigarette smoke might in some way protect the brain against the loss, or the effects of loss, of dopamine-secreting cells in the basal ganglia. A third theory is that there is a pre-Parkinsonian personality associated with a particular balance of dopamine and other neurotransmitters in the brain which is characterized by a distaste for, or a lack of susceptibility to, the supposed pleasures of tobacco smoke. The evidence at present for this personality theory is not convincing but the assumption is that the pre-Parkinsonian personality, if it exists, is likely to be of the somewhat rigid and orderly type that is exemplified by Mr. Newman's profession and lifestyle. It could also be postulated that if Mr. Newman had smoked, he may well have died from lung cancer, heart disease, stroke, or emphysema years before he would have developed Parkinson's disease, thereby explaining the possible increased association of non-smokers with Parkinson's disease.

Prognosis in Parkinson's disease has been greatly altered with the introduction of levodopa therapy. Although levodopa therapy may not have greatly increased the average total duration of survival in Parkinson's disease, there is little doubt that it has reduced the length of time spent in a disabled state. This aspect is usually of greatest interest to older individuals with the disease. There is a wide variation in the rate of progression and responsiveness to treatment. In reviewing Mr. Newman's history, symptoms attributable to Parkinson's disease can be traced back as far as five years. The fact that there is little disability at the time of Mr. Newman's presentation to Dr. Campbell suggests that his is one of the more slowly progressive forms of the disease. An initial presentation with tremor rather than rigidity and akinesia has also been found in some series to be a favorable feature (Hoehn and Yahr 1967). The absence of dementia is also a good sign. There is always hope that the disease may not progress. In Mr. Newman's case, it is reasonable to suggest that it is likely to be years rather than months before the disease interferes in any major way with his lifestyle.

In conversation with patients and families, discussion of possible drug treatments (the facts and uncertainties) is appropriate. There is a wide range of drugs available for the treatment of Parkinson's disease. Four main drugs are in common use. The oldest are the anticholinergics. These can be effective for the treatment of tremor but tend to have a high frequency of side-effects in older people. In order to exert their anti-Parkinsonian effect they need to penetrate the blood–brain barrier. Unfortunately, their anticholinergic effect partially blocks the cholinergic system that is involved in memory and orientation so that it can produce delirium and hallucinations. This problem is more likely to occur in those patients who show evidence of cognitive impairment before treatment is initiated.

Levodopa is the essential component of the second group of drugs. Levodopa is the amino acid precursor of dopamine and is usually given in combination with a peripheral dopa decarboxylase inhibitor in order to prevent metabolic breakdown in the bloodstream. It works by providing increased levels of the substrate for cells that produce the neurotransmitter dopamine. Levodopa's side-effects include delirium, dystonic movements, and a variety of symptoms including nausea and anxiety. A few patients, however, seem to enjoy the psychological effects of levodopa and may increase the dosage of the drug inappropriately.

The third group of drugs that are useful in Parkinson's disease are those that act as dopaminergic agonists on the postsynaptic cells of the basal ganglia. Pergolide and bromocriptine are frequently used agents of this group. Opinions differ about the suitability of these drugs for older patients and whether they are best used early or late in the course of the disease. There is a high incidence of delirium and other side-effects when using the dopamine agonists with elderly patients. Most geriatricians restrict the use of these agents to patients showing signs of reduced effectiveness of levodopa or as a means of smoothing out the effects of levodopa.

The fourth group of drugs are those that inhibit the enzyme monoamine oxidase. The only such drug presently in common use is selegiline, a selective monoamine oxidase B inhibitor. It delays the breakdown of dopamine in the basal ganglia and may increase the synaptic concentration of dopamine by inhibiting its reuptake by the presynaptic neurons. Selegiline can be used to reduce the dose of levodopa and to smooth out the action of intermittent doses. Selegiline has also been thought to have other actions, possibly including interference with the metabolism of monoamines or of some unidentified exogenous or endogenous substance into neurotoxic products, which then might result in a decrease in the rate of loss of dopaminergic activity in the basal ganglia. The balance of evidence is now against any such effect on the natural history of the disease, but the drug has a low frequency of side-effects and is effective in reducing some of the symptoms of early disease (Shoulson et al. 1989; Schulzer et al. 1992). There has, however, been one report suggesting that selegiline treatment might be associated with increased mortality (Lee et al. 1995).

Neurosurgical approaches to the treatment of Parkinson's disease are rarely appropriate for older patients.

Dr. Campbell discusses with Mr. and Mrs. Newman the need for treatment of Parkinson's disease to be tailored to a patient's requirements in co-operation with a physician. Mr. Newman, after considering the treatment options, is concerned that starting levodopa therapy before it is absolutely necessary might 'tax the body system'. He decides that it would be best to keep the drug in reserve until it becomes really necessary. Dr. Campbell explains that there has indeed been anxiety in the past that increasing the metabolic activity of the remaining basal ganglia cells by loading them with dopamine precursor might hasten their degeneration but that there is no experimental evidence to support this. She went on to mention that there is some evidence that patients who start levodopa early may do better in the long run but that this may simply reflect the benefits of enhanced activity levels. After more discussion with Mr. Newman, he remains unwilling to start levodopa until it becomes necessary. Dr. Campbell wisely realizes that the evidence regarding beginning levodopa therapy early is not strong enough to justify going against Mr. Newman's wishes. Given his detailed and meticulous nature, it is important to allow him to have as much control as possible over managing his own illness. With that in mind, Dr. Campbell outlines the importance of remaining physically active and strongly encourages him to maintain a positive approach to life. She also provides specific counseling about the situations and movements that are likely to result in falls. These include turning while standing or walking, beginning to walk after getting up from a chair, or getting up from a bed. Symptoms are also likely to be more prominent if he gets cold. Mr. Newman thanks Dr. Campbell for her suggestions and her careful approach to his problems. At the end of the visit, Dr. Campbell makes arrangements for regular follow-up appointments.   Some six months later, Mr. Newman admits that his Parkinsonism is causing him problems. Family members have hinted that he is slowing up. His feelings of stiffness and aches in the muscles and joints have become worse. He has particular difficulty getting started in the mornings. Mr. Newman notes that he has become more prone to stumble but by recognizing his difficulties and taking appropriate care he has had only one fall. He also recognizes that his voice is weaker and is disturbed by an apparent increase in saliva production. He feels that the time has come to try levodopa therapy.

Mrs. Newman asks Dr. Campbell if she can have a word with her while her husband is changing for his examination. She finds it quite difficult to speak but admits that she is embarrassed by her husband's appearance and demeanor. She feels that her friends suspect that he is becoming senile. At night, his snoring and increasing immobility is disturbing her sleep but she feels that to move into separate beds after 50 years of marriage would be upsetting for both of them.

Beginning levodopa therapy requires consideration of two different approaches. One can aim for normal function around the clock. Alternatively, the prescription of the drug can be kept to a minimum by aiming only to have its effects at times that Mr. Newman feels he particularly needs it. While there is no general reason for preferring one strategy over the other, one prescribes continuous therapy if there is evidence of complications (e.g., aspiration due to impaired swallowing, during the untreated phases). In the case of Mr. Newman, the information Mrs. Newman provides suggests that around-the-clock prescribing regimen is indicated. Since there is great individual variability in the response to levodopa, the prescribing strategy should be to start with minimal doses and to increase, as necessary, the size of the dose and the frequency with which it is taken. Patients and their families need to be warned that although some immediate benefit from levodopa is usually apparent, the maximum effects are seen only after some weeks of therapy.

There are several preparations of levodopa combined with a peripheral decarboxylase inhibitor available. These differ in the decarboxylase inhibitor used, in the ratio of decarboxylase inhibitor to levodopa, and whether or not the tablets are formulated as sustained release preparations. A dispersible combination is also available for patients with severe dysphagia. As mentioned, one of the most widely used preparations combines levodopa with carbidopa as the decarboxylase inhibitor (Sinemet). It is available in carbidopa/levodopa ratios (in milligrams) of 1:10 and 1:4. The total daily dose of carbidopa needed to produce complete inhibition of extracerebral dopa-decarboxylase is 75 mg. In patients who begin with low daily doses of levodopa, therefore, it is logical to use the 1:4 preparations prior to switching to the 1:10 ratio preparations when higher dosage levels of levodopa are required. In some cases, where the pattern of levodopa therapy that is essential for the control of a patient's Parkinsonism leads to nausea, domperidone may be used as adjunctive treatment.

Mr. Newman is begun on 100 mg of levodopa (with 25 mg of carbidopa) three times per 24 hours. Dr. Campbell recommends that the doses be taken after meals to reduce the risk of nausea and vomiting. This can be modified depending on the needs and tolerance of Mr. Newman. He is encouraged by Dr. Campbell to find, by trial and error, the best timing of the medication. After a week or two, Mr. Newman finds that he can tolerate the first daily dose of Sinemet if he takes it with a cup of tea and a snack in bed at 7.00 a.m. before he gets up. He takes the second dose after his midday meal and his last dose with a warm milky drink on retiring. This gives him the greatest effects of the levodopa during the day but also ensures that the drug is acting through the night to improve his mobility during sleep and to provide some residual effects on first waking.

Mr. Newman does well on this drug regimen for nearly a year. A problem arises, however, when he finds that he has 'seized up' one evening while having supper with some friends. His first warning of difficulty was that he was eating his supper much more slowly than everyone else and was forced to leave some of his food in order to keep up with the other guests. He then found he had difficulty in keeping his voice audible and in handling his cutlery. Dr. Campbell recognizes that the symptoms may reflect a low period in the level of levodopa in the brain. She prescribes an extra tablet containing 50 mg of levodopa with 12.5 mg of carbidopa for late afternoon or early evening, to boost

levels during the evening hours or at other times when problems can be anticipated.

On a visit several months later, the Newmans report that this modified drug regimen initially worked well but in recent weeks it has become clear that Mr. Newman's condition is deteriorating. He has begun to consider giving up his work for the church. Mrs. Newman mentions that it is most unlike her husband to ever consider giving in to circumstances. She wonders if he may be depressed.

Depression is common in Parkinson's disease, affecting up to 50% of patients at some stage. The diagnosis of depression can be complicated (Hammond-Tooke and Pollock 1992). The mask-like facies and retarded motor function of the Parkinsonian patient can give a false impression of depression. Despite the possibility of a mistaken diagnosis of depression, a high index of suspicion is appropriate when caring for persons with Parkinson's disease. Although depression may be an understandable reaction to the problems of late Parkinsonism, depression can also occur very early in the disease. In some case series, patients with Parkinson's disease have a previous history of severe depression more often than controls, suggesting there may be a biochemical basis for a specific association between the two conditions. Not only is depression common during the course of Parkinson's disease but there is often a greater history of depression before the onset of Parkinson's disease in the patients compared to the controls (Mayeux et al. 1984). One possible mechanism lies in the reduced serotonin concentrations that are found in the forebrain of patients with Parkinson's disease. Serotonin-enhancing drugs, such as imipramine, offer a rational approach to the treatment of depression in Parkinsonian patients. Imipramine and related tricyclic drugs, because of their anticholinergic effects, have the potential to induce delirium and other undesirable effects such as constipation or urinary retention. They also can cause postural hypotension and have the potential to cause cardiac arrhythmias. Despite these concerns, they may, in addition to relieving depression, improve the problem of saliva retention.

Dr. Campbell speaks with Mr. Newman at length about the possibility of depression. She finds no organic features of depression and suggests that he is sad at the loss of his functional independence. Given this, she feels that pharmacological treatment for depression would not be indicated at this time, although she needs to remain vigilant about the possibility that depression will develop. Dr. Campbell noted that as Mr. Newman is on selegiline, care would have to be taken to avoid the 'serotonin syndrome' produced by an interaction between selegiline and antidepressant medication, particularly of the serotonin selective reuptake inhibition type (SSRI) [Williams and Lowenthal 1995]. Dr. Campbell suggests that Mr. Newman consider what can be done to improve his life situation and his attitude towards his activities. He recognizes that it is sensible to give up some responsibilities before he becomes unable to discharge them efficiently.

Some discussion with Mrs. Newman, and through her, with Mrs. Newman's minister, leads to the happy device of Mr. Newman resigning from his position as church treasurer but his appointment to a permanent seat on the church finance committee as Treasurer Emeritus. Mr. Newman is pleased at the result, being relieved of some duties he finds difficult but able to feel valued and involved with his friends.

Unfortunately, Mr. Newman notes further increase in the general intensity of Parkinsonian features, particularly bradykinesia. He also is suffering wide fluctuations in functional ability during a 24-hour period. Dr. Campbell suggests that Mr. Newman keep a diary in which the day is divided into hourly segments. With the help of his wife, he records on a scale of 1 to 5 his subjective assessment of his function during his waking hours. The diary also notes the times when Mr. Newman takes his levodopa/carbidopa. An inspection of the diary after a few days of records finds no clear evidence that his 'on-off' problems reflect an end of dose relationship between the severity of Parkinsonism and the likely tissue levels of levodopa.

The treatment options include increasing the frequency of levodopa therapy, providing increased background levels of levodopa, or adding another drug. As bradykinesia rather than tremor is now Mr. Newman's main problem, an anticholinergic drug is not indicated. The dopaminergic agonists are best not considered at this time. They are often not tolerated by older patients and it is wise to avoid them until absolutely necessary. The information from Mr. Newman's diary does not show any clear pattern to the 'off' periods he records so there seems to be no rational basis for interpolating additional doses of levodopa/carbidopa. The therapeutic choice, therefore, is to try giving all or a proportion of the levodopa therapy in the form of a continuous release preparation [Rodnitsky 1992].

Dr. Campbell is well aware that care must be taken in changing to a controlled release form of levodopa/carbidopa because of the possibility of toxicity due to overlapping dosages. The initial changeover is planned with an eight–hour period free from therapy (most conveniently overnight) followed by a twice daily dosage of a continuous preparation of levodopa with decarboxylase inhibitor. Dr. Campbell prescribes for Mr. Newman a continuous release dose of 200 mg levodopa with 50 mg of carbidopa every twelve hours. Additional doses of the ordinary preparations can be added to provide extra effect at times of day when symptoms are worse. In a few days, Mr. Newman phones Dr. Campbell and reports that he is having some problems with dizziness when he stands up as well as an upset stomach. At the levels of dosage he is taking, it is not surprising that Mr. Newman is experiencing some side-effects of levodopa therapy, particularly nausea and postural hypotension. The latter is worrisome because of the increased risk of falls. The need to balance the benefits of the drug against the side-effects is a common therapeutic problem in the later (and sometimes the early) stages of Parkinson's disease.

Over the next twelve months, Mrs. Newman becomes worried over her husband's loss of weight.

Weight loss is a recognized feature of Parkinson's disease. It is thought to have two causes. The first is the reduced food intake secondary to bradykinesia and swallowing difficulties. The second is the rise in metabolic rate due to increased muscle tone and, when prominent, tremor. Other possibilities for weight loss need to be considered. Thyrotoxicosis and diabetes are easily excluded biochemically (bearing in mind that levodopa therapy may interfere with the diagnosis of thyroid disease). The possibility of tuberculosis also needs to be considered and skin testing and a chest radiograph may be appropriate. A difficult question, in the absence of specific or localizing features of disease, is how far invasive or uncomfortable investigations for occult malignant disease should be pursued in investigating weight loss. This is a matter for clinical judgement and negotiation with the patient. As in most situations in geriatric medicine, the issue centers on the potential benefit to the patient from the various possible outcomes of investigations and their relative probabilities of providing a diagnosis.

Mr. Newman is, at this stage, quite severely disabled by his Parkinsonism and is not a good candidate for major surgery should, for example, a carcinoma of the colon be discovered. In discussion, he states that he would not be willing to undergo elective surgery. Mr. Newman says that he is now content to live day by day, and feels no particular need to pursue invasive investigations at this time. Dr. Campbell agrees to keep the situation under review.

A few months later, Mr. Newman's swallowing difficulties, which had been well controlled by the levodopa, are now clearly manifest at meal times. He is troubled by the accumulation of saliva. Dr. Campbell recognizes the problem with swallowing saliva in Parkinson's disease as potentially socially disabling. It is also difficult to treat. She elects a trial of propantheline, an anticholinergic drug which does not cross the blood–brain barrier, in an effort to reduce saliva secretion. Mr. Newman is instructed about side-effects, including thrush and other forms of stomatitis, and the need for regular rinsing with mouth washes. A happy effect of the propantheline is that it provides Mr. Newman some benefit from troubles with urinary urgency.

Mrs. Newman calls Dr. Campbell a short time later. She says that her husband's mobility is steadily decreasing, despite what Dr. Campbell has explained are maximum doses of levodopa/carbidopa. She also is concerned that her husband's memory is slipping, although he is good at covering up lapses. Mrs. Newman suspects that he is not always sure of where he is or what is happening. Dr. Campbell, at the next regular visit, notes that Mr. Newman becomes quite annoyed when she attempts to carry out a formal mental status test. But prior to that failed effort, she is aware that he evades questions about his children's ages, the names of grandchildren, or even his home telephone number. She feels it is likely that Mr. Newman does have some mental impairment.

Cognitive loss in Parkinson's disease can be multifactorial. Although there is some controversy, the general view is that there is an increased incidence of mental impairment in Parkinson's disease that is not extirely explicable by coexistent cortical Lewy body disease, cerebrovascular disease, or Alzheimer's disease. The issue is clouded by the concept of bradyphrenia in which the Parkinsonian patient exhibits slowness of mentation but without formal evidence of defects in memory or reasoning power. Individuals with cognitive impairment are at high risk for developing delirium. Delirium can occur in Parkinson's disease and is a particularly common side-effect of anti-Parkinsonian medications.

Dr. Campbell realizes that there are few therapeutic options left for Mr. Newman. His physical function has now deteriorated to the point where the addition of a dopaminergic agonist drug seems to offer some slight hope of improvement. Because of his cognitive impairment, however, the risk of a delirious reaction is high. After discussion with Mr. and Mrs. Newman, a single trial dose is given one evening. Mr. Newman reports that it has made him anxious and he felt muddle-headed. No further dopaminergic medication is prescribed.

Mrs. Newman has employed some regular daily help in caring for her husband, and, with the advice of an occupational therapist, has obtained suitable modifications to furniture and fittings to assist her husband's mobility. He has also been provided with modified clothing to ease with dressing. Despite these efforts, Mr. Newman is now largely confined to one room of their home.

One morning, in attempting to get up from his chair without assistance, Mr. Newman falls and suffers a fracture of the proximal femur. He is in such pain that surgical fixation is indicated and pinning is carried out under spinal anesthesia, in the hope of reducing the risk of post-operative complications. Despite careful titration of medications, Mr. Newman is confused after the operation and develops a chest infection, probably due in part to his problems in swallowing. An active program of antibiotics and physical therapy is pursued for a week but Mr. Newman does poorly, remaining immobile and intermittently confused. While apparently in a lucid state, Mr. Newman tells a nurse that his life is no longer worth living and that he hopes the doctors will not keep him alive.

There is no previous advance directive.

Careful thought needs to be given to setting goals for the care of individuals with advanced Parkinson's disease and to consideration of potential new therapies (Siegfried and Lippitz 1994). Ideally, these goals can be set in collaboration with the patient. An advance directive can provide some guidance although there always remains the potential problem of whether a previously stated expression of wishes regarding care genuinely represents the wishes of an individual who is now demented. Likewise, although family members must be consulted, there is abundant evidence that even close relatives often fail to identify an older person's wishes accurately. It is also inappropriate for a physician to place the burden of decision making on the relatives. In Mr. Newman's case, where there is evidence that he does not wish further therapy, his physician must ultimately take the responsibility for decision making regarding further therapy.

Dr. Campbell, in discussion with Mr. Newman's family, reviews his case. It is unlikely that he will recover to independent functioning. The recent fracture and pinning have been a large setback and, if he survives the hospitalization, will probably require nursing-home care. Dr. Campbell has attempted to discuss care issues with Mr. Newman. Because of the severe Parkinson's disease, it is extremely difficult to communicate with him. He is also intermittently confused. Mr. Newman's wife and children make it clear that they feel that he would not want any sort of aggressive therapy to continue his life given his current condition. They also believe, given his strong religious faith and independent personality, that he is ready to die and that a prolonged period as an invalid would not be what he would want. It is their unanimous feeling that he has suffered enough.

Dr. Campbell speaks with members of the nursing and medical staff regarding Mr. Newman's care. Based on these conversations, her best estimate of what Mr. Newman would want, and her knowledge that Mr. Newman has a poor prognosis, she writes an order that Mr. Newman is not to be resuscitated in the event of cardiac or respiratory arrest and that the prime goal of his current care is to be made as comfortable as possible.

Questions for further reflection

1. What features on history and physical exam are diagnostic of Parkinson's disease? What other diagnoses should be considered in a patient who has features suggestive of Parkinson's disease?

2. What types of medications are useful in the management of Parkinson's disease?

3. You have made the diagnosis of Parkinson's disease in one of your patients. How would you tell her this diagnosis and

what sort of other information would you want to discuss with her?

References

Cummings, J.L. (1992). Parkinson's disease and parkinsonism. In *Movement disorders in neurology and neuropsychiatry*, (ed. A.B. Joseph and R.R. Young), pp. 195–203. Blackwell, Boston.

Godwin-Austen, R.B., Lees, P.N., Marmot, M.G., and Stern, G.M. (1982). Smoking and Parkinson's disease. *Journal of Neurology, Neurosurgery, and Psychiatry*, **45**, 577–81.

Hammond-Tooke, G.D. and Pollock, M. (1992). Depression, dementia and Parkinson's disease. In *Movement disorders in neurology and neuropsychiatry*, (ed. A.B. Joseph and R.R. Young), pp. 221–9. Blackwell, Boston.

Hoehn, M.M. and Yahr, M.D. (1967). Parkinsonism: onset, progression and mortality. *Neurology*, **29**, 1209–14.

Jenner, P., Schapira, A.H.V., and Marsden, C.D. (1992). New insights into the cause of Parkinson's disease. *Neurology*, **43**, 2241–50.

Lee, A.J. on behalf of the Parkinson's Disease Research Group of the United Kingdom (1995). Comparison of effects and mortality data of levodopa levodopa combined with selegiline in patients with early, mild Parkinson's disease, *British Medical Journal*, **311**: 1602–7.

Mayeux, R., Williams, J.B.W., Stern, Y., and Cote, L. (1984). Depression and Parkinson's disease. In *Advances in neurology*, (ed. R.G. Hassler and J.F. Christ), pp. 241–50. Raven, New York.

Rodnitsky, R.L. (1992). The use of Sinemet CR in the management of mild to moderate Parkinson's disease. *Neurology*, **42**(suppl.), 44–50.

Schulzer, M., Mak, E., and Calne, D.B. (1992). The antiparkinson efficacy of deprenyl derives from transient improvement that is likely to be symptomatic. *Annals of Neurology*, **32**, 795–8.

Shoulson, I. and the Parkinson Study Group (1989). Effect of deprenyl on the progression of disability in early Parkinson's disease. *New England Journal of Medicine*, **321**, 1364–71.

Siegfried, J. and Lippitz, B. (1994). Bilateral chronic electrostimulation of ventroposterolateral pallidum: a new therapeutic approach for alleviating all parkinsonian symptoms. *Neurosurgery*, **35**, 1126–30.

Williams, L.S. and Lowenthal, D.T. (1995). Clinical problem solving in geriatric medicine: Obstacles to rehabilitation. *Journal of the American Geriatrics Society*, **43**, 179–83.

Young, R.R. (1992). Extrapyramidal syndromes sometimes mistaken for Parkinson's disease. In *Movement disorders in neurology and neuropsychiatry*, (ed. A.B. Joseph and R.R. Young), pp. 257–61. Blackwell, Boston.

6

Depression in the elderly

Andrew Satlin

Frank Hollis, at age 69, had been a patient of Dr. Seth Goldstein's for twenty years. When Dr. Goldstein first met Mr. Hollis, he was a 49-year-old, healthy and vigorous executive in a large shoe-manufacturing company that he and his brother had built from a modest business started by their father. At that time Mr. Hollis and his wife had three children: a 23-year-old son who was beginning medical school, a 20-year-old daughter in her sophomore year of college, and a 15-year-old son in high school. Mr. Hollis had first come to see Dr. Goldstein at age 49 because of a bad case of bronchitis associated with his two-pack a day smoking habit.

Over the ensuing years, Mr. Hollis sustained pneumonia twice. The pneumonias responded quickly to oral antibiotics. After the second bout of pneumonia, Mr. Hollis initially seemed to recover well with disappearance of cough and fever and a clearing of his chest X-ray. Surprisingly, however, Mr. Hollis remained sleepless and fatigued, his appetite did not return, and, to the amazement of his wife and family, he showed no desire to return to work. Over the next month he began spending more time in bed, often remaining in pajamas until late in the day. Mrs. Hollis urged her husband to speak with Dr. Goldstein. In conversation with Mr. Hollis, Dr. Goldstein learned that he was having brief bursts of severe free-floating anxiety, had stopped taking an interest in his children, and had begun to question if life was worth living. Dr. Goldstein suspected that Mr. Hollis had developed a major depressive disorder and prescribed nortriptyline. After four weeks at a dose of 100 mg a day, Mr. Hollis gradually came to feel like his old self and he returned to work with enthusiasm. Dr. Goldstein had a difficult time explaining to his patient why he needed to remain on the antidepressant medication for at least six months. Mr. Hollis, however, finally agreed but he kept Dr. Goldstein to his promise to taper off the medication over one week as soon as the six months of treatment ended.

At age 57, Mr. Hollis was noted to have hypertension which Dr. Goldstein treated with propranolol. As he had so many times previously, Dr. Goldstein urged Mr. Hollis to quit smoking. Mr. Hollis did manage to cut back to one pack a day. Mrs. Hollis also urged her husband to reduce his hours at work. If anything, however, by the time he turned 60, Mr. Hollis was working more rather than less. He had worked hard his entire life, enjoyed the details of his business enormously, had no other lasting interests, and had no intention of retiring at age 65 or in the foreseeable future. Mr. Hollis' brother, however, had developed a passion for sailing, and when he reached age 65 he transferred his own interest and responsibilities in the company to his two sons. Almost immediately, Mr. Hollis found himself in conflict with his two nephews. Disputes about the running of the business threatened to lead to its collapse. Within a year of his brother's retirement, Mr. Hollis was forced out of the business.

Mr. Hollis took the opportunity of this unexpected retirement to begin the foreign travel that he and his wife had put off for many years. Instead of deepening their relationship, however, the increased contact led to bickering between the two about minor matters and quarrels over how much time they should spend visiting their children. By this time, their oldest son was a successful endocrinologist living 1500 miles away. Their daughter was married, raising a family, and working part-time in a suburb 30 miles away from her parents. Their youngest son had come to an accommodation with his cousins and was working in the family shoe business. After several years of retirement, Mr. Hollis, at age 69, sought to start a new company on his own. Unfortunately, the planned company failed to generate enough investment interest to be viable. The business folded before it began. One week after closing the books, Mr. Hollis experienced chest pain and shortness of breath. He had suffered a mild myocardial infarction, was admitted to the hospital, and, in the intensive care unit, was noted to have several brief periods of ventricular tachycardia. Six weeks post discharge, after a good physical recovery and a brief period in which he seemed to his family to be mildly though oddly elated and brimming with new ideas, the unmistakable signs and symptoms of depression again began to emerge. Dr. Goldstein, considering Mr. Hollis' recent cardiovascular problems, questioned if nortriptyline would again be a good choice for his patient. Dr. Goldstein asked for a consultation from Dr. Michael Stevens. A few months before, Dr. Stevens, a geriatric psychiatrist, had joined the large multi-disciplinary practice where Dr. Goldstein worked. They had shared several patients. Dr. Goldstein called Mr. Hollis and urged him to make an appointment with Dr. Stevens.

Why, at age 69, did Dr. Goldstein ask Mr. Hollis to see a geriatric psychiatrist? When does someone become 'geriatric'? There is no age at which a person suddenly becomes 'old,' and there is no one age at which the assistance of a geriatric specialist may be beneficial. In the medical care of patients, the evaluation and treatment of common conditions, such as hypertension, chronic obstructive pulmonary disease, and myocardial infarction, are the staple of internists. These problems often present in middle age. At the time that these conditions first occur, patients are typically not very different physiologically from patients in their thirties or forties and there are usually few complicating factors present in their overall management. For many persons in their seventies and eighties, however, a complex set of chronic medical conditions may be superimposed on the physiological changes in organ systems that can accompany aging. This combination of processes can affect the presentation and treatment of subsequent acute illnesses, including psychiatric illness, and a clinician with expertise in aging may be more sensitive to the subtleties of diagnosis and therapeutics with older patients.

Psychologically, the issues associated with the aging process may begin much earlier than the physiological changes commonly associated with advancing years. Feelings about growing old may emerge with an awareness that one has reached one's vocational peak, or at the time that the last child has left home, or when the first serious medical illness occurs. Psychiatric illnesses that emerge in the context of these changes are the particular interest of the geriatric psychiatrist. Most severe psychiatric illnesses, such as schizophrenia, major affective disorders, and obsessive-compulsive disorder usually begin early in life, as depression did for Mr. Hollis. When these illnesses persist or recur past middle age, they may coexist with medical disorders that can exacerbate psychiatric illness or affect the pharmacokinetics of psychotherapeutic agents. A geriatric psychiatric consultation can be invaluable in the management of these problems. Dr. Goldstein is making an appropriate use of consultation when he brings Dr. Stevens into the care of Mr. Hollis at this time.

The epidemiology of late-life depression has received much attention recently in response to the finding from the large Epidemiological Catchment Area (ECA) survey that the six-month prevalence of major depression in the elderly (defined as those above 65) was significantly lower than that in the population as a whole (Myers *et al*. 1984). This surprising result appears to contradict the commonly held belief that depression is more common in older people. Several factors probably help to explain the discrepancy. First, the ECA study was a community survey that did not evaluate persons in institutions such as hospitals or nursing homes, where rates of depression are invariably higher. Second, any depressive disorder that was possibly associated with bereavement was excluded, thereby excluding from the study many elderly persons in whom this is a common precipitant of depression.

A third factor is that the ECA study specifically sought prevalence rates of major depressive disorder, and many elderly with clinically significant depressive symptoms may not fit easily into this diagnostic category. The diagnosis of major depression requires the presence of a depressed mood plus at least four other vegetative symptoms of depression, including loss of interest and of the ability to feel pleasure; impaired concentration; guilty thoughts; suicidal ideation; and disturbances of sleep, appetite, energy, and the level of arousal. These vegetative symptoms are common in depression in younger individuals, but may be unreliable markers of depression in the elderly. For example, elderly persons with depression may not report guilt or changes in concentration. They may also have changes in sleep or energy that might be attributed to medical illness rather than invoking a diagnosis of depression. It is less likely, therefore, that major depression would be diagnosed in this group.

Clinically significant depressive disorders that do not meet the criteria for major depression are very common in the elderly. In a community study that examined the rates of depressive disorders in the elderly associated with medical illness, or cognitive change, or that was deemed clinically significant without meeting the rigorous criteria for major depression, the rate of depressive syndromes was much higher than the 1–2% percent found for major depression in the ECA study, with approximately 15% of those above age 65 who were surveyed having a significant depressive disorder (Blazer and Williams 1980).

Because elderly depressed patients are less likely to fulfill the established criteria for major depression, the presentation of depression in the elderly is commonly said to be 'masked'. It must be kept in mind, however, that the criteria were established based on studies of younger patients, and may not be appropriate for diagnosing depression in the elderly. It would be more accurate to say that the symptoms of depression are many and varied, and that different age groups can present with different constellations of symptoms in the depressive cluster. Among some very elderly nursing-home residents, for example, depressed mood is not a consistent complaint. More common, however, is apathy, loss of interest in activities, and a persistently negative attitude toward the people and events in their environment. Even in the absence of a depressed mood, these symptoms may improve with antidepressant treatment. Conversely, difficulty sleeping, or loss of appetite, are non-specific markers for a large number of potential physiological or medical disturbances in the very old, and may not predict response to antidepressants as reliably as in younger patients.

At age 69 years, and still in generally good health despite his recent myocardial infarction, Mr. Hollis presented with symptoms that were typical of depression. The diagnosis was not particularly difficult. Dr. Goldstein recognized that his patient was starting to look similar to the way that he did when he treated him for his first depression 15 years before. The history of a previous depressive episode, of course, would have aided in the diagnosis later in life even if the later presentation was not as clear or had been altered by concomitant medical illness or other factors. Depression is often a cyclical, recurrent illness. Among all patients who develop a first episode of major depression, about half will go on to have a second episode, and most of these persons will experience additional episodes. Often, as patients age, the episodes of depression become more frequent, with shorter periods of remission between them.

A number of factors may be associated with the development, persistence, and recurrence of depressive disorders in late life, as is also the case with their response to treatment. Medical illness is an important risk factor for depression in the elderly. Mr. Hollis, even though he appeared to be making a good recovery from the physical symptoms, became depressed shortly after his myocardial infarction. While depressive syndromes are found in 15% of the elderly living in the community (Blazer and Williams 1980), these same depressive syndromes are found in as many as 25–33% of patients with medical illnesses requiring out-patient or hospital treatment. Many of the medical illnesses associated with depression are common in the elderly. These include cardiovascular disease, cancer, renal failure, and endocrine abnormalities such as hypothyroidism and hypercortisolism (Cushing's disease). Metabolic disorders that can cause depression include electrolyte imbalances, hepatic encephalopathy, uremia, and diminished oxygen transport as can occur in lung disease, congestive heart failure, or severe anemia.

Prior to seeing Mr. Hollis in consultation, Dr. Stevens would have reviewed his medical record, including laboratory tests that could exclude acute medical conditions as possible causes of depression. The presence of active medical illness at the time of presentation for depression, or the development of a new medical illness over the period of follow-up, is also a predictor of a less favorable one-year outcome. Major depressive disorder is a risk factor for mortality over a 12- to 18-month period. This increased risk may be accounted for in part by the increased rates of serious underlying medical disease, particularly cardiovascular disease, found in these patients; it may also be due to other, unknown factors (Parmelee et al. 1992).

The frequent occurrence of multiple medical problems in the older person usually leads to treatment with a variety of medications. Some of these may cause depression, including antihypertensives (alpha-methyldopa, guanethidine, and the beta blockers), histamine blockers like cimetidine, immunosuppressives, and steroids. In Mr. Hollis's case, the use of propranolol may have contributed to his depression. Older persons may use alcohol or over-the-counter sleep medications in an attempt to improve the fragmented, delayed, or non-restorative sleep that can be experienced as part of the sleep changes that accompany aging. Unfortunately, use of these medications or alcohol may have only transient benefit and may, because they suppress the deep stages of sleep, actually produce chronic insomnia with long-term use. In evaluating an older person for depression, it is a good idea to ask the patient to bring to the first visit all the medications that he or she is currently using, including non-prescription drugs.

Psychological factors often play a role in the development of depression in the elderly. Bernice Neugarten has suggested that those elderly people who have successfully negotiated the stresses that accompany the transitions from childhood to adulthood and to middle age are generally well able to handle the expected changes that occur with aging (Neugarten 1979). Unexpected changes, however, are more likely to be precipitants for depression. Voluntary retirement is not typically an important risk factor for depression. In Mr. Hollis' case, however, his forced retirement may have been a significant stressor. For other elderly, retirement may be precipitated by illness, or by the need to suddenly care for an ailing spouse. Similarly, marital separation or divorce, although uncommon events in late life, are associated with higher rates of depression than widowhood. Dissatisfaction with one's social network, rather than mere contraction of social contact or physical isolation, is another risk factor for depression.

Mr. Hollis arrived at the consultation visit with Dr. Stevens accompanied by his daughter. Dr. Stevens greeted both of them in the waiting room, and then said that he would like to begin by interviewing Mr. Hollis alone, but that he would leave time to talk to his daughter later. In the office, he explained that Dr. Goldstein had provided him with a great deal of information, but that he wanted to hear about Mr. Hollis' symptoms in his own words. Mr. Hollis began slowly: 'Well, doctor, I really have never felt well since the heart attack. After the first few days in the intensive care unit, I tried hard to throw myself back into life. I wanted to return to activity. But my dad died of a heart attack at 65. Here I am at 69. Hell, I'm living on borrowed time. I just can't keep on pushing myself. I'm getting old. There is nothing I want to do. I've lost my appetite, I'm a big burden. Now here I am to see another doctor and I have my daughter all frightened and pulled her away from her kids. She had to get a baby sitter in to take care of her old man.'

Dr. Stevens listened carefully to Mr. Hollis. He asked him about his business dealings and inquired into his marriage. Mr. Hollis mentioned that he had been badly hurt by his recent business failure and that there had been friction and disappointment with his wife. He went on to complain that he was also concerned about his memory, that he was having trouble remembering facts and details, and that his concentration was so bad he was unable to even concentrate enough to read the headlines in the newspaper.

Dr. Stevens asked a few more questions: 'Mr. Hollis, you mentioned that you felt you couldn't keep pushing yourself, that your life seems over. You do not seem to have much hope for the future. Do you ever feel so bad that you feel life is not worth living?'

'Well, Doctor, it does seem bleak. I do feel down most of the time.'

'Mr. Hollis, do you ever wish that your life was over? Do you find yourself thinking about death?'

'Yes, Dr. Stevens, I think about that a lot.'

'Have you ever thought about taking your own life?'

Mr. Hollis shifted a bit on his seat. He paused a moment and then responded: 'I would never kill myself. I wish I would just go in my sleep, that would be fine. But I could never deliberately harm myself. That would be a terrible thing to do to my family.'

Dr. Stevens asked Mr. Hollis if he could follow-up on his concerns about his memory. He had noted that Mr. Hollis, although his vocabulary was excellent and his speech fluent, spoke slowly and seemed to stop talking in mid-thought. Dr. Stevens explained that he was going to ask him a series of questions that he routinely used to test the memory and thinking of his patients. Dr. Stevens went through the Mini-Mental State Examination. Mr. Hollis was off by two days on the date, but gave the day of the week correctly and was otherwise fully oriented. He seemed to need encouragement to answer the questions, sometimes prefacing an attempt by saying that he was not sure that he could answer correctly. Despite his concerns, however, Mr. Hollis could recall three unrelated words at three minutes after a distracting delay, could name common objects, could follow a three-step oral command as well as a written command, and could write a complete sentence and copy a figure of two intersecting pentagons. When asked to write a complete sentence, Mr. Hollis wrote: 'I hope you can make me feel better.'

Dr. Stevens told Mr. Hollis that he thought Mr. Hollis

was again suffering from a depressive episode, similar to the one that he had had after his pneumonia several years before. He noted that such recurrences were common in the context of another serious illness. Dr. Stevens reassured Mr. Hollis that his sense of poor concentration and memory were also likely to be due to his depression and that his thinking should improve as his depression lifted. If there were persistent memory problems, then they could discuss how to evaluate them further. Dr. Stevens asked permission to call Dr. Goldstein and discuss Mr. Hollis' visit with him. He explained that he wanted to suggest discontinuing the propranolol to Dr. Goldstein but that the choice of a new antihypertensive medication would be up to Dr. Goldstein. Dr. Stevens also told Mr. Hollis that he was prescribing for him a new antidepressant medication called paroxetine. He discussed with him that although he was helped by nortriptyline in the past, the paroxetine would probably have fewer side-effects and seemed a safer choice, given the recent heart attack.

Dr. Stevens asked Mr. Hollis if his daughter could join them. Mr. Hollis agreed and, at Dr. Stevens's request, remained in the room while they all talked together. Mr. Hollis's daughter noted that the family had tried everything they could think of to try and cheer up their father, but he just became more withdrawn despite their best efforts. Mrs. Hollis was so discouraged about her husband she could not bring herself to come to the appointment. Dr. Stevens explained that he felt it likely that Mr. Hollis was suffering from a depression, that it was a treatable illness, that the apparent personality changes were common in depression, and that, with treatment, he would recover over time.

As with medical encounters in other settings, the psychiatric evaluation of the elderly patient usually involves one or more family members. The presence or absence of family may be significant pieces of information for the physician. The presence of Mr. Hollis' daughter, rather than his wife, hinted at the stress caused in the family by Mr. Hollis' depression, but also indicated that the family had the resources to compensate, at least in part. Dr. Stevens had enough past history from Dr. Goldstein to enable him to begin the evaluation by meeting with the patient alone. In seeing the patient privately, Dr. Stevens helped to foster in Mr. Hollis some sense of self-confidence and independence. Mr. Hollis recognized that the doctor viewed him as someone who was capable of understanding the situation and could address his problems directly with him. At the same time, Dr. Stevens also showed that he valued the concerns of the family by leaving open the possibility that a family member might have important information that would aid in the evaluation. By inviting Mr. Hollis' daughter into the office after speaking with her father, Dr. Stevens stressed the importance of the family's co-operation in the treatment. This also provided the opportunity for some education about the nature of depression that could be heard both by the patient and a family member.

The expressions of hopelessness by the patient, including his expectation that he was near the end of his life, prompted Dr. Stevens to inquire carefully about suicidal ideation and intent. Questioning about suicide is a routine part of any adequate clinical evaluation of a depressed patient. Careful questioning is especially crucial in the elderly, because their rates of suicide are higher and they are less likely to communicate their intention to commit suicide compared to younger patients. Even though the elderly comprise about 12% of the population in the United States, this group accounts for about 17 to 25% of all suicides (McIntosh 1992). They also are more likely to select more lethal methods. Although only one out of every twenty suicide attempts by persons under 40 is successful, one in four is successful among those over the age of 60. In addition to depression, risk factors for suicide include male sex, white race, medical illness, alcohol abuse, and social isolation.

Many clinicians are reluctant to ask about suicide and find it difficult to question patients about self-destructive desires and plans. As Dr. Stevens did with Mr. Hollis, approaching the topic in stages that build progressively on the responses of the patient is an appropriate and relatively easy way to discuss suicidality. Most people experiencing a severe depression have already at least vaguely considered the idea of ending their life. Raising this issue in the interview will not be planting the idea in the patient's mind. Openly discussing suicide may provide a patient who is frightened by self-destructive impulses the opportunity to discuss these concerns. It may help to alleviate their despair. A thorough assessment of suicidality is essential to management and follow-up. If Mr. Hollis had indicated that he had considered specific means to end his life, Dr. Stevens would have proceeded to ask about plans and inquire if the patient thought he might actually carry them out. A clear intention to commit suicide, without the ability to agree to contact the physician if the intention becomes imminent, would warrant hospitalization for the patient's safety. If the patient did not agree to voluntary hospitalization, such a situation would constitute appropriate grounds for an involuntary commitment.

In evaluating any older person with depression, a cognitive assessment is an essential component of the workup. Dr. Stevens used the Mini-Mental State Examination (MMSE) as a screening instrument (see MMSE in Chapter 3 on 'cognitive decline in the elderly). This scale assesses most major areas of cognitive function and can be completed in less than fifteen minutes. Scores range from zero to thirty, with the lower scores indicating more impairment (Folstein et al. 1975). Frequently in the elderly, depression will be accompanied by memory loss or other cognitive dysfunction. This symptom complex has sometimes been called 'pseudodementia' because the cognitive losses that sometimes accompany depression will resemble the dementing illness that afflicts older individuals with degenerative brain disorders. The term 'pseudodementia', however, is not apt. The 'dementia of depression' (a preferable term) is as real and disabling as that caused by a presumed neurodegenerative condition.

In reaching a decision about the appropriate treatment, Dr. Stevens carefully considered the patient's past psychiatric history as well as his current medical condition. In his bout with depression fifteen years ago, Mr. Hollis had responded to nortriptyline, a tricyclic antidepressant. Generally, prior response to an antidepressant is the best predictor of future success. Mr. Hollis' history of a recent myocardial infarction, however, led Dr. Stevens to consider alternatives. The tricyclic antidepressants are chemically related to class I antiarrhythmics. In the past, this similarity supported the belief that they might be especially suitable for depressed patients with cardiac arrhythmias. But, surprisingly, the results of the Cardiac Arrhythmia Suppression Trial revealed that patients with ischemic heart disease who were prophylactically maintained on class I antiarrhythmics after myocardial infarction had higher mortality rates than similar patients on placebo. It appears that the mechanism of excess mortality involves a drug-induced increased risk of ventricular fibrillation in ischemic myocardium. Analogously, persons with ischemic heart disease who are treated with tricyclic agents for depression might be at increased risk for sudden death (Glassman et al. 1993).

Does this reasoning imply that the tricyclic antidepressants should not be used in the elderly? This conclusion is not warranted for two reasons. First, most older persons do not have ischemic heart disease, even though there is a high prevalence of this condition with advancing age. Second, not treating depression carries with it the risk of increased deaths from cardiovascular causes. Any assessment of treatment must weigh the risk–benefit ratio. In such an assessment, consideration should be given to possible alternatives to tricyclic antidepressants. All of the newer antidepressants, including bupropion, venlafaxine, and the selective serotonin uptake inhibitors (SSRIs) fluoxetine, sertraline, and paroxetine have been found to be effective antidepressants for elderly patients (Nemeroff 1994). Unfortunately, the question of relative efficacy of these newer agents compared with tricyclic antidepressants is unsettled. Tricyclic antidepressants have proven efficacy in the management of severe depression requiring hospitalization.

The data are not available to give clear guidelines as to whether to choose a tricyclic agent or a newer antidepressant when balancing severity of depression and the risks of ischemic heart disease. Several rules of thumb, based on a combination of clinical experience and findings from the literature, can, however, give reasonable guidance. For older persons with mild to moderate depression who have no evidence of ischemic heart disease, start with a tricyclic antidepressant such as nortriptyline or desipramine. For patients with a mild to moderate depression with ischemic heart disease, start with buproprion, venlafaxine, or one of the SSRIs. For patients with severe depression and ischemic heart disease, consider using a tricyclic, but with close monitoring of the electrocardiogram. A 12-lead electrocardiogram should be repeated every four to five days after each dose increase, to check for broadening of the QRS complex or QT interval. Patients with the combination of severe depression and severe ischemic heart disease, or with severe depression occurring less than six months after a myocardial infarction, present particularly difficult management issues. In such cases, admission to the hospital and treatment with electroconvulsive therapy under close cardiac monitoring might be a safer and more effective approach.

Dr. Stevens made an appropriate choice in selecting one of the SSRIs, paroxetine, given his moderate depression and recent myocardial infarction. The SSRIs also have the benefit, compared to the tricyclics, of low anticholinergic side-effects. Unfortunately, they are more likely to cause anorexia, nausea, diarrhea, and insomnia. Bupropion would have been another good choice for Mr. Hollis. Bupropion is the only one of the newer agents that has been studied specifically in patients with pre-existing cardiac disease. At standard doses over three weeks, patients had a low rate of orthostatic hypotension, no conduction problems, no change in pulse, and no exacerbation of ventricular arrhythmias (Roose et al. 1991). Bupropion, however, like the SSRIs, may cause gastrointestinal side-effects and insomnia. The differences in side-effect profile among the different agents may be crucial in affecting compliance. Several points should be stressed in discussing drug therapy with patients. Careful discussion of the side-effects to be expected and the possible courses of action if these side effects develop can help a patient to continue on therapy. The patient should also be told of the need to wait at least several weeks before expecting any therapeutic benefit. The presence of a family member for these discussions often helps to improve retention of the information, and allows the family to provide support while everyone waits for the medication to take effect.

Mr. Hollis continued to see Dr. Stevens weekly for the next month, as the dose of paroxetine was gradually increased. Four weeks after initiating treatment, many of his symptoms were objectively better and his family noted significant improvement. At this point Mr. Hollis acknowledged their observations but he felt minimal improvement in his mood. Dr. Stevens, noting the partial response, elected to make a further dose increase. Two weeks later Mr. Hollis reported that he finally felt better. A repeat Mini-Mental State Examination now resulted in a perfect score and Mr. Hollis no longer complained of poor memory or concentration.

Dr. Stevens encouraged Mr. Hollis to become involved with outside activities and suggested that he get in touch with the Council on Aging in his town. The Council had a need for volunteer drivers to deliver hot meals through the Meals-on-Wheels program to homebound elderly. Mr. Hollis agreed to volunteer for four half-days a week. He and his wife also enrolled in a music appreciation course offered in the extension program of the local university.

As with younger patients, antidepressant therapy requires at least several weeks to achieve efficacy, and the delay in response may be longer for elderly patients. An understanding of this fact is essential for compliance in the initial stages of treatment when all that the patient may experience from the drug are its side-effects. Often, objective reductions in vegetative symptoms precede the patient's own awareness of an improvement in mood.

Psychosocial interventions, including supportive psychotherapy and family counseling, are important adjuncts to the use of psychotropic medication in the treatment of depression. Efforts to get the patient involved in structured social or vocational activities frequently must be delayed. Although it is appropriate to encourage the patient to become involved in outings and recreational activities, it is not useful to force a person to participate when he or she is clearly too depressed, psychomotor retarded, without energy, or delusional. Depression is an illness that prevents function in its acute stages. Any premature attempt to overcome this inability to function may lead to a greater sense of failure and frustration. As recovery begins, participation in activities will foster a sense of mastery and will increase confidence. After resolution of the acute phase of the illness, ongoing recreational and vocational activities may be helpful in preventing relapse.

The next eight years were good ones for Mr. Hollis. He enjoyed his volunteer work. He and his wife took other classes together, and resumed some traveling, mostly to visit the children and grandchildren. About a year after beginning the paroxetine, Mr. Hollis, under Dr. Stevens' direction, tapered and discontinued the medication without any recurrence of his depressive symptoms. Mr. Hollis continued to see Dr. Goldstein, who was still unable to get him to stop smoking entirely.

At age 77, Mr. Hollis found himself unsteady on his feet one morning when he got out of bed. After a few hours, he seemed to be back to his usual self, but over the next few months he occasionally thought he detected some dizziness, or sometimes clumsiness in his hands when shaving. Over the next two years, additional similar episodes occurred. Individually, none of the episodes struck Mr. Hollis as particularly troublesome; he even managed to hide some of them from his wife. Mrs. Hollis, however, was aware of other changes in her husband that troubled her a great deal. Sometimes he had difficulty remembering the way when he was driving to deliver a meal. He began to repeat the same story he had already told her earlier in the day. Particularly upsetting for her was the fact that Mr. Hollis was frequently more irritable. At times he would yell at her about the dinners she prepared, then might suddenly appear tearful, but only for a few minutes. His mood seemed to change rapidly, and without clear provocation. One day, while on his way to deliver lunches, Mr. Hollis had a minor car accident. When approached by a police officer, Mr. Hollis could not find his registration nor explain to the officer the presence of the wrapped lunches in the seat next to him. Mrs. Hollis called the Council on Aging the next day to say that her husband could no longer serve as a volunteer.

Two years later, at age 79, Mr. Hollis awoke one day and fell when he tried to get out of bed. His right leg was weak and numb. Mrs. Hollis called the ambulance and he was brought to the hospital where it was clear that he had had a thrombotic stroke involving the left middle cerebral artery. Recovery was slow for Mr. Hollis and he was transferred to a rehabilitation hospital for further therapy. At the rehabilitation hospital, the staff noted Mr. Hollis to be lethargic, without motivation to co-operate with physical therapy, eating little, and often weepy. A psychiatric consultation was arranged.

The psychiatric consultant spent only fifteen minutes with Mr. Hollis. The stroke had left the patient's speech somewhat garbled, although word choice and fluency were not significantly affected. Mr. Hollis, however, mostly moaned throughout the interview, and when asked direct questions tended to answer 'I can't,' or 'I don't know,' even when he was asked if he felt sad. Mr. Hollis appeared fatigued, and after a short time with the psychiatrist turned around in bed and ignored the doctor. Discussing Mr. Hollis with the staff, the consultant learned that Mr. Hollis slept poorly at night, had no interest in food, and refused to work with the physical therapist. A magnetic resonance imaging scan done just before transfer to the rehabilitation

hospital clearly showed the left frontal stroke and also several 2–3 mm cavitated lesions, consistent with old lacunar infarctions, in the region of the thalamus and the basal ganglia.

The psychiatrist prescribed methylphenidate, 5 mg twice a day. Over the next week, Mr. Hollis became somewhat less weepy and more alert. He regained his appetite and would intermittently appear interested in the television. As he regained some strength in his leg, he made some attempts to assist in transferring from bed to chair. He was put back on paroxetine, and after three weeks, taken off the methylphenidate. Gradually, Mr. Hollis returned to a nearly euthymic state but with continued mood lability and brief episodes of crying that he was at a loss to explain. After twelve weeks, he was transferred to a nursing home for long-term care.

Depression is a common sequel of cortical thromboembolic strokes (Robinson *et al.* 1983), with a 30–50% prevalence after a major hemispheric stroke. Patients with functional disability similar to that of a major stroke, but due to orthopedic or other problems, have only a 10% prevalence of depression, indicating that the depression due to stroke is likely to be based on physiologic changes associated with the brain lesion itself. Further evidence for this hypothesis is the relationship between the location of the stroke lesion and the development of depression. Most studies suggest an increased incidence of depression with lesions of the left hemisphere, particularly in the frontal region. Right hemispheric lesions may be more likely to produce depression if they are posterior, but the evidence for this is not as strong. The period of risk is up to two years after the stroke.

Several studies indicate that post-stroke depression responds well to antidepressant therapy. Although only the tricyclic antidepressant nortriptyline, trazodone, and stimulants such as methylphenidate have support in the literature, it is likely that other antidepressants may be equally efficacious. The major concern with the use of these agents in patients who are post stroke is their potential effects on blood pressure and cardiac function. The newer agents may hold particular promise for their lack of deleterious effects on the vasculature or the heart. One older class of antidepressants, the monoamine oxidase inhibitors, can cause more orthostatic hypotension than the tricyclics, and probably are best avoided in older post-stroke patients.

Depression also frequently accompanies dementia. The relationship between depression and dementia is complex. As mentioned, depression can produce a reversible dementia, especially among the elderly. In persons with degenerative dementias, such as Alzheimer's disease, the rates of depression appear highest (about 25%) in the early stages of disease, suggesting that the depression may represent a psychological reaction to the loss of cognitive ability, independence, and function.

Depression may be even more common in patients with multi-infarct dementia than in those with senile dementias of the Alzheimer type. The course of Mr. Hollis' decline prior to his major thrombotic stroke, marked by episodes of transient neurological symptoms that largely resolved but left some residual dysfunction, suggests the possibility of multi-infarct disease. This disorder is characterized by a 'stepwise' deterioration, and frequently is accompanied by changes in personality and lability of mood. The distinction between a true depressive syndrome and pathological crying or mood lability due to the subcortical lesions of multi-infarct disease may be difficult to make. Mr. Hollis' brief episodes of weepiness may be part of this so-called 'pseudobulbar' clinical presentation, but his past history and present vegetative symptoms suggest a recurrence of his depressive disorder. Mr. Hollis, however, unlike his earlier presentations, is neither aware of his depressed mood or able to communicate it. The complexity of the clinical picture may not have clear therapeutic implications, as antidepressant medication may be of some benefit in uncontrollable mood lability from any cause.

At the nursing home, Mr. Hollis made a good initial adjustment. His family found it distressing that he often did not recognize them. At other times, however, he greeted his wife by name, and they were able to reminisce about their long life together. Mr. Hollis was confined to a wheelchair because of right-side weakness. Mrs. Hollis would try to push him up and down the halls, until her arthritic knees forced her to stop. With time, Mr. Hollis gradually declined further in his cognitive ability, until communication was nearly impossible. Over the ensuing four years, Mrs. Hollis visited less and less often, although she felt increasingly guilty as a result. She lost interest in the television soap operas that she had loved to watch for many years. Slowly, she developed other signs of depression herself. She denied any problem. But it was not long before she, too, was accompanied by her daughter to Dr. Stevens' office.

At age 83, Mr. Hollis was noted by the staff at the nursing home to have a sudden change in his behavior. He had not been able to communicate his needs to the staff for the last year, but had generally sat quietly in his wheelchair in the hallway for most of the day, idly watching the staff and other patients walk by. Now, over the span of a few weeks, he again lost interest in food, and would angrily push away the tray or spit out food. He seemed to have lost the ability to put on his clothes, and suddenly required total assistance for all activities of daily living. Between meals he would yell or shriek loudly, and had to be put in his room to avoid disturbing others.

At times he would try to strike at other residents as they passed by in the hall.

The nursing home physician, in an effort to control these distressing behaviors, prescribed haloperidol, 0.5 mg bid. Over the next two weeks, there was a slight reduction in the striking out, but the nurses felt that this might be due to some stiffness in his arms. The other symptoms were mostly unchanged. At times, Mr. Hollis appeared more restless, with rapid shuffling of his feet even while he was restrained in a chair. The shuffling worsened with an increase in the haloperidol to 1.0 mg bid. A psychiatric consultation was requested. The consultant noted the rapid, unexplained regression in Mr. Hollis' behavior. In discussions with the nursing staff, he identified no change in the patient's external environment in the last few weeks: no change in room or roommate, no change in staff, nor changes in the frequency of visits from family members. Further, a careful physical examination and laboratory tests did not reveal any new physical disorder. Mr. Hollis was not constipated and there was no reason to suspect that he was in pain. The consultant reviewed Mr. Hollis' previous history, and suspected that depression might have recurred, despite the lack of the typical symptoms that the patient had exhibited in the past. The consultant chose to restart the patient's antidepressant medication. Over the next month, enough of the new symptoms had remitted on paroxetine to allow Mr. Hollis to resume his usual place in the hall during the day. He began to eat again, but now required the nurses to feed him as he seemed unable to use a fork or spoon.

Elderly patients like Mr. Hollis who are severely demented and unable to report, or even to experience, depressive symptoms may develop unrecognized depression and fail to receive adequate treatment. There is some evidence that degeneration of subcortical cholinergic neurons in severe Alzheimer's disease may reduce the clinical expression of depression. A similar pathology in other dementing illnesses may occur. Thus, depression in the later stages of dementia may be more common than suspected. In order not to miss the diagnosis, a high degree of suspicion must be maintained in any demented patient. Anxiety, tension, anger, irritability, restlessness, or hyperactivity may be as common presenting features in the depressed demented patient as withdrawal, hypoactivity, or unresponsiveness to external stimulation. Depression must also be considered in the differential diagnosis of any demented patient who develops an acute or rapid regression in behavior or functioning, or who develops agitated behavior or psychosis. Unfortunately, if depression is not suspected, clinicians may instead prescribe antipsychotic medications to treat the agitation. Although some benefit may be seen, the full clinical syndrome usually will not resolve, and the patient may even become worse, with increased restlessness. Akathisia is the term that describes this motor hyperactivity in patients on antipsychotic medications and, as in Mr. Hollis' case, is manifest by rapid foot shuffling or an inability to stay still. If this side effect is not recognized, an iatrogenic vicious cycle may be the result, with increased antipsychotic medication used to treat the restlessness, only to make the symptom worse.

In the evaluation of any demented patient who has an abrupt change in mood, behavior, or functioning, one must first exclude a treatable medical cause for the deterioration, or identify an environmental change that could have been the immediate precipitant. If neither a medical nor environmental cause is identified that could be the cause for the clinical change, a presumptive diagnosis of depression may be appropriate and an empirical trial of antidepressant medication is warranted. Even in the late stages of a dementia, such treatment can be helpful and the excess disability due to the superimposed depression may be reduced. The resulting improvement can have a dramatic effect on the well-being of the patient, and on the ability of the nursing home staff to manage him or her.

Mr. Hollis's depression illustrates a number of important issues in the diagnosis and management of this disorder in the elderly: the changes in symptomatic presentation as medical and neurologic diseases are superimposed on the psychiatric illness; the role of the geriatric psychiatrist; the impact on the family; considerations in the selection of appropriate pharmacologic therapy; and the interaction of dementia and depression.

Although it is unfortunate that Mr. Hollis' stroke resulted in nursing home placement, and that his depressive disorder recurred several times in late life, between the ages of 69 and 83 he also had several long periods (years) of recovery of good function with excellent quality of life. During the 14 years that were covered in this chapter, recognition of his depression and thoughtful approaches to its treatment substantially improved the level of functioning and well-being for the patient and his family.

Questions for further reflection

1. Is depression more common in the elderly than in younger persons?

2. Which of the following pose the greatest suicide risk: a 35-year-old woman with a recent divorce, a 78-year-old woman with a chronic depression who lives with her daughter, an 80-year-old man who lives alone and has been widowed for several years?

3. Outline a reasonable approach to pharmacologic therapy of depression in an older man with moderate ischemic heart disease who is suffering from depression.

References

Blazer, D. and Williams, C.D. (1980). Epidemiology of dysphoria and depression in an elderly population. *American Journal of Psychiatry*, **137**, 439–44.

Folstein, M.F., Folstein, S.E., and McHugh, P.R. (1975). 'Mini-Mental State': a practical method for grading the cognitive state of patients for the clinician. *Journal of Psychiatric Research*, **12**, 189–98.

Glassman, A.H., Roose, S.P., and Bigger, J.T. (1993). The safety of tricyclic antidepressants in cardiac patients: risk–benefit ratio reconsidered (Commentary). *Journal of the American Medical Association*, **260**, 2673–5.

McIntosh, J.L. (1992). Epidemiology of suicide in the elderly. In *Suicide and the older adult*, (ed. A.A. Leenars, R.W. Maris, J.L. McIntosh, and J. Richman), pp. 15–35. Guilford Press, New York.

Myers, J.K., Weissman, M.M., Tischler, G.L., Holcer, C.E., and Lief, P.J. (1984). Six-month prevalence of psychiatric disorders in three communities. *Archives of General Psychiatry*, **41**, 959–70.

Nemeroff, C.B. (1994). Evolutionary trends in the pharmacotherapeutic management of depression. *Journal of Clinical Psychiatry*, **55**, 3–15; 16–17.

Neugarten, B.N. (1979). Time, age, and the life cycle. *American Journal of Psychiatry*, **136**, 887–93.

Parmelee, P.A., Katz, I.R., and Lawton, M.P. (1992). Depression and mortality among institutionalized aged. *Journal of Gerontology*, **47**, 3–10.

Robinson, R.G., Starr, L.B., Kubos, K.L., and Price, T.R. (1983). A two-year longitudinal study of post-stroke mood disorders: Findings during the initial evaluation. *Stroke*, **14**, 736–41.

Roose, S.P., Dalack, G.W., Glassman, A.H., Woodring, S., Walsh, B.T., and Giardiana, E.G.V. (1991). Cardiovascular effects of buproprion in depressed patients with heart disease. *American Journal of Psychiatry*, **148**, 512–16.

7

Congestive heart failure in the elderly

Daniel E. Forman and Jeanne Y. Wei

Frances Kyriakis considered herself a fortunate person, especially when it came to her health. She was 84 years old, and except for high blood pressure, never had significant medical problems. Tonight everything changed. She had felt well all day and went to sleep at 10 p.m., as usual, but at 2 a.m. she woke up suddenly feeling severely short of breath. She sat bolt upright in bed, propping her pillows behind her to help ease her breathing. For half an hour she sat in her bed and struggled to breath. She considered calling for help, but she kept hoping that the episode would pass. Instead, her symptoms got worse. By 2.30 a.m. she could no longer catch her breath. Looking at herself in the bedside mirror, she thought to herself, 'Frances, you're in trouble . . . You're sweaty, you can't breathe, you're coughing, and it feels like your heart is doing flip-flops.' She felt scared and miserable. Next to the mirror was a picture of her late husband Jim, who had died five years earlier. 'Darling, I might be on my way to join you . . .' She reached for the phone and called for an ambulance. 'Help me!' she gasped, 'I can't breathe! I think I'm dying!'

The ambulance came quickly. The emergency technicians put a facemask with 100 per cent oxygen on Mrs. Kyriakis and immediately began her transport to the emergency unit of the hospital. At first Mrs. Kyriakis felt confused and disoriented in the ambulance and she didn't like the sensation of the plastic face mask and cold air on her face. She kept pulling it off as the ambulance raced along.

It was 3 a.m. by the time Mrs. Kyriakis arrived in the emergency unit. Dr. Maria Rodriguez was on duty that night. As Mrs. Kyriakis' stretcher passed in front of her, Dr. Rodriguez didn't like what she saw. Mrs. Kyriakis breaths were rapid and shallow. Her face was gray and bathed in sweat, and she looked very tired. After introducing herself, Dr. Rodriguez immediately began asking Mrs. Kyriakis questions intended to clarify the history and diagnose Mrs. Kyriakis' breathing distress. Inability to converse because of breathlessness and/or confusion from hypoxia can help identify patients requiring acute intubation for respiratory support.

Mrs. Kyriakis communicated clearly. She had revived a bit since breathing some of the oxygen, though her distress and fatigue were obvious. In the first moments they talked,

Dr. Rodriguez explained the need for her oxygen mask, and Mrs. Kyriakis agreed to wear it. Mrs. Kyriakis briefly described her rocky night, and she explained that for many years her only daily medication was 25 mg of hydrochlorothiazide for hypertension. She had no prior history of myocardial infarction or angina pectoris.

Dr. Rodriguez and a nurse quickly checked Mrs. Kyriakis' vital signs. Temperature was 101°F, blood pressure 80/40 mmHg, pulse rate was irregularly irregular at about 120 beats per minute, and breathing was shallow at 24 breaths per minute. A pulse oximeter on Mrs. Kyriakis' finger indicated an oxygen saturation of 86% while Mrs. Kyriakis was wearing the facemask. Immediately, Dr. Rodriguez asked the nurse to start an intravenous line, and send blood for a complete blood count, electrolytes, urea nitrogen, creatinine, and cardiac enzymes. Another nurse connected Mrs. Kyriakis to a heart monitor.

Glancing at the heart rhythm on the monitor, Dr. Rodriguez diagnosed atrial fibrillation with a rapid ventricular response. Then, quickly examining Mrs. Kyriakis, Dr. Rodriguez saw that her neck veins were distended to the angle of the jaw, even as she sat upright. Carotid upstrokes were brisk. Chest exam was difficult as auscultation revealed diminished breath sounds. Listening carefully, Dr. Rodriguez could only distinguish basilar crackles and scattered wheezes. Mrs. Kyriakis' heart sounds were also diminished. The heart rhythm was irregularly irregular, and heart sounds S1 and S2 seemed normal. Dr. Rodriguez also detected a loud holosystolic murmur radiating from the apex to the axilla. The abdomen was soft and without any obvious masses. Her legs were slightly swollen and the skin pitted slightly as Dr. Rodriguez pressed gently but firmly on the shins. However, Mrs. Kyriakis commented that her legs have been similarly swollen for at least three years.

Even as she proceeded through her exam, Dr. Rodriguez asked the nurses to complete an electrocardiogram, and then to place a catheter in Mrs. Kyriakis' bladder, making sure to save a urine sample for urinalysis and possible culture. She also asked that radiology be called for an emergency chest film. She specified that the X-ray had to be done at bedside as Mrs. Kyriakis was too unstable to travel to the radiology suite.

Immediately, Dr. Rodriguez pulled on some gloves and an eye-shield and then reached for an arterial blood gas kit that was on a nearby supply cart. She found Mrs. Kyriakis' radial pulse thready and rapid, but easily palpable. Quickly, she performed an Allen test to check for collateral circulation before puncturing the radial artery. Her sterile prep took only moments, and she watched with concern as the dark red, pulsating arterial blood filled the syringe. She asked a nurse to press firmly on the puncture site with gauze as she walked to the adjacent doctor's station where she could arrange for the blood gas sample to be transported to the lab, and then paused to think about how best to treat Mrs. Kyriakis.

Mrs. Kyriakis is suffering from congestive heart failure. The absence of impressive râles may seem surprising, but râles are less specific or sensitive signs of heart failure in older compared to younger adults. Even Mrs. Kyriakis' peripheral edema is not specific for heart failure, and may merely be venous insufficiency, which is common among the elderly. Appropriately, Dr. Rodriguez's assessment did not hinge on her rales or peripheral edema. She was more concerned about her patient's marked respiratory distress and progressing fatigue, as well as Mrs. Kyriakis' dark (probably hypoxic) arterial blood, rapid atrial fibrillation, hypotension, holosystolic murmur, and fever. Dr. Rodriguez had to determine how to relieve Mrs. Kyriakis' respiratory problems without additionally depressing her hemodynamic state. There was very little time for reflection as Mrs. Kyriakis was near death.

Mrs. Kyriakis' scenario is typical of elderly adults. While heart failure is relatively rare among younger adults, with a prevalence of less than less than 3% in adults aged under 65 years, it climbs rapidly in advanced age. Prevalence of heart failure is greater than 20% in adults aged over 80 years. Furthermore, while prevalence of other cardiovascular diseases is decreasing overall, heart failure continues to rise among older adults. It remains the most common reason for which older adults are hospitalized (Luchi et al. 1991; Lye 1993; Wei 1994).

Before considering optimal management strategies, a knowledge of the underlying pathophysiology is essential. Certain age-related changes in cardiovascular function are inevitable and increase the susceptibility of older adults to develop heart failure. Even very healthy individuals such as Mrs. Kyriakis are predisposed to heart failure, especially when an acute stress, such as an infection, overwhelms the older person's heart, due to its limited homeostatic reserve capacity.

Mrs. Kyriakis is literally drowning as the alveolar spaces in her lungs are filling with fluid. She is susceptible to such precipitous pulmonic congestion because the physiologic demands of her high fever and increased heart rate surpass her heart's forward pumping capacity, and the blood accumulates rapidly in the pulmonary vasculature.

To understand why congestive heart failure is so common among older adults, three interrelated concepts require review: (1) age-related changes in the heart and circulation predispose to heart failure; (2) cardiac ischemia is common among the elderly, (3) diastolic heart failure is common among the elderly.

Typical cardiovascular changes associated with aging

With typical aging, a number of cardiovascular functional changes are common. Changes that occur in: (1) the vasculature, (2) the heart itself, and (3) systemic neurohormonal activity, tend to compound upon one another so as to diminish cardiovascular reserve and lower the threshold of instability.

Age-related vascular changes

Arteries stiffen with age. The intimal and medial layers of the walls of the senescent arteries tend to thicken throughout the body. There are increasing accumulations of tightly cross-linked collagen, fibrosis, and calcium within the arterial walls. In the media, the smooth muscle cells tend to undergo hypertrophy and proliferation, adding to the vessel wall thickness and rigidity.

Such age-associated arterial stiffening has important physiologic implications. The great central vessels (aorta and large arteries) in younger adults are highly elastic, a property that confers efficient pumping advantages. These large central arteries in young adults distend to accept a bolus of blood from the contracting ventricle (converting kinetic to potential energy) and then the arteries locally recoil during diastole (converting potential to kinetic energy); blood is thereby moved (pulsated) by the central arteries into the periphery. In contrast, arteries lose their elasticity with age. The heart in the older person must, therefore, supply all the force that is needed to push blood into the periphery. In addition, more work is required to push blood through the stiff, inelastic, nondistensible arteries. With these typical age-changes in the arteries, aortic pulse pressures widen. Systolic pressures increase as blood traverses the rigid arteries and diastolic pressures fall since the rigid senescent arteries lack the compliance necessary to maintain the intravascular pressures during diastole.

The effects of arterial rigidity are further compounded by the tendency of pulse waves to reflect from the rigid peripheral arterial walls and bifurcations back into the central vasculature, creating high aortic impedance in elderly adults. The central arterial impedance rises disproportionately to the increases in peripheral arterial pressure. Even normotensive elderly adults have significantly elevated aortic impedance, resulting in associated increases in the physiologic burdens on the heart (Vaitkevicius et al. 1993; Wei 1992).

Aging is also associated with altered morphology of the vascular endothelium, in addition to changes in media. The endothelial cells in young adults are homogeneous in size and alignment. In contrast, endothelium in older adults is progressively more heterogeneous in size, shape, and orientation. Consequently, the blood flow is less laminar and the vessel walls are more vulnerable to subendothelial dissection, lipid deposition, and atherosclerosis.

The aging vasculature also has intrinsically altered endothelial cell-mediated vasodilatory functional characteristics. The endothelium usually plays a key role in the ability of the vessels to respond to physiologic demands. In young adults, endothelial cell-mediated dilation of the large arteries results in important afterload reduction during physiologic stresses, facilitating the increases in cardiac output and ejection fraction. Similarly, coronary artery dilation increases blood flow in response to higher cardiac work demands. Such endothelial cell-mediated vasodilation is diminished in elderly adults, in both large central arteries as well as coronary arteries. Therefore, the associated susceptibility to cardiac ischemia due to insufficient vasodilation is increased in older adults.

Age-related myocardial changes

The age-associated changes in the vasculature are interrelated to alterations in the myocardium. As the central and peripheral arteries stiffen with age, the heart of the older person must pump against ever greater resistances, or afterload pressures. The normal physiologic response to increased afterload is myocyte hypertrophy with increased myofibrillar and other protein synthesis. Another part of the normal physiologic response is myocyte cell death.

Progressive cell death of myocytes during aging is associated with increased fibrosis and collagen deposition. Moreover, aging is associated with increased cross-linking of collagen fibers as well as calcium and amyloid deposition in the interstitium within the myocardial walls. Typically, therefore, the heart in the older person is modestly hypertrophied and becomes progressively stiffened by the cumulative myocyte and connective tissue changes.

In addition to the likely effects of hypertrophy and ventricular stiffening on cardiac performance, primary changes occur within the myocytes of an older person's heart that have important functional implications. Myocyte relaxation depends on the dissociation of ionized calcium from the myofilaments and the uptake of ionized calcium from the cytoplasm into the sarcoplasmic reticulum. Calcium is actively pumped into the sarcoplasmic reticulum. This allows actin and myosin myofilaments to disengage progressively and the ventricle to relax. However, the rate of calcium sequestration by the sarcoplasmic reticulum slows with age, due in part to age-related decreases in enzymes of mitochondrial oxidative phosphorylation and of the sarcoplasmic reticulum (Odiet et al. in press; Wei 1994). As a consequence, there is a decreased rate as well as capacity for intracellular ionized calcium handling. Myocyte relaxation progressively slows in the left and right ventricles with advancing age, even in the absence of significant ventricular hypertrophy (as measured by increased wall thickness or heart weight).

The combined effects of vascular and cardiac hypertrophy, stiffening, and impaired intracellular calcium handling confer important cardiovascular functional differences between young adult and older individuals. In young adults, most ventricular diastolic filling occurs early in diastole: blood is literally 'pulled' from the atria into the ventricular chambers as the compliant ventricles relax and dilate. In an elderly adult, however, ventricular relaxation usually occurs more slowly and sometimes incompletely, while at the same time the ventricles become stiff and noncompliant as a result of hypertrophy and fibrosis. Hence, early diastolic filling tends to be progressively decreased with advancing age. Instead, ventricular filling in older individuals depends increasingly on late diastolic filling, when atrial contraction pushes blood from the atria into the stiff, non-compliant ventricles.

Because the majority of ventricular filling occurs during early diastole in the hearts of younger adults, they tend to tolerate fast heart rhythms quite well. Even at high heart rates, blood rapidly enters the ventricle during early diastole in young adults, as the myocytes swiftly relax. In the older person's heart, however, early diastolic filling is minimal and at fast heart rates there is incomplete ventricular relaxation. This is because with increasing heart rate, the progressive decrease in the R-R interval often occurs predominantly through a shortening of diastole (by >90%), not systole. Therefore, at high heart rates in the older person, the late diastolic ventricular filling becomes progressively compromised. The ventricle becomes underfilled and cardiac output is markedly diminished. In the older person, atrial fibrillation is particularly threatening to hemodynamic stability because with age, coordinated atrial contraction becomes increasingly essential for effective ventricular filling. Exercise conditioning, certain antihypertensive agents, and long-term afterload reduction have been demonstrated to delay or partly reverse these age changes (Forman *et al.* 1992; Manning *et al.* 1991; Eysmann *et al.* 1995).

The number of pacemaker cells decreases progressively with age while the amount of fibrosis around the heart's sinus node and conduction pathways progressively increases. As a result, elderly adults are more susceptible to atrial and ventricular arrhythmias. Progressive atrial dilatation and left ventricular hypertrophy with aging further increase the propensity for these arrhythmias, and the incidence of atrial fibrillation is especially common with advancing age.

Neurohormonal changes with aging

Changes in the beta-adrenergic system tend to affect typical cardiovascular performance among elderly adults. There is progressively diminished beta-adrenergic responsiveness (both agonistic and antagonistic) with age. Manifestations include reduced maximal heart rate, diminished inotropy, increased maximal end-systolic volumes, and blunted adrenergically mediated vasodilation (Cody 1993; Lakatta 1993). One implication of such neurohormonal changes is that peripheral vascular tone may be slightly higher in older persons. This amplifies the physiologic burden on the heart of the older person, which is already forced to cope with stiffened arteries and elevated central vascular impedance. The age-associated adrenergic changes also contribute to lower maximal cardiac outputs in elderly adults, since maximum heart rates decline and minimal end-systolic volumes rise, resulting in lesser increases in heart rate and stroke volume during adrenergic stimulation.

As she struggles to breath, Mrs. Kyriakis is coping with the effects of many age-related cardiovascular changes. She has a history of hypertension, which is one typical manifestation of peripheral arterial stiffening with aging. Even though her blood pressure has been well-controlled on medication, the central vascular impedance of the aorta remains high. The pumping burden on Mrs. Kyriakis' heart is increased. It is probable that Mrs. Kyriakis' heart is hypertrophied and stiff. In addition, her coronary arteries very likely are rigid, thickened, poorly dilating, and may be even significantly atherosclerotic. She is likely to have diminished coronary flow reserve, contributing to an increased susceptibility to myocardial ischemia, especially as her heart beats rapidly in the context of a fever. Finally, Mrs. Kyriakis is not likely to be able to tolerate the atrial fibrillation. Her heart's ventricular hypertrophy and susceptibility to ischemia make it especially vulnerable to the development of filling

abnormalities and make it critically dependent on late diastolic filling from the atrial contraction. Loss of this coordinated atrial pumping function therefore often precipitates rapid hemodynamic instability and compromise in older persons.

Despite a history of good health and/or no cardiovascular risk factors, aging itself represents a risk for cardiovascular disease. There is a continuum between 'typical aging' and the development of disease that becomes more difficult to differentiate as one reaches advanced age. The sedentary lifestyle of most elderly adults and the lifelong dietary habits (including high sodium and sugar intake) typical of Western societies may further exacerbate many of the aging changes that increase the risk of cardiovascular instability.

The increased susceptibility of the elderly person's heart to ischemia

The combined effects of vascular and myocardial changes render elderly adults particularly susceptible to the development of cardiac ischemia. Ischemia refers to a relative lack of blood supply for the metabolic needs of the cells. In contrast to 'necrosis', which implies cell death, ischemia is reversible if the ischemic cells receive a prompt return of the blood flow.

Ischemia is commonly conceptualized as arterial blockage that prevents adequate blood supply to meet normal cellular needs. Ischemia can also occur if metabolic demands are high and the maximum blood flow through the open arteries is insufficient. Both age-related vascular and myocardial changes predispose elderly adults to the development of such insufficient oxygenation.

Stiff coronary arteries in older persons have limited vasodilatory capacity, reducing their potential for increased coronary flow during periods of increased metabolic demand. In addition, the aorta and major arteries stiffen with age, which translates into higher afterload pressures. Consequently, cardiac metabolic requirements for adequate pumping are increased with age.

The previously mentioned age-related changes in the vascular endothelial cells further amplify the elderly adults' predisposition to the development of supply ischemia. These changes predispose the elderly person to a high incidence of atherosclerosis with aging. Once again, a continuum may be seen between typical aging and disease: coronary arteries in older persons are more prone to develop stenotic lesions. In addition to the susceptibility to develop demand ischemia, maximum blood supply to the older person's heart is also commonly reduced.

Ischemia in hearts of older persons may increase also because the myocardium is commonly hypertrophied. Perfusion of blood from the epicardial coronary arteries to the inner myocardial regions or endocardium depends on blood traversing the thickness of the heart wall during diastole. As myocytes undergo hypertrophy in response to increased afterload pressures, blood must perfuse a greater distance through the thickened myocardium to reach the endocardium (Litwin and Grossman 1993; Lorrell 1991). Ventricular hypertrophy in young adults is associated with cardiac microvasculature proliferation that serves to maintain perfusion through the thickened walls. There is, however, relatively less microcapillary proliferation around hypertrophied myocytes in older persons. The diminished microcapillary proliferation leaves the endocardial myocytes more vulnerable to the development of inadequate perfusion (i.e., ischemia).

Finally, myocardial perfusion usually occurs during diastole, when the heart muscle is sufficiently relaxed for blood to travel through the myocardium. The age-related delay in myocellular relaxation and the increased myocardial stiffness interfere with such diastolic perfusion. Tachyarrhythmias further exacerbate relative ventricular underperfusion. This is because as heart rate increases, the shortened R-R interval does not result in much shortening of systole but rather, the diastolic interval is shortened disproportionately; this results in incomplete ventricular relaxation and insufficient diastolic time for adequate perfusion to occur in older persons.

Since ventricular relaxation is energy-dependent, ischemia due to insufficient perfusion is associated with further impaired relaxation and that creates a vicious cycle. Subsequent perfusion of blood across the myocardium is hampered by the already incomplete myocyte relaxation. Ischemia progressively worsens, with escalating myocardial stiffness. Furthermore, the acutely stiffening ischemic myocardium not only exacerbates diastolic perfusion abnormalities, but it causes the heart to become ever more resistant to diastolic filling. Blood cannot enter the stiff and non-compliant chamber. Cardiac output deteriorates precipitously as there is insufficient blood in the ventricular chamber to pump forward to the periphery.

Mrs. Kyriakis is suffering from such a cycle of cardiac demand ischemia and the spiraling cycle of impaired ventricular filling. Her fever and elevated heart rate create high metabolic demands, for which there is insufficient coronary blood flow through her rigid and poorly vasodilating coronary vessels. Occult coronary artery disease may also be present, adding to inadequate blood flow. She probably also has left ventricular hypertrophy which exacerbates her propensity for ischemia. Her ischemic heart becomes increasingly stiff, which not only increases her ischemia, but further impairs ventricular filling. Her arterial pressure falls, as there is insufficient blood entering the ventricle and therefore little blood to pump subsequently.

Systolic versus diastolic heart failure

The function of the heart – to pump oxygenated blood to the body – depends on the two portions of the cardiac cycle: systole and diastole. Although the actual pumping of blood from the ventricle occurs with contraction of the ventricle in systole, the filling of the ventricle in diastole is also critical for adequate cardiac function. When the heart cannot meet the needs of the body for oxygen-rich blood, then the heart is said to be failing. Failure can occur because of dysfunction in systole, diastole, or both.

Systolic failure generally describes abnormalities of contractile function. Clinical symptoms of systolic heart failure evolve behind the weakened left and/or right ventricles as blood accumulates in the pulmonic and systemic vasculature, respectively.

Among the elderly, impaired systolic failure is common. Coronary artery disease is one major etiology, since the progressive endothelial changes predispose the elderly to develop atherosclerotic disease (i.e., progressive myocyte death and related contractile impairment). Hypertension is another common etiology of systolic failure among the elderly as systolic function commonly deteriorates in end-stage hypertensive cardiomyopathy. Other common etiologies of systolic impairment in the elderly include valvular heart disease, diabetes mellitus, other cardiomyopathies, and thyroid disease.

As the systolic performance of the heart deteriorates in elderly adults, heart failure symptoms tend to evolve slowly. Gradually worsening exercise tolerance and weakness occur from decreased cardiac output and diminished peripheral muscle function. Such decreased peripheral muscle function in most elderly heart failure patients tends to be exacerbated by the typically sedentary lifestyles of most older adults as well as peripheral muscle deconditioning with which sedentary lifestyles are associated. Progressive peripheral fluid accumulation and hypoxia are also common heart failure sequelae among older adults.

Heart failure can also occur if there is an impairment in ventricular filling in hearts with normal systolic performance (i.e., diastolic heart failure). This is also common among elderly adults. In fact, approximately 50% of persons aged 80 years and over with heart failure have preserved or nearly normal systolic function (Luchi et al. 1991; Lye 1993; Wei 1992).

Diastolic heart failure may occur when the ventricle is too stiff to distend normally to allow adequate ventricular filling. Since blood cannot properly fill the stiff ventricles, it tends to accumulate in the pulmonic and systemic vasculature. Diastolic heart failure often arises suddenly when the vicious cycle of ischemia described above is manifest in a rapidly stiffening left ventricle with blood precipitously accumulating in the pulmonic vasculature. It is commonly associated with rapid heart rates, because they add to cardiac work demands, foreshorten late diastolic ventricular filling, and launch the cycle of progressively worsening ischemia and dysfunction. Atrial fibrillation is particularly destabilizing since the late diastolic filling mechanism of atrial contraction, a critical component of normal function in older persons, is disrupted.

Diastolic and systolic failure are frequently linked. As systolic failure evolves, ventricular pressures at end-systole steadily increase because of the inadequate emptying. This leads to elements of diastolic filling impairment. Similarly, diastolic failure is often conjoined with aspects of systolic failure: impaired diastolic filling leads to decreased cardiac output as there is insufficient blood to pump. Since the uptake of calcium by the sarcoplasmic reticulum is acutely sensitive to ischemic effects, diastolic heart failure often precedes systolic heart failure during myocardial ischemia.

Mrs. Kyriakis probably suffers from predominantly diastolic heart failure. The metabolic demands of her fever have precipitated ischemia and a progressively stiffening heart. Blood is accumulating in her lungs and her blood pressure has fallen as there is little ventricular volume to sustain an adequate cardiac output.

Moreover, it is important to remember that pathology of the heart rarely occurs in isolation among elderly adults. A series of age-related physiologic changes in other vital organs add to the vulnerability of the elderly to develop instability and add to the propensity for developing heart failure. The aging kidney, for example, has fewer nephrons and renal blood flow and glomerular filtration are decreased. Therefore, older adults are less able to mobilize excess intravascular or extravascular fluid, further exacerbating their susceptibility to the development of congestive heart failure.

Hypertension, diabetes mellitus, and coronary artery disease are other common disease states among the elderly that are associated with increased risk for heart failure. Some researchers now describe such disease states as advanced points in the continuum of normal aging, which leave the older person's heart ever more susceptible to failure. Similarly, thyroid disease, infections, and valvular heart disease are common conditions among the elderly that add to the risk for heart failure.

Therefore, for Mrs. Kyriakis and other older individuals, the notion of robust health is illusory. The normal processes of aging result in multiple physiologic alterations that leave the older person vulnerable to any insult of a fragile homeostatic balance. Even a relatively minor problem such as a urinary tract infection or a respiratory illness can tip a seemingly robust elderly adult into life-threatening heart failure and hemodynamic compromise.

One of the challenges for geriatrics and cardiovascular research is to differentiate the irreversible changes of aging from those that are modifiable and preventable. Hypertension and resultant left ventricular hypertrophy can be avoided and perhaps partly reversed in many cases, by appropriate exercise, lifestyle, and dietary changes, as well as medication. Likewise, coronary artery disease can also be modified by efforts at exercise, diet, and cholesterol lowering in early and mid adulthood. Some of the changes that are so commonly seen with aging that predispose the older person to the development of heart failure may be delayed and or avoided by a healthy lifestyle that begins when a person is still relatively young.

Dr. Rodriguez considered the possible causes of Mrs. Kyriakis' troubles and the optimal treatment. The symptoms had come on suddenly, unlike the gradual progression of symptoms typical in systolic failure. Although Dr. Rodriguez could not be positive, it seemed likely that most of Mrs. Kyriakis' current problems could be attributed to diastolic dysfunction. Her fever was a key contributor to tachycardia and to demand ischemia that resulted in ventricular stiffness and filling dysfunction. Rapid atrial fibrillation further exacerbated the ventricular filling abnormalities, leading to diminished cardiac output and hypotension.

Immediately, Dr. Rodriguez surmised that the fever had resulted from an infection. She prioritized therapy with antipyretic medications and broad-spectrum antibiotics after culturing the blood, urine, and sputum. The chest X-ray was also crucial in identifying the source and best treatment for her fever. Dr. Rodriguez considered the possibility of sepsis contributing to hypotension, and blood cultures were completed. She checked Mrs. Kyriakis' hematocrit to exclude bleeding, a less-likely cause of her hypotension. Although her quick history had revealed no evidence of previous myocardial infarction or angina, Dr. Rodriguez considered coronary artery disease in her differential diagnosis and ordered serial cardiac enzymes and ECGs. She considered that Mrs. Kyriakis might have thyroid disease and ordered thyroid function tests, although she knew these studies would take several days to complete.

Dr. Rodriguez scanned the electrocardiogram the nurse handed her, and confirmed that the rhythm was atrial fibrillation with a left bundle branch block configuration. Left bundle branch block can indicate a significant myocardial infarction, but they also occur commonly in elderly adults because of 'normal aging' effects on progressive fibrosis. Dr. Rodriguez considered whether the block was an old finding or the consequence of new ischemia or an infarct. Fortunately, an old tracing was available which indicated that it was also present a year earlier, making the diagnosis of acute infarction less likely.

Dr. Rodriguez received the report of the blood gas: pH 7.30, pCO_2 of 50, and pO_2 of 50. But in the five minutes since the arterial blood test was completed Mrs. Kyriakis' energy and color had improved. She was now complying with her facemask and supplemental oxygen. Her breathing rate was slightly slower and auscultated sounds of air movement in her lungs had improved. The pulse oximeter on Mrs. Kyriakis' finger showed 92% saturation. Consideration of intubation was postponed.

The chest film also returned. Placing it on the view screen, Dr. Rodriguez saw the findings typical of congestive heart failure: redistribution of the blood vessels to the upper zones of the lung and haziness of the pulmonary fields with interstitial and alveolar filling. No obvious pneumonia was present, but Dr. Rodriguez could not exclude this possibility in the context of so many other chest film abnormalities.

Diagnosis

When, as in Mrs. Kyriakis' case, an individual presents in frank pulmonary edema, the diagnosis of congestive heart failure is not difficult. In other less florid cases of heart failure, the diagnosis can be subtle (Hausdorff *et al.* 1994). Differentiation of systolic and diastolic failure is important because the treatment strategies may be different.

A full discussion of the many etiologies of heart failure is beyond the scope of this chapter and the reader is referred to a standard text of geriatric medicine or cardiology (Wei 1994). It is, however, important to consider, for example, in a case of systolic failure whether the heart dysfunction is due to ischemia and infarction or if it is related to a problem with the heart muscle, such as a viral myocarditis. Although coronary artery disease with resultant ischemia and infarction are common causes of systolic heart failure, it is important to consider other etiologies since management may differ.

One crucial consideration is whether the presence of valvular heart disease might be a factor in a patient's heart failure. The loud holosystolic murmur that Dr. Rodriguez detected during Mrs. Kyriakis's exam may represent a valvular abnormality. A holosystolic murmur could also be a manifestation of ischemia: energy-deficient left ventricular papillary muscles might contract abnormally, resulting in severe transient mitral regurgitation. Ischemic mitral regurgitation may also

be an important contributor to hemodynamic instability since forward output can be acutely compromised by the significant papillary muscle dysfunction-related mitral insufficiency.

Commonly, systolic heart failure evolves gradually with progressive shortness of breath and gradual worsening of exercise intolerance. In the elderly, however, symptoms may be subtle. Progressive somnolence, decreased alertness or altered mental status, and fatigue may be the only complaints. Unfortunately, some practitioners might inaccurately ascribe such atypical symptoms to dementia, depression, or 'aging', thereby delaying prompt and effective therapy.

Clinical signs of systolic heart failure often include an S3 gallop, jugular venous distention, and hepato-jugular reflux. While in younger persons inspiratory crackles and peripheral edema are reliable signs of heart failure, they are less specific for heart failure in elderly adults. Assessment of ventricular function is especially useful in clarifying whether heart failure is the result of systolic and/or diastolic abnormalities. Echocardiography and radionuclide ventriculograms provide useful estimates of ejection fraction and wall motion abnormalities. Echocardiography may be additionally useful in characterizing valve function and ventricular wall thickness.

In contrast to systolic heart failure, diastolic failure tends to manifest more abruptly. As with Mrs. Kyriakis, diastolic heart failure is often associated with sudden, marked shortness of breath and jugular venous distention. The precipitous onset of symptoms, especially in the setting of coronary ischemia (as manifest by complaints of chest pain and/or changes in the electrocardiogram) or obvious clinical stress (such as fever and tachyarrhythmias), are suggestive of diastolic dysfunction. Acute papillary muscle dysfunction may be a component in this acute process, with acute mitral regurgitation adding to the symptoms and hemodynamic instability.

Dr. Rodriguez's provisional diagnosis of diastolic failure in the case of Mrs. Kyriakis makes clinical sense. Mrs. Kyriakis' history of hypertension indicates likely left ventricular hypertrophy. The combination of a fever, atrial fibrillation, left ventricular hypertrophy and the symptoms of acute shortness of breath associated with hypotension all are consistent with diastolic dysfunction. However, Dr. Rodriguez cannot be completely sure that Mrs. Kyriakis does not have at least a partial component of systolic failure in her presentation. After she is stabilized, an assessment of ventricular function will be helpful to guide long-term management. It is important to emphasize that neither echocardiography nor radioventriculograms can clearly diagnose diastolic dysfunction. Their primary clinical utility is to demonstrate or exclude systolic dysfunction by determining the ejection fraction and the presence or absence of wall motion abnormalities.

As Mrs. Kyriakis was gasping for breath before her, Dr. Rodriguez recognized the need to do several things simultaneously: improve oxygenation, slow the heart rate, improve ventricular filling, restore blood pressure, and treat ischemia. She had already ordered supplemental oxygen to be given via a facemask. Now she ordered an 0.3 mg tablet of sublingual nitroglycerin and she ordered 10 mg of intravenous furosemide. She also asked that the patient be given an acetaminophen suppository to reduce her fever as she considered other measures to slow the rapid heart rate.

Given Mrs. Kyriakis' systolic blood pressure of 80 mmHg, Dr. Rodriguez considered cardioversion for her atrial fibrillation. Applying an electrical shock to convert atrial fibrillation to sinus rhythm may provide essential stabilization since ventricular filling would be improved. She asked that the defibrillator be charged. However, Dr. Rodriguez decided there was an opportunity first to attempt pharmacological treatment of the tachyarrhythmia. She elected to use a small amount of intravenous diltiazem. She also considered morphine administration but chose to wait for now.

Treatment

The treatment of congestive heart failure varies depending on the clinical condition of the patient, the presence of systolic or diastolic failure, and the underlying condition that has led to the heart failure. Dr. Rodriguez's acute management of Mrs. Kyriakis is based on the interrelated goals to adequately oxygenate her blood, improve her cardiac function, and remove fluid from her lungs. To achieve these goals, and not risk doing more harm than good, Dr. Rodriguez must proceed rapidly but carefully, basing her therapeutic decisions on her knowledge of the likely pathophysiology.

Dr. Rodriguez knows that a person in pulmonary edema is not effectively exchanging oxygen and carbon dioxide. Not only does the edema fluid in the alveoli block gas exchange, but mismatches in ventilation and perfusion contribute to poor oxygenation and increased carbon dioxide. Individuals in pulmonary edema need supplemental oxygen. Frequently, in an emergency room setting, supplemental oxygen is given via facemask. A variety of amounts of oxygen can be delivered. Measurement of arterial blood gases can be very useful in determining the effectiveness of the supplemental oxygen in improving the arterial oxygen content as well as determining if hypercarbia or acidosis is present. Not uncommonly, individuals

in pulmonary edema may require endotracheal intubation and mechanical ventilation because of inadequate oxygenation or increased amounts of carbon dioxide. Mrs. Kyriakis' pO_2 of 50, pCO_2 of 50, and pH of 7.30 indicate that she may well require intubation if she does not improve quickly. The low pO_2 is a clear sign of inadequate oxygenation. The high pCO_2 and low pH reveal that Mrs. Kyriakis is in acute respiratory failure: she is not removing carbon dioxide from her circulation rapidly enough and there is a resultant respiratory acidosis. It is appropriate that Dr. Rodriguez order supplemental oxygen via facemask. Still, she will have to recheck a blood gas in a short time to see if the arterial oxygen content is improved but also to make sure that Mrs. Kyriakis is not further retaining carbon dioxide and becoming more acidotic. Typically, carbon dioxide retention makes people more somnolent and weak. The fact Mrs. Kyriakis shows improved alertness and breathing after supplemental oxygenation is initiated suggests that carbon dioxide retention is not occurring.

Dr. Rodriguez orders that the acetaminophen be given promptly in order to reduce Mrs. Kyriakis' fever. Fever is a stimulus for tachycardia and increased cardiac metabolic demand. Acetaminophen works well to at least partially suppress the fever and interrupt the spiral of ischemic diastolic dysfunction. In some situations where the patient has a very high fever, consideration should be given to the use of a cooling blanket or cold compresses.

Fever and hypotension also raise the possibility of sepsis. Blood cultures, sputum cultures, urinalysis and urine cultures are all completed. Likewise, a blood count sample has been sent. An intravenous antibiotic has also been promptly started.

Dr. Rodriguez realizes that a key to treating Mrs. Kyriakis is to slow her heart rate and, ideally, to restore sinus rhythm. With a slower heart rate there will be more time for ventricular filling, decreased myocardial oxygen consumption, and improved myocardial function. Even if sinus rhythm cannot be restored immediately, a reduction in the ventricular response rate will likely lead to improved blood pressure and decreased pulmonary congestion. The therapeutic decision is how best to slow the cardiac rate. The use of an electrical shock to restore a normal rhythm is used emergently in the setting of a non-sinus tachycardia (i.e. ventricular tachycardia or atrial tachycardia, such as rapid atrial fibrillation, rapid atrial flutter, or another supraventricular tachycardia) when the patient is experiencing chest pain, pulmonary edema, or hypotension. Mrs. Kyriakis is clearly hypotensive and in pulmonary edema. She is not describing chest pain, but elderly adults more often develop symptoms of shortness of breath than chest pain when stricken with cardiac ischemia. The low threshold for ischemic ventricular stiffening precipitates diastolic dysfunction and pulmonic congestion before other symptoms develop.

There are also a variety of potent drugs available to slow cardiac rate and Dr. Rodriguez opts to try this pharmacologic therapy first. In current practice, short-acting intravenous beta-blocking agents or intravenous calcium channel-blocking agents are frequently used. One normally chooses a drug based on its mechanism of action, the likely side-effects, and the ability to treat these side-effects. One should not switch indiscriminately from one class of drug to another without considering drug interactions and synergistic effects.

Intravenous cardiac glycosides, such as digoxin and ouabain, are common agents for rate control. Recently, many physicians have come to feel that their narrow window of safety, slow onset of action, and limited capabilities as rate-controlling agents renders them second line drugs in emergent situations such as Mrs. Kyriakis' instability. Alternative options include beta-blocking drugs and calcium channel-blockers.

Intravenous esmolol is a useful beta-blocker for the control of ventricular rate in individuals with rapid supraventricular tachycardias. It can be titrated precisely to afford maximum benefits and minimal side-effects. However, many emergency room physicians are not completely comfortable with mixing and titrating esmolol, and therefore small intravenous doses of metoprolol may be preferable. In general, beta-blockers merit certain clinical precautions. They can result in hypotension and they can promote bronchospasm in individuals with bronchospastic disease. Acute administration of beta-blockers can also reduce inotropic action of the heart and actually worsen heart failure in some patients with systolic dysfunction.

Two calcium channel-blocking agents, intravenous verapamil and diltiazem, may be useful in the setting of a rapid supraventricular arrhythmia. Caution is essential since both can cause hypotension. Like beta-blockers, both are also negatively inotropic, with verapamil having a more negative inotropic effect than diltiazem. Beta-blockers and calcium channel-blockers can also induce heart block, particularly among elderly adults in whom sick sinus syndrome conduction disease is common. Appropriate precautions with atropine and/or external cardiac pacing capabilities are mandated during their administration.

In a patient with rapid atrial fibrillation or rapid atrial flutter, one should attempt to exclude a history of previous arrhythmias associated with a concealed bypass tract, as is the case with Wolff–Parkinson–White or Long–Ganong–Levine syndromes. In these individuals, the use of a calcium channel-blocking agent or digoxin may facilitate conduction along the bypass tract with further clinical deterioration. Beta-blockers or other antiarrhythmic drugs would be preferable in those situations.

In addition to rate and fever control, Dr. Rodriguez also wants to palliate the acute cardiac fluid instability. She orders a small dose of intravenous furosemide, a potent loop diuretic. Furosemide, and the other diuretics in its class, act by blocking chloride absorption in the loop of Henle. This results, in most cases, in a prompt and impressive increase in urine flow. It is beneficial in reducing Mrs. Kyriakis' pulmonary congestion and in re-establishing fluid equilibrium. However, furosemide and other powerful diuretics can lead to an excessive reduction in preload, a potentially critical loss to a stiffened ischemic heart that is intrinsically resistant to filling. Overdiuresis can, therefore, worsen heart failure and further compromise ventricular filling, thereby causing further decreases in cardiac output and blood pressure. For this reason, Dr. Rodriguez begins with a relatively low dose (10 mg) of furosemide. She recognizes that the removal of a small amount of excess intravascular fluid could lead to a marked improvement in oxygenation and resultant improvement in cardiac function. The use of furosemide entails risk but this is mitigated by the low dose and the potential benefits.

Dr. Rodriguez also ordered a small dose of sublingual nitroglycerin. Nitrates are vasodilators and are useful particularly in the management of acute heart failure. The vasodilation decreases both the amount of blood entering the heart (preload reduction) as well as reduces the afterload. Nitroglycerin also dilates the coronary arteries and can relieve ischemia, leading to improved myocardial performance which may be particularly useful in modifying acute ventricular ischemic stiffness. Just as with diuretic therapy, one must be cautious with nitrates since reducing preload can potentially compromise ventricular filling and further reduce the already low blood pressure. Any further blood pressure reductions may undermine renal or cerebral perfusion, or exacerbate ischemia in other vital organs.

The risks of nitroglycerin are balanced in Mrs. Kyriakis by the benefit of decreasing coronary ischemia, improving myocardial blood flow, and improving diastolic function. Dr. Rodriguez elects to use a small dose of sublingual nitroglycerin initially because she needs to carefully monitor the blood pressure and can add more tablets as needed. Because it is difficult to stop the effect of the sublingual nitroglycerin once the tablet has dissolved, it is prudent to begin with the smallest dose available to avoid instability. She can also add intravenous nitroglycerin as needed. Intravenous nitroglycerin, however, is extremely potent, requires careful monitoring, precise administration, and the presence of personnel skilled in its use. A large or rapid bolus may be harmful. Other physicians in this setting might also choose to use a topical nitrate preparation in which the nitroglycerin is part of an ointment that can be applied to the skin. If hypotension worsens, it can be wiped off.

Intravenous morphine is a potent vasodilator and has substantial value in treating heart failure patients by shifting preload volume away from the heart. It is particularly beneficial for acute systolic failure. Its benefits in diastolic heart failure are also substantial, but those benefits may at times be outweighed by respiratory or hemodynamic risks in certain settings. Since Mrs. Kyriakis is already in respiratory failure and she is now receiving nitrates and diuretics, Dr. Rodriguez decides to wait a few minutes before making the decision about morphine.

After administering an acetaminophen suppository, Dr. Rodriguez began intravenous diltiazem therapy for heart rate control. To prevent ill effects from heart block, she made sure that atropine was also nearby as she administered a low dose of diltiazem (15 mg) in a slow intravenous bolus over 2 minutes. Shortly after finishing the slow bolus she noticed on the monitor that Mrs. Kyriakis' rate had slowed from 120 to about 90 beats per minute. Dr. Rodriguez quickly took Mrs. Kyriakis' blood pressure and found it to be mildly improved at 90/50 mmHg. She asked that the nurses prepare a continuous intravenous infusion of diltiazem. Acute myocardial infarction remains a possibility, and Mrs. Kyriakis is asked to chew one aspirin (325 mg) tablet for its beneficial antiplatelet aggregating effects.

After another half-hour, Dr. Rodriguez was gratified at the improvement in Mrs. Kyriakis' condition. She remained in atrial fibrillation, but her heart rate was now staying in the low 90s with a blood pressure close to 100 mmHg systolic as the low dose diltiazem drip continued. Mrs. Kyriakis felt better and told the doctor her breathing was easier. Her fever had improved since receiving acetaminophen and ampicillin with a beta-lactamase inhibitor (sulbactam). About 150 cc of cloudy yellow urine had drained from her bladder. A repeat blood gas revealed a pO_2 of 95, a pCO_2 of 40, and a pH of 7.35.

Dr. Rodriguez took a sample of urine to the laboratory and placed it in the centrifuge. She examined the sediment under the microscope and saw sheets of white cells. She took a drop of unspun urine and let it dry on a microscope slide, then carefully stained it. It revealed gram-negative rods. As Dr. Rodriguez documented the urinalysis data in Mrs. Kyriakis's chart, her blood test results returned. White count was 14 000 with a normal hemoglobin, hematocrit, and platelet count. Glucose was moderately elevated at 180 mg/dl, but electrolytes and renal function were normal (blood, urea, nitrogen/ creatinine at 18/1.0). Dr. Rodriguez concluded that Mrs. Kyriakis had a urinary tract infection with fever that lead to tachycardia, atrial fibrillation, and resultant diastolic heart failure.

Mrs. Kyriakis listened carefully as Dr. Rodriguez explained to her what she thought had happened and what was planned. Mrs. Kyriakis was relieved to hear that she probably had not had a heart attack but that there would be further tests to be sure. She understood that she was in an abnormal rhythm but that it was not dangerous so long as it did not go too fast, and that she probably had an infection in her urine that might have been the cause of all the breathing problems.

Dr. Rodriguez began to write transfer orders for Mrs. Kyriakis to go from the emergency room to a hospital bed. Antibiotics would be continued for her urinary tract infection and to minimize possible sepsis. Antipyretics would be continued every 4 hours as needed. Cardiac monitoring was continued along with a diltiazem drip. Cardiac enzymes and electrocardiograms would be checked every 8 hours over the next 24 hours to rule out a myocardial infarction. A low-dose aspirin tablet would be given each day to minimize risks and possible sequelae of supply ischemia. Furthermore, after checking to see that her stool was negative for occult blood, a low bolus of heparin (5000 Units) with slow intravenous drip (700 U/h) would be started for antiischemic as well as thromboembolic benefits while she remained in atrial fibrillation. Dr. Rodriguez looked up at the clock with mild surprise and realized that a lot had happened over the last hour!

Management of systolic failure

Prognosis for systolic heart failure is grim. The Framingham Heart Studies data show that over 60% of older men and 40% of women of all ages died within five years of onset of systolic heart failure symptoms (Kannel 1989). Mortality is even greater for older adults. Therapeutic goals for heart failure include improved function as well as reduced mortality. Treatment of acute systolic failure must address the needs of a dilated, poorly contracting heart. Weakened hearts initially dilate to accrue mechanical work advantages through configurational changes, but the heart fails subsequently when dilation is too great and/or systemic volumes or demands overwhelm the weak heart's limited pumping capacity.

Traditional therapy for systolic heart failure centered on redistribution of fluid away from the overly burdened heart as well as on improving the heart's contractile performance. More recently, it has been appreciated that modifying the renin–angiotensin system also plays a pivotal role in heart failure management.

With systolic or diastolic heart failure, there is insufficient cardiac output to meet the body's metabolic needs. In an effort to regulate cardiac output and fluid balance there are specific neurohumoral responses. Renin is released from the juxtaglomerular apparatus, cells located near the renal glomeruli. Norepinephrine is released from the adrenal medulla. These hormones initially stimulate improved contraction (inotropy), myocardial relaxation (lusitropy), and hemodynamic stability via retention of fluid. Unfortunately, over time, the benefits ebb with an ultimate cost of progressive cardiac strain, fluid overload, and other factors that contribute to long-term decline. Several clinical trials have indicated that modifying the neurohumoral response reduces mortality in systolic heart failure.

Diuretics remain a mainstay of acute systolic heart failure therapy since intravascular volumes are generally reduced and the heart can more easily accommodate the lower pumping demands. Diuretics, however, do not reduce mortality. It is now thought that the high mortality relates to the neurohormonal stimulation that persists and even worsens with diuretic therapy.

Still, diuretics provide symptomatic improvement, especially for acute systolic failure. They must be used with caution in older patients as the side-effects are notorious. Hyponatremia, hypokalemia, hyperglycemia, and increased serum uric acid and gout are all common. Furthermore, the brisk diuresis caused by agents such as furosemide can cause or aggravate urinary incontinence, especially in those individuals who cannot get to the toilet quickly. Many patients are reluctant to complain of urinary complications and will end up not taking their diuretic medication. In general, small doses are used initially to minimize complications, titrating the dose upward as needed.

Digoxin's use as an inotropic agent for systolic failure remains somewhat controversial. Many studies indicate that digoxin improves the functional capacity of heart failure patients. Digoxin also has benefit in slowing ventricular response rate with atrial fibrillation. However, digoxin has a narrow therapeutic window and side effects are common, especially in elderly adults whose renal clearance of the drug is reduced. Serum digoxin levels must therefore be carefully monitored in older patients and are best maintained in the mid-to-low normal range. High serum digoxin levels do not increase inotropic effect, and increase the potential for toxic effects including atrial and ventricular arrhythmias, as well as nausea, anorexia, and vision changes.

Vasodilators represent another class of therapeutic agents that are effective therapy for systolic failure, again with the goal to minimize cardiac blood volumes. Vasodilators can act on the venous circulation, the arterial circulation, or both, depending on the agent. Vasodilators such as morphine, nitrates, and nitroprusside are effective in acute heart failure since they rapidly reduce preload and afterload volumes, thereby improving pumping efficacy. Dobutamine also lowers afterload with arteriolar dilation. It is an effective medication for a severely weakened heart in acute failure since it helps the heart to pump harder while it reduces afterload pressures.

In contrast to diastolic failure, with systolic heart failure the heart tends to fill with minimal resistance. Therefore, in systolic failure therapy, preload reduction with nitrates, morphine, and diuretics proceeds with less hemodynamic risk. Total ventricular filling is preserved despite the reduced preload pressures.

Angiotensin-converting enzyme (ACE) inhibitors are useful for systolic failure. They are arteriolar vasodilating agents. Earliest trials of ACE inhibitors showed mortality and functional benefits that were attributed entirely to their lowered afterload effects. Subsequent trials demonstrated that other purely arteriolar vasodilator regimens, such as prazosin and hydralazine, although beneficial, did not provide similar mortality benefits. The greatest mortality benefit of ACE inhibitors may relate, therefore, to their unique ability to modify the neurohormonal response.

Use of ACE inhibitors has become the first line therapy for chronic systolic heart failure. Use of combined hydralazine and nitrates has also been associated with mortality and functional benefits, but ACE inhibitors are the preferred medications, probably in part because of their important neurohormonal effects. The benefits of ACE inhibitors for acute systolic failure are unproven. They may even increase mortality when administration for acute failure is associated with secondary hypotension. In stable asymptomatic adults with weakened hearts, use of ACE inhibitors has been shown to significantly forestall the onset of congestive symptoms and to yield trends of reduced mortality. The major studies showing ACE inhibitor benefits in mortality enrolled only younger adults and used very high doses of ACE inhibitors that would not be well tolerated by most elderly persons. For now, many clinicians empirically extrapolate from the heart failure trials in younger adults and simply choose lower doses of ACE inhibitors for their elderly systolic heart failure patients. Ongoing research should help to clarify which agents in which doses create optimal benefits for older adults.

Because of their peripheral vasodilating effects, many clinicians assume calcium channel-blockers are also useful for chronic systolic heart failure therapy. Unfortunately, the negative inotropic effects of the first generation of calcium channel-blockers (verapamil, diltiazem, and even nifedipine) apparently outweigh their beneficial vasodilating effects. The newer second generation calcium channel-blockers, including amlodipine and felodipine, may provide vasodilatation with less reduction of contractility. These medications are currently under investigation for their efficacy.

During the morning of her first day of hospitalization, Mrs. Kyriakis underwent echocardiography. Left ventricular systolic function was normal and the ventricle was concentrically thickened, as is often the case in elderly hypertensive adults. Only mild mitral regurgitation was present and the left atrial size was normal. The echocardiogram helped to confirm the hypothesis that diastolic heart failure was her main problem. It also provided more evidence that suggested that the loud murmur heard when Mrs. Kyriakis came to the emergency room was a consequence of papillary muscle dysfunction due to ischemia. This valvular dysfunction had improved after the cycle of ischemia was successfully interrupted.

Serial creatine phosphokinase (CPK) enzymes drawn over 24 hours showed no evidence of heart damage, and serial electrocardiograms showed persistent left bundle branch block even when sinus rhythm was restored.

After 24 hours, her heart rhythm spontaneously converted to sinus rhythm. Heparin was stopped. Still, diltiazem was continued with oral dosing. Diltiazem provides effective antihypertensive benefits, potential rate control in a patient with a propensity for atrial fibrillation, and benefits in reducing left ventricular hypertrophy.

Mrs. Kyriakis's urine culture revealed the presence of *Escherichia coli* sensitive to ampicillin. Treatment with oral ampicillin was planned for the next 10 days. A follow-up appointment was scheduled for Mrs. Kyriakis with her primary care physician.

Mrs. Kyriakis was discharged to home in good spirits after two days in the hospital. She looked strong and felt well. There was little to indicate how ill and precariously close to death she had been two days earlier.

Mrs. Kyriakis' care after the acute emergency was relatively straightforward. Diltiazem was chosen as it had worked in the acute setting. It is a good antihypertensive agent, useful in the management of potentially recurrent supraventricular arrhythmias, and helps to reduce left ventricular ischemia. The provoking event of urinary tract infection required a course of antibiotics and out-patient follow-up.

The echocardiogram provided important information for management. There was no indication for digoxin in her medical regimen since she did not suffer from systolic failure and diltiazem had provided sufficient rate-controlling benefits. Likewise, the echocardiogram helped to exclude other causes of failure such as significant valvular disease, which would require other management strategies.

If Mrs. Kyriakis had not converted spontaneously to normal sinus rhythm, management decisions for atrial fibrillation would have become more complicated. Since her atrial fibrillation was of very recent onset and the left atrium was normal in size, many clinicians might have attempted to convert her to sinus rhythm with antiarrhythmic therapy (e.g., quinidine or sotalol)

or electric cardioversion. Although other options might include letting the atrial fibrillation persist with continued anticoagulation using coumadin to prevent cerebral embolism, chronic atrial fibrillation is known to increase morbidity and mortality, so conversion to sinus rhythm is almost always worth considering. The treatment of atrial fibrillation is an important and common problem that is often encountered in caring for older persons (Disch *et al.* 1994; Wei 1994).

Questions for further reflection

1. Describe how systolic failure and diastolic failure might be clinically differentiated.
2. What role does myocardial handling of calcium play in the development of diastolic dysfunction?
3. Outline how you would care for an older person in acute pulmonary edema. What special concerns are appropriate with regard to nitrates, diuretics, and the use of morphine?

References

Cody, R.J. (1993). Physiological changes due to age: Implications for drug therapy of congestive heart failure. *Drugs and Aging*, 3, 320–34.

Disch, D.L., Greenberg, M.L., Holzberger, P.T., Malenka, D.J., and Birkmeyer, J.D. (1994). Managing chronic atrial fibrillation: A Markov decision analysis comparing Warfarin, Quinidine, and low-dose Amiodarone. *Annals of Internal Medicine*, 120, 449–57.

Eysmann, S.B., Douglas, P.S., Katz, S.E., Sarkarati, M., and Wei J.Y. (1995). Left ventricular mass and diastolic filling patterns in quadriplegia and implications for effects of normal aging on the heart. *American Journal of Cardiology*, 75, 201–3.

Forman, D.E., Manning, W.J., Hauser, R., Gervino, E., Evans, W., and Wei, J.Y. (1992). Enhanced left ventricular diastolic filling associated with long-term endurance training. *Journal of Gerontology*, 47, M56–8.

Hausdorff, J.M., Forman, D.E., Ladin, Z., Goldberger, A.L., Rigney, D.R., and Wei, J.Y. (1994). Increased walking variability in elderly persons with congestive heart failure. *Journal of the American Geriatrics Society*, 42, 1056–61.

Kannel W.B. (1989). Epidemiological aspects of heart failure. *Cardiology Clinics*, 7, 1–9.

Lakatta, E.G. (1993). Deficient neuroendocrine regulation of the cardiovascular system with advancing age in healthy humans. *Circulation*, 82, 631–6.

Litwin, S.E. and Grossman, W. (1993). Diastolic dysfunction as a cause of congestive heart failure. *Journal of the American College of Cardiology*, 22, 49A-57.

Lorrell, B.H. (1991). Significance of diastolic dysfunction of the heart. *Annual Review of Medicine*, 42, 411–36.

Luchi, R.J., Taffet, G.E., and Teasdale, T.A. (1991). Congestive heart failure in the elderly. *Journal of the American Geriatrics Society*, 39, 810–25.

Lye, M.D.W. (1993). Chronic cardiac failure in the elderly. In *Textbook of geriatric medicine and gerontology*, (ed. J.C. Brocklehurst, R.C. Tallis, and H.M. Fillit), pp. 108–205. Churchill Livingstone, London.

Manning, W.J., Shannon, R.P., Santinga, J.A., Parker, J.A., Gervino, E.V., Carl, L.V., and Wei, J.Y. (1991). Reversal of changes in left ventricular diastolic filling associated with normal aging using diltiazem. *American Journal of Cardiology*, **67**, 894–6.

Odiet, J.A., Boerrigter, M.E., and Wei, J.Y. (1995). Carnitine palmitoyl transferase-I activity in the young and old mouse heart. *Mechanisms of Ageing and Development*, **79**, 127–36.

Vaitkevicius, P.V., Fleg, J.L., Engel, J.H., O'Connor, F.C., Wright, J.G., Lakatta, L.E., *et al.* (1993). Effects of age and aerobic capacity on arterial stiffness in healthy adults. *Circulation*, **88**, 1456–62.

Wei, J.Y. (1992). Age and the cardiovascular system. *New England Journal of Medicine*, **327**, 1735–9.

Wei, J.Y. (1994). Disorders of the Heart. In *Principles of geriatric medicine and gerontology*, (3rd edn) (ed. W. Hazzard), pp. 517–32. McGraw-Hill, New York.

8

Thyroid disorders in the elderly

Harold N. Rosen

The beeper went off, seemingly for the thousandth time that morning. The coronary care unit rotation had been demanding, although a good learning experience. Renee Ginot, a medical intern, looked down to stop the annoying paging signal and check the number to which she had been paged. 'Hmmm . . . this is an outside call.' She walked off to the phone at the nurses' station.

'This is Dr. Ginot, how can I help you.'

'Dr. Ginot, this is Tillie Benson.'

'Yes, Mrs. Benson, what's up?' Dr. Ginot remembered Mrs. Benson, an 82-year-old woman, from her visit to the out-patient clinic two months ago.

'Doctor, for the last two days, my breathing has been real hard. Just to walk around the house seems too much. I am all right if I sit still, but I am worried.'

'Mrs. Benson, I am seeing patients in clinic this afternoon. Could you come in about 1 p.m. and I'll squeeze you in?'

'Yes, I'll get a ride from my neighbor. Thank you, Doctor.'

After a hurried lunch, Dr. Ginot reviewed Mrs. Benson's chart before seeing her. Her past medical history was remarkable for three major problems. First, Mrs. Benson underwent a left modified radical mastectomy seven years ago for stage I breast cancer. She has done well and has been free of recurrence. Second, three years ago Mrs. Benson had suffered an anterior wall myocardial infarction complicated by mild congestive heart failure. After the infarct, Mrs. Benson's stress test was normal and her cardiac echocardiogram revealed a decreased ejection fraction of 35% with anterior wall hypokinesis and normal valves. The congestive heart failure had not been a major problem on a regimen of digoxin 0.125 mg/day and enalapril 10 mg daily. Her digoxin level was stable at 1.2 ng/ml. Mrs. Benson's third significant problem was a mild memory loss of gradual onset. Dr. Ginot had first met Mrs. Benson when she came to the clinic two months ago with this complaint. Physical exam was normal as were routine serum chemistries, cell blood count, thyroid hormone, vitamin B_{12} and folate levels. A computerized tomogram (CT) of the brain, done with intravenous contrast, showed only slight atrophy. Dr. Ginot was concerned about the possibility of early dementia of the Alzheimer type and had planned to see Mrs. Benson again in six months.

Dr. Ginot greeted Mrs. Benson and took her history. Usually, Mrs. Benson could walk for three blocks before getting short of breath, but now going from the kitchen to the living room resulted in shortness of breath and forced her to sit down. Mrs. Benson also complained that she felt quite tired and that she was more comfortable lying down with several pillows to prop her up. She denied any swelling in her legs, chest pain, fever, wheezing, or cough. Dr. Ginot was alarmed by Mrs. Benson's physical exam. She had an irregularly irregular pulse of 130, blood pressure of 160/80 and a respiratory rate of 30. She was afebrile. Her lung exam was remarkable for râles heard bilaterally and a cardiac exam without murmur but revealing the irregularly irregular and rapid rhythm. Dr. Benson brought in the electrocardiogram machine and prepared Mrs. Benson for the tracing. The ECG revealed atrial fibrillation with a ventricular response of about 130, normal axis, Q waves in the anterior precordial leads and some ST flattening in the inferior and lateral leads. Dr. Ginot quickly compared the current tracing with a previous one in the chart. The atrial fibrillation was new as was the inferolateral ST flattening. Dr. Ginot called the attending physician who supervised the resident's clinic. He concurred with Dr. Ginot that Mrs. Benson needed to be admitted to the hospital and helped to make the arrangements for Mrs. Benson to be transferred to the coronary care unit. Dr. Ginot smiled ironically when she realized she was giving herself another admission, not something an intern usually wanted!

Mrs. Benson's clinical presentation seems fairly straightforward. She was well until she became short of breath and found to be in new onset atrial fibrillation with a rapid ventricular response. Most likely, Mrs. Benson went into rapid atrial fibrillation which precipitated an exacerbation of congestive heart failure (CHF). The clinical question was why Mrs. Benson's heart rhythm suddenly changed from sinus to atrial fibrillation. Unfortunately, as Dr. Ginot was aware, another explanation was possible: Mrs. Benson may have suffered an acute ischemic event that caused both the CHF and atrial fibrillation. The immediate clinical priorities were to control the rate of Mrs. Benson's atrial fibrillation and treat her CHF.

Over the next few hours, Mrs. Benson improved. With oxygen, furosemide, topical nitroglycerin ointment, and small intravenous doses of digoxin, Mrs. Benson's heart rate slowed a bit to 110, her breathing became easier, and she urinated some 600 cc. Admission laboratories were normal and showed no evidence for infection. The chest X-ray revealed moderate CHF without focal infiltrates and also revealed mild tracheal deviation to the left. Dr. Ginot was surprised to find that the digoxin level returned low at 0.3 ng/ml. An initial blood sample was negative for enzymatic evidence of myocardial infarction and a repeat electrocardiogram revealed resolution of the inferolateral ST flattening with slowing of the rate. Dr. Ginot considered the possible causes of Mrs. Benson's atrial fibrillation. Her echocardiogram from before showed no evidence for valvular heart disease and her cardiac exam was without murmurs, making it unlikely that mitral valve disease was a cause. There was little evidence for a pulmonary embolus, although it was always a necessary consideration in the differential diagnosis of new atrial fibrillation. Hyperthyroidism remained a possibility, although Dr. Ginot had obtained normal thyroid function tests two months ago at the time of Mrs. Benson's evaluation for memory loss.

There are a number of possible causes of hyperthyroidism. Hyperthyroidism is usually caused by Graves' disease, where the thyroid is stimulated by abnormal immunoglobulins to secrete excessive amounts of thyroid hormone. Another frequent cause of hyperthyroidism is multi-nodular goiter, where a thyroid gland is enlarged by the presence of numerous thyroid nodules, which can begin to secrete thyroid hormone autonomously. As a cause of hyperthyroidism, multinodular goiter becomes more common with increasing age. A third cause of hyperthyroidism can occur after a patient, especially a patient with an abnormal thyroid gland, has received iodide, either in a cough medication (such as elixir of organidin) or after a diagnostic test which involves the use of radiopaque iodide as intravenous contrast. Thyroiditis can also lead to hyperthyroidism when the inflamed gland releases thyroid hormone in an uncontrolled fashion. In patients with thyroiditis, the thyroid gland may be painful but it often is not. Some individuals may be hyperthyroid as a consequence of excessive ingestion of exogenous thyroid hormone (thyrotoxicosis factitia). Although this can happen when a person is prescribed too large a dose of thyroid hormone, it also occurs with individuals who surreptitiously ingest thyroid medication. This last possibility may be difficult to diagnose. A final, and rare, cause of hyperthyroidism is when a pituitary adenoma produces excessive amounts of thyroid-stimulating hormone (TSH).

Dr. Ginot wonders if hyperthyroidism could be the problem that precipitated atrial fibrillation. Her initial thought was to dismiss excess thyroid hormone as the cause of atrial fibrillation given the lack of signs and symptoms classically associated with hyperthyroidism. There was no history of heat intolerance, insomnia, anxiety, tremor, diarrhea, goiter, or eye findings, such as proptosis or lid lag. After thinking for a moment, however, Dr. Ginot recalled that studies have shown that the older patient with hyperthyroidism frequently will not have the signs and symptoms that are seen in younger persons. She remembered a rule of geriatric medicine: new complaints referable to a given system may be due to illness in another system [Greenspan and Resnick 1994]. Just as a urinary tract infection could make its presence known by severe confusion in a person with dementia, hyperthyroidism may initially manifest itself as atrial fibrillation in a patient whose heart is vulnerable.

Dr. Ginot wondered if it made sense to consider hyperthyroidism as a possible diagnosis given the normal thyroid function tests from two months ago. Why should Mrs. Benson suddenly become hyperthyroid? Dr. Ginot recalled, however, that Mrs. Benson had received contrast with her CT scan two months ago. Mrs. Benson had some focal findings on her neurologic exam. Given the past history of breast cancer, Dr. Ginot wanted to exclude the possibility of metastatic breast cancer to the brain and had ordered the CT scan with and without contrast. Perhaps Mrs. Benson could have iodide-induced thyrotoxicosis? It is more common in people who have abnormal thyroid glands. Dr. Ginot recalled that Mrs. Benson's admission chest X-ray showed some tracheal deviation, suggesting, among a variety of possibilities, a substernal goiter.

Another finding perplexed Dr. Ginot. Why should Mrs. Benson have a low serum digoxin level when it had been stable over the last few years? Of course, it could be, given her recent memory problems, that she had forgotten to take her medication. Mrs. Benson, however, was vehement in her insistence that she took her digoxin faithfully. The other possibility was that hyperthyroidism may increase digoxin requirements, either by increasing the volume of distribution or increasing the clearance rate [Croxson and Ibbertson 1975]. Dr. Ginot called the laboratory and asked that they measure total T_4, total T_3, a T_3 resin uptake, and a TSH from the serum specimen drawn on admission.

What tests should be ordered when checking thyroid function? A measurement of serum TSH alone usually will suffice. If the TSH level were normal, it would virtually exclude the possibility of hyperthyroidism, because significant hyperthyroidism will suppress TSH secretion by the pituitary. A low TSH level, however, can also occur in euthyroid patients with serious non-thyroidal illness. A low TSH alone is an unreliable indicator of hyperthyroidism in sick patients. Measurement of both TSH and thyroid hormone levels is indicated in hospitalized patients with a serious possibility of hyperthyroidism.

In measuring thyroid hormone levels, an estimate of the free, or unbound, amount of hormone is crucial.

Most circulating thyroid hormone is bound to proteins in the blood, the vast majority being thyroid-binding globulin (TBG) and a smaller amount being albumin and transthyretin. A small fraction of the total amount of thyroid hormone is not bound to these proteins. This is the active (free) thyroid hormone.

A variety of methods for measuring thyroid hormone levels are available. While a full discussion of the various methods is beyond the scope of this chapter, some basic principles will be discussed here. One can measure free T_4 (FT_4) directly. Another way to estimate the amount of free thyroid hormone is to measure the total thyroid hormone level (which measures both bound and unbound hormone), and then measure in the serum the thyroid hormone binding capacity. Knowledge of the thyroid hormone-binding capacity and the total amount of thyroid hormone allows calculation of a free T_4 index (FT_4I). Dr. Ginot ordered a measurement of total T_4 and T_3 resin uptake (T_3RU). The T_3RU test is a measure of thyroid hormone binding capacity. With the total T_4 and the T_3RU, the lab will be able to report a FT_4I for Mrs. Benson. (A T_3RU should not be confused with a reverse T_3 measurement, which is a completely different test that is rarely indicated.)

Dr. Ginot, along with ordering the total T_4 and T_3RU, also ordered a measurement of serum triiodothyronine (T_3) because hyperthyroid patients occasionally have elevations of T_3 without elevations of T_4. This condition is referred to as T_3 toxicosis. A low TSH raises the differential diagnosis of hyperthyroidism versus a euthyroid person with serious illness. An individual with a normal FT_4I and suppressed TSH may be suffering from T_3 toxicosis or be euthyroid and sick. A measurement of T_3 will be low in euthyroid sick syndrome and high in T_3 toxicosis. In Mrs. Benson's case, acutely ill in the coronary care unit, the simultaneous measurement of TSH, T_4, and T_3 is wise, in order to establish a diagnosis rapidly and initiate appropriate treatment if needed.

Dr. Ginot is concerned that Mrs. Benson's atrial fibrillation is caused by the recent onset of hyperthyroidism secondary to the use of iodide in the CT contrast medium. Dr. Ginot knows that the earliest time she will know the results of the thyroid function tests (TFTs) is still 24 hours away. She could empirically treat for hyperthyroidism, but wonders if 24 hours of empiric antithyroid treatment could be harmful if it turns out that Mrs. Benson is not hyperthyroid.

It is extremely unlikely that 24 hours of antithyroid therapy will cause any significant negative effect. Should it turn out that Mrs. Benson is euthyroid, she could have, in time, a transient decrease in thyroid hormone that would reverse on its own. If there is substantial clinical suspicion for hyperthyroidism and there is some urgency to treat, then thyroid hormone levels should be drawn and antithyroid therapy begun while awaiting the results.

Beginning immediate treatment, however, may affect the ability to perform a radioactive iodine uptake test. In this procedure, a patient is given a tracer dose of radioactive iodine and the percentage taken up by the thyroid gland is measured. Radioactive iodine uptake can by useful in the differential diagnosis of hyperthyroidism. The measurement is elevated in Graves' disease, multi-nodular goiter, and a TSH-producing pituitary adenoma. Radioactive iodine uptake is low in hyperthyroid patients with thyroiditis and iodide-induced thyrotoxicosis. As the treatment for most of these conditions does not differ greatly, and given that Mrs. Benson is quite ill, it is reasonable in her case to forego the radioactive iodine uptake.

After obtaining a consultation from the endocrinology service, Dr. Ginot decides to treat Mrs. Benson empirically, pending the results of the TFTs. She orders propylthiouracil (PTU) 300 mg by mouth every 6 hours. In addition, Mrs. Benson receives further intravenous digoxin for rate control. Intravenous sodium iodide is also ordered.

Two drugs, propythiouracil and methimazole, are commonly used as antithyroid medications in hyperthyroid patients. Both act to prevent the production of thyroid hormone. Under most circumstances, it is preferable to treat patients with methimazole because it may be given once daily, a factor that greatly improves compliance (Roti *et al.* 1989). When patients are acutely ill, however, most endocrinologists recommend administering propythiouracil (PTU). PTU not only inhibits production of thyroid hormone but it blocks the conversion of T_4 to the more active T_3 (Abuid and Larsen 1974). Both PTU and methimazole reduce thyroid hormone levels slowly. Although they inhibit the production of new thyroid hormone, at the time that the medication is begun, the thyroid gland is already filled with thyroid hormone that may be released into the circulation. For hyperthyroid emergencies, therefore, another agent, sodium iodide, may need to be given intravenously to prevent the release of preformed thyroid hormone. It may seem contradictory to give iodide to a hyperthyroid patient, as it can be the cause of hyperthyroidism. Large doses of sodium iodide were given to Mrs. Benson after the second dose of PTU had been administered. Once PTU has been given, the iodide will not be taken up for the production of new hormone.

Treatment with beta-blocking agents is another therapeutic option in the management of patients with hyperthyroidism. Beta blockers decrease the heart rate as well as other symptoms in hyperthyroidism. They are very useful for symptomatic outpatients with hyperthyroidism and normal cardiac function (Woeber 1992). They can, however, be risky in patients with impaired systolic cardiac function. For Mrs. Benson, one could consider the use of beta blockers to reduce her heart rate, but she would need careful monitoring to detect possible deterioration from worsening heart failure.

By midnight, Mrs. Benson was greatly improved. She was no longer short of breath and her heart rate was down to 100/min, with a respiratory rate of 18/min. The next day, repeat blood drawn for cardiac enzymes failed to reveal any evidence for myocardial infarction. Thyroid function tests returned later in the afternoon of the day after admission: serum T_4 was 20 µg/dl (normal 4–11 µg/dl), FT_4I 22 (normal 5–12), T_3 583 ng/dl (normal 90–180 ng/dl), and TSH was less than 0.1 mIU/l (normal 0.3–5 mIU/l). A thyroid ultrasound was ordered, revealing substantial substernal enlargement of the thyroid with many areas of nodules and inhomogeneity.

The diagnosis of hyperthyroidism is confirmed, probably due to prior iodide administration to a woman with a multi-nodular goiter. The thyroid ultrasound was ordered to provide more diagnostic information regarding Mrs. Benson's thyroid. The tracheal deviation noted on the admission chest X-ray suggests thyroid enlargement. The thyroid ultrasound provided information as to whether the thyroid has variations in density or the presence of nodules and cysts. An enlarged homogeneous gland in a hyperthyroid person is suggestive of Graves' disease. An inhomogenously enlarged thyroid gland is consistent with multi-nodular goiter. The ultrasound also confirmed that the tracheal deviation was due to thyroid enlargement and not to some other cause, such as adenopathy.

Could the diagnosis of iodide-induced hyperthyroidism be wrong? It is possible that Mrs. Benson could have spontaneously developed hyperthyroidism because of autonomous hormone production from her nodular goiter? It is quite unlikely that this would develop so precipitously over the two months since the normal TFTs were obtained. Likewise, one cannot definitively reject Graves' disease but the time course is too rapid. Other laboratory tests could be obtained, normally positive in Graves' disease, to exclude this diagnosis. The treatment with methimazole is appropriate in all these instances. There is nothing in Mrs. Benson's history to suggest surreptitious thyroid hormone ingestion. A final possibility is that Mrs. Benson could have thyroiditis. Again, this is unlikely and quite rare in Mrs. Benson's age group. In addition, thyroiditis is usually painful, and Mrs. Benson did not have a tender thyroid.

Mrs. Benson improved quickly and was ready for discharge after a few days. While in the coronary care unit, heparin was begun which was switched to coumadin for anticoagulation to lower the risk of embolization with atrial fibrillation [Woeber 1992]. It was decided to defer any attempt at cardioversion while Mrs. Benson remained hyperthyroid.

Although there are no data on the success rate of cardioversion in restoring normal sinusrhythm in hyperthyroid individuals, it is reasonable to wait until the hyperthyroid state resolves before attempting cardioversion. One study of the natural history of atrial fibrillation found that if spontaneous reversion to sinus rhythm occurred, it did so within weeks of the resolution of the hyperthyroidism (Woeber 1992). If the patient can tolerate atrial fibrillation, then one should delay elective cardioversion for three months until after the TFTs have normalized. After that time, if the person has not spontaneously converted to sinus rhythm, chemical or electrical cardioversion may be attempted.

Mrs. Benson was discharged on methimazole 40 mg orally daily. She had been switched from PTU, which was given four times a day, to the once daily methimazole in order to make it easier to follow Dr. Ginot's plan. Mrs. Benson was also warned to report any sore throat or fever. About 0.5% of patients taking antithyroid drugs may suffer from agranulocytosis as a consequence of the medication [Cooper 1984]. Any sign of infection in a patient on methimazole or PTU requires evaluation with a cell blood count and differential. In addition, Mrs. Benson was instructed to double her usual digoxin dose to 0.25 mg daily, with weekly monitoring of digoxin levels. The digoxin dose would need readjustment downward as the hyperthyroidism remits. It was planned that after a few months of control of the hyperthyroidism, definitive treatment would be undertaken with therapeutic doses of radioactive iodine. Given Mrs. Benson's complaints of difficulty with memory on her office visit two months ago, Dr. Ginot arranged for a visiting nurse to monitor Mrs. Benson's compliance with her new medical regimen and to assist with obtaining prothrombin and digoxin levels, as well as follow-up thyroid function tests.

Just after saying goodbye to Mrs. Benson, Dr. Ginot heard her beeper go off again. Dr. Ginot read the far too familiar number off her beeper: 'Oh no. The emergency room. That's another admission.'

Dr. Ginot calls the emergency room to speak with the senior resident: 'Dave, this is Renee, have you got a patient for me?'

'Oh, Renee, yes, there's a really nice 75-year-old lady down here by the name of Sarah Consoletti. She's got unstable angina. Want to come down and make her acquaintance?'

'I'll be there in a couple of minutes.'

Dr. Ginot introduced herself to Mrs. Consoletti and

took her history, supplemented by a look through her old medical record. She had a six-year history of angina with complaints of chest pain and pressure on exertion. In the past, stress thallium testing showed a reversible inferior wall perfusion defect with 1 mm of ST segment depression in leads II, III, at a high level of exercise. Until recently her angina has been very well controlled on atenolol 50 mg daily. Mrs. Consoletti had been walking about a mile daily without any symptoms. She would only rarely take nitroglycerin, usually just once every month or two, for angina precipitated by unusually intense physical activity. In the past three or four weeks, however, Mrs. Consoletti has noted gradually increasing frequency of angina at lesser levels of exertion. At her physician's advice, she doubled her atenolol without improvement. This morning she had chest pressure while still in bed. It resolved after a single nitroglycerin but she was concerned about this episode of chest discomfort at rest and called her physician. He instructed her to come to the hospital.

Mrs. Consoletti's current medications included atenolol 100 mg daily and one enteric-coated aspirin daily. Past medical history, other than cardiac history, was remarkable for a cholecystectomy some 20 years ago. Review of systems was positive for some anxiety and some difficulty sleeping. On physical exam, Dr. Ginot noted that the thyroid gland was enlarged to about twice the normal size. It was rough and firm in texture and without definite nodules. She felt no nodes. The remainder of Mrs. Consoletti's exam was normal. Cardiac exam revealed normal rate and rhythm, no murmurs or gallops. Lungs were clear to percussion and auscultation.

Mrs. Consoletti gave a good history for unstable angina with a change in the frequency and pattern of her chest discomfort. Unstable angina is usually related to the progressive occlusion of a coronary artery. Rarely, however, the increasing angina may be caused by increased demand with the coronary anatomy remaining stable, or it may be caused by coronary spasm. Anemia, fever, and hyperthyroidism are examples of conditions that increase cardiac demand and may lead to a patient presentation of worsening angina. Hyperthyroidism may also be associated with coronary spasm (Wei *et al.* 1979).

Dr. Ginot, with the memory of Mrs. Benson's admission very fresh in her mind, wondered if the cardiac problem in Mrs. Consoletti's case could be related to her hyperthyroidism. She has complaints associated with hyperthyroidism, such as anxiety and insomnia, as well as having a goiter on exam. Mrs. Consoletti was not tachycardic, but that could be because of the beta blocker or because of her age. Dr. Ginot asked the emergency room nurse to add a total T_4, T_3RU, and TSH on the labs already drawn. Routine labs revealed a normal cell blood count, chemistry profile, urinalysis, and chest X-ray. The electrocardiogram was within normal limits. Dr. Ginot started Mrs. Consoletti on

intravenous nitroglycerin and heparin for unstable angina. She called the cardiology fellow about Mrs. Consoletti's admission and mentioned her concern about the possibility of hyperthyroidism. Because the patient was now pain-free and receiving intravenous nitroglycerin and heparin, Dr. Ginot felt that it was reasonable to await the return of TFTs before beginning antithyroid treatment.

The next day, Mrs. Consoletti's TFTs returned: T_4 2.0 µg/dl (normal 4–11 µg/dl), FT^{4}I 2.2 (normal 5–12), and TSH 56 mIU/l (normal 0.3–5.0 mIU/l).

Much to Dr. Ginot's surprise, the thyroid function tests revealed Mrs. Consoletti to be significantly hypothyroid rather than hyperthyroid, as she reasonably suspected on the basis of her symptoms. This somewhat embarrassing situation is not uncommon. TFTs are often drawn because of a complaint consistent with hyperthyroidism and the patient turns out to be hypothyroid. The signs and symptoms of thyroid dysfunction are neither sensitive nor specific in older persons (Greenspan and Resnick 1994). Likewise, evidence of a goiter with diffuse thyroid enlargement on exam does not tell the examiner whether a person is hyperthyroid, hypothyroid, or euthyroid. Thyroid size and function may vary independently.

What is the cause of Mrs. Consoletti's hypothyroidism? Hypothyroidism can be primary, due to failure of the thyroid gland, or secondary, due to pituitary failure. Primary hypothyroidism is much more common. Distinguishing between primary and secondary hypothyroidism is important because secondary hypothyroidism may be accompanied by other pituitary hormone deficiencies. The key clue to distinguishing primary from secondary hypothyroidism is the serum TSH: high levels document thyroid gland failure with an intact pituitary. In Mrs. Consoletti's case, she clearly has primary hypothyroidism, with a TSH of 56 mIU/l. There is no particular need to further characterize the exact reason why Mrs. Consoletti's thyroid gland has failed to produce an adequate amount of thyroid hormone.

Dr. Ginot decides to treat Mrs. Consoletti's hypothyroidism and writes an order for thyroxine, 0.1 mg daily. Later that evening, just as she gets ready to go home, Dr. Ginot is paged by Elizabeth Rawlins, the nurse caring for Ms. Consoletti. Ms. Rawlins is one of the more experienced nurses, having worked at the hospital for many years. She tells Dr. Ginot that she recalls that years ago before effective antianginal therapy was available, patients were made hypothyroid using radioactive iodine in an effort to control their angina. Because hypothyroidism decreases the severity of angina, Ms. Rawlins wonders if it is advisable to give thyroxine to a hypothyroid patient with unstable angina.

Ms. Rawlins' experience served her well. In the 1950s,

it was shown that euthyroid patients with severe angina improved after they were made hypothyroid by radioactive iodine (Blumgart et al. 1955). The theory behind this therapy was that hypothyroid patients have a lower cardiac output and therefore have less demand on their hearts. Because of the more effective antianginal therapy currently available, hypothyroidism is no longer deliberately induced in persons with angina. It is, however, unwise to begin thyroxine therapy in patients with unstable angina while they are awaiting revascularization.

Dr. Ginot held the order for thyroxine. Mrs. Consoletti underwent cardiac catheterization the next morning. The study revealed severe stenoses involving the left main, left circumflex, left anterior descending, and posterior descending coronary arteries. The stenoses were mainly proximal with good distal vessels.

Mrs. Consoletti's cardiologist felt that coronary artery bypass surgery was the best approach and that the vessels involved made angioplasty a poor choice.

Dr. Ginot wondered what to do about Mrs. Consoletti's hypothyroidism given the plans for coronary bypass surgery. Given her severe coronary artery disease, it would be best to wait until after revascularization. That would have been easy if all that was needed was angioplasty. The thyroxine could have been held until the procedure. But what about bypass surgery? Will the hypothyroidism jeopardize her ability to tolerate surgery? Is it better to treat the hypothyroidism first even if it is risks destabilizing her coronary artery disease?

Dr. Ginot's questions are pertinent and they have been studied and reported in the medical literature. Some studies find that post-operative morbidity due to gastrointestinal complications and congestive heart failure may be higher in hypothyroid than in euthyroid patients (Ladenson et al. 1984). Despite this, the consensus is that unstable cardiac patients are better off waiting until after surgery to begin treatment with thyroxine (Mandel et al. 1993).

Mrs. Consoletti underwent successful coronary artery bypass grafting. Her recovery was uneventful except for some post-operative confusion which cleared spontaneously. After five days she was walking with assistance and tolerating a full liquid diet. Mrs. Consoletti then received her first dose of thyroxine, 50 µg orally, daily.

Choosing the initial dose of thyroxine requires care and consideration. The mean dose for thyroxine replacement is 1.6 µg/kg/day (Mandel et al. 1993). This dose is an average, and many patients will require higher or lower doses. In caring for older persons, a crucial point to keep in mind is that thyroxine requirements usually decrease with age, with some studies showing a decrease by as much as one-third in the elderly, so that for older patients the average thyroxine requirement

is probably lower than 1.6 µg/kg/day (Davis et al. 1984). Furthermore, many elderly patients have mild hypothyroidism and require only partial replacement with thyroxine. A final consideration is that some older persons without overt cardiac symptoms may manifest evidence of cardiac disease once thyroxine is begun. Therefore, it is standard practice to begin thyroxine replacement very slowly, at 12.5–25 µg/day, monitoring for cardiac symptoms. The dose can be adjusted according to symptoms and hormone levels every six weeks. If the dose needs to be increased, small increments of 12.5–25 µg are used. Mrs. Consoletti was started on a slightly higher dose of 50 µg/day as she was quite hypothyroid and, given her just completed coronary bypass surgery, her risk of cardiac ischemia was relatively low.

The day prior to Mrs. Consoletti's planned discharge she experienced incisional pain in her legs and was given acetaminophen with codeine. She then complained of nausea, vomiting, and abdominal distension which persisted for two days. The clinical impression was an ileus secondary to opiate use in the post-operative setting. There was no evidence for cardiac ischemia.

Mrs. Consoletti's gastrointestinal difficulties could be due to a post-operative ileus compounded by the use of codeine. The incidence of post-operative ileus, however, is higher in hypothyroid than in euthyroid patients. It would be sensible to continue with Mrs. Consoletti's thyroxine replacement therapy, but she cannot tolerate oral medications. Thyroxine is available as an intravenous preparation, with the appropriate intravenous dose being approximately 80% of the oral dose.

In addition to intravenous fluids and nasogastric drainage, Mrs. Consoletti was given 40 µg of thyroxine intravenously for three days, at which time she was again able to tolerate oral medications. After another day, Mrs. Consoletti was discharged, eating well and without vomiting. Her TSH was checked six weeks after discharge and remained high. The thyroxine dose was increased to 75 µg/day. At this dose of thyroxine, Mrs. Consoletti's TFTs and TSH normalized. After a few months, Mrs. Consoletti reported that she felt better than she had in years.

Mrs. Consoletti did well over the next three years. She then visited Dr. Ginot, now in private practice and an attending physician, and mentioned some fatigue and constipation. Laboratory results revealed that the serum FT_4I was slightly low with an elevated serum TSH, consistent with under-replacement of thyroxine. Mrs. Consoletti was adamant that she had taken her thyroxine as prescribed. Dr. Ginot wondered why her long-standing dose of thyroxine should now be inadequate.

There are a variety of reasons for an increasing thyroxine requirement. The dose of thyroxine may need to be increased if, at the time of diagnosis, the thyroid was making some hormone, and is now making less. Other potential causes for increasing thyroxine requirements include gastrointestinal malabsorption syndromes which interfere with thyroxine absorption or severe nephrotic syndrome which may increase urinary thyroxine clearance. Switching among different pharmaceutical preparations of thyroxine can also cause alterations in the level of thyroid hormone because of differences in bioavailability among the various products. Medications may also affect thyroid requirements. Concomitant administration of various medications, such as iron sulfate, sucralfate, or cholestyramine, may interfere with thyroxine absorption. Antiseizure drugs or rifampin can also lower serum thyroid hormone levels by increasing the hepatic clearance of thyroxine.

Dr. Ginot checked her records and saw that she had always prescribed the same brand of thyroxine for Mrs. Consoletti rather than allow substitution with a different formulation. She then questioned Mrs. Consoletti, keeping in mind the various possible causes of decreased thyroid hormone levels in a patient who has been previously doing well on a stable dose of thyroxine. Mrs. Consoletti admitted to having begun taking some over-the-counter iron supplements during the past year. She thought that as she got older she needed more iron. Dr. Ginot suggested that the seeming increase in thyroxine requirement may be due to the iron supplement interfering with absorption. Dr. Ginot explained that there was no need for an older woman to take iron supplements in the absence of iron deficiency, and that Mrs. Consoletti should stop the iron, continue on her old dose of 75 μg/day and have her TFTs rechecked in two months time. Mrs. Consoletti's repeat TFT's and TSH two months later showed normalization. She also told Dr. Ginot that she felt better and that her constipation had resolved.

Questions for further reflection

1. What are possible causes for hyperthyroidism?
2. Why are the elderly particularly vulnerable to cardiac manifestations of thyroid disorders?
3. Discuss the indications and dangers for anti-thyroid therapy and thyroid replacement therapy.

References

Abuid, J. and Larsen, P.R. (1974). Triiodothyronine and thyroxine in hyperthyroidism: Comparison of the acute changes during therapy with antithyroid agents. *Journal of Clinical Investigation*, **54**, 201–8.

Blumgart, H.L., Freedberg, A.S., and Kurland, G.S. (1955). Treatment of incapacitated euthyroid cardiac patients with radioactive iodine. *Journal of the American Medical Association*, **157**, 1–4.

Cooper, D.S. (1984). Antithyroid drugs. *New England Journal of Medicine*, **311**, 1153–62.

Croxson, M.S. and Ibbertson, H.K. (1975) Serum digoxin in patients with thyroid disease. *British Medical Journal*, **3**, 566–8.

Davis, F.B., LaMantia, R.S., Spaulding, S.W., Wehmann, R.E., and Davis, P.J. (1994) Estimation of a physiologic replacement dose of levothyroxine in elderly patients with hypothyroidism. *Archives of Internal Medicine*, **144**, 1752–4.

Greenspan, S.L. and Resnick, S.M. (1994). Geriatric endocrinology. In *Basic and clinical endocrinology*, (4th edn), (ed. F.S. Greenspan, and J.D. Baxter), pp. 729–46. Appleton & Lange, Norwalk, CT.

Ladenson, P.W., Levin, A.A., Ridgway, E.C., and Daniels, G.H. (1984). Complications of surgery in hypothyroid patients. *American Journal of Medicine*, **77**, 261–6.

Mandel, S.J., Brent, G.A., and Larsen, P.R. (1993). Levothyroxine therapy in patients with thyroid disease. *Annals of Internal Medicine*, **119**, 492–502.

Roti, E., Gardini, E., Minelli, R., Salvi, M., Robuschi, G., and Braverman, L.E. (1989). Methimazole and serum thyroid concentrations in hyperthyroid patients: Effects of single and multiple daily doses. *Annals of Internal Medicine*, **111**, 181–2.

Wei, J.Y., Genecin, A., Greene, H.L., and Achuff, S.C. (1979). Coronary artery spasm with ventricular fibrillation during thyrotoxicosis: response to attaining euthyroid state. *American Journal of Cardiology*, **43**, 335–9.

Woeber, K.A. (1992). Thyrotoxicosis and the heart. *New England Journal of Medicine*, **327**, 94–8.

9

Diabetes mellitus in the elderly

Linda A. Morrow and Kenneth L. Minaker

Dr. Eric Dierdorf's next patient was a 75-year-old woman who was seeing him for the first time. Perusing the nurse's notes, he noted that she had made the appointment because of a lack of energy and pain in her feet. Dr. Dierdorf paused for a moment to think. A 75-year-old woman with a lack of energy and pain in her feet—there were so many possibilities. He briefly considered some of the more common causes of a lack of energy—anemia, depression, congestive heart failure, hypothyroidism—and then he thought about common things that might cause pain in the feet—bunions, heel spurs, arthritis, gout. Or could the problem be something that did both? He went in to see his patient, Mrs. Miriam Cohen.

Mrs. Cohen was an obese woman who looked a bit dejected. 'Dr. Dierdorf, I've had terrible pain in my feet for several months It seems to be getting worse. I thought it might go away on its own, but now I'm having trouble sleeping and trouble walking. My daughter finally insisted that I see a doctor about this.' Taking a careful history, Dr. Dierdorf discovered that Mrs. Cohen's pain had a burning type quality, that it was present all the time, and that there were no precipitating or alleviating factors. It extended from her toes to her ankles on both feet. Mrs. Cohen had also noticed some numbness in her toes.

Dr. Dierdorf ran through the differential diagnosis for peripheral neuropathy in his head. The list was long and included alcoholism, vitamin B_{12} deficiency, diabetes mellitus, drug-induced neuropathy, and even lead poisoning. Dr. Dierdorf began questioning Mrs. Cohen to try to narrow the list, but he realized that one disease was the likely culprit—diabetes mellitus.

Diabetes mellitus is a very common disease among older adults. The National Health and Nutrition Examination Survey II revealed that disorders of carbohydrate metabolism (diabetes mellitus and impaired glucose tolerance) affected more than 40% of individuals over the age of 65 (Harris *et al.* 1987). Over 18% of those elderly individuals who participated in the survey had diabetes and, strikingly, only half of those who had diabetes were previously aware of the diagnosis. (Although this statistic would be alarming enough if it were true only for older adults, the survey revealed that half of all American who had diabetes were not aware of their diagnosis.) In addition to those older individuals with diabetes, another 23% had impaired glucose tolerance, a condition that places them at increased risk for coronary artery disease.

Mrs. Cohen's answers pointed to the diagnosis of diabetes mellitus. She had no history of alcoholism or anemia. The only medication that she was taking was an occasional acetaminophen for arthritis and a laxative when she was constipated. She tried to eat a good diet, but commented that she had not really felt hungry lately. In fact, she thought she might have even lost a few pounds. She had noticed that she was getting up a couple of times a night to go to the bathroom, but she did not feel much more thirsty than usual. And yes, she had noticed that her vision was somewhat blurry. She'd been thinking that she needed to have her prescription for her lenses changed. No, she had never been told that she had diabetes. As far as she knew, all of her previous blood sugars had been normal, although she could not remember that she had ever been tested for diabetes. Her fatigue had been of gradual onset, worsening over the past several months. It seemed to be worse at the end of the day, but there was no associated joint pain. She had not had any chest pressure or shortness of breath nor had she noticed any skin rashes. There had been no particular bouts of constipation and no complaints of dryness of her skin or hair. She just did not seem to have the energy that she had had even six months ago.

Dr. Dierdorf moved on to Mrs. Cohen's past medical history. She had several urinary tract infections as a younger woman but none in the past decade or so. She had a hysterectomy in her fifties and her gallbladder was removed around that time also, but otherwise she had been fairly healthy. Mrs. Cohen admitted to Dr. Dierdorf that she really did not like to go to the doctor and, in fact, she had not had a check-up in years. Dr. Dierdorf asked about her family history. Yes, 'sugar ran in the family.' Mrs. Cohen's mother and her older brother had had diabetes mellitus, but both of them were dead now. Her mother had died in her seventies following several strokes and her brother had died three years ago following a heart attack. Mrs. Cohen was not sure how old her brother and mother were when they were diagnosed, but she did recall that her mother had developed diabetes while hospitalized with her first stroke.

Diabetes mellitus occurs in two forms: insulin-dependent diabetes mellitus (IDDM); and non-insulin-dependent diabetes mellitus (NIDDM). While the clinical manifestations of these two types of diabetes are similar, the etiologies are quite different, so much so that they can almost be thought of as two different diseases. Insulin-dependent diabetes mellitus (so named because the patient will die without exogenous insulin) is also known as type I diabetes mellitus. Other names previously used include juvenile-onset diabetes mellitus, because most individuals with this form of diabetes develop the disease in childhood, and ketosis-prone diabetes mellitus, because these individuals are prone to developing diabetic ketoacidosis. Insulin-dependent diabetes mellitus is thought to result from autoimmune mediated destruction of pancreatic islet cells, perhaps triggered in susceptible individuals by viral infections. Patients with IDDM produce no insulin. Non-insulin-dependent diabetes mellitus, a disease with a strong genetic component, appears to result from a combination of peripheral insulin resistance in insulin utilizing tissues, and impaired pancreatic islet cell function (although insulin production is still present). Controversy remains regarding which defect is dominant. While insulin-dependent diabetes mellitus is perhaps the more dramatic disease, 90–95% of individuals with diabetes mellitus have non-insulin-dependent diabetes mellitus. Interestingly, about 5% of older adults who develop diabetes mellitus develop the insulin-dependent form, although features of their illness are somewhat different from that of insulin-dependent diabetes mellitus in younger adults. For example, older adults with IDDM do not have an increased frequency of antibodies to islet cells, although the HLA-DR3 subtype is more common (Kilvert et al. 1986).

Dr. Dierdorf began the physical exam. Mrs. Cohen had been scheduled as an acute patient, not for a complete evaluation, and he needed to focus his exam. He concentrated on those parts that would be pertinent to her complaints. Her weight was 184 pounds, her height 65 inches, blood pressure 165/94 mmHg, and her resting heart rate was 70/min. She was moderately obese and did not appear to be in acute distress. The examination of her head, eyes, ears, nose, and throat was cursory, but did not demonstrate any remarkable abnormalities. Examination of Mrs. Cohen's neck revealed a soft right carotid bruit that he almost missed (reminding him to be a little more careful), but her thyroid felt normal. Heart and lung exams were normal except for the presence of a fourth heart sound. Her extremities revealed trace pitting edema bilaterally around the ankles, but her feet

were well-shaped, without deformities or evidence of inflamed joints. Dr. Dierdorf, looking for ulcerations, carefully inspected the bottoms of Mrs. Cohen's feet and in between each toe. He was happy not to find any ulcerations but he did note the presence of thick calluses on both feet. His screening neurologic exam was normal until he reached the lower extremities. He found markedly decreased vibratory sensation in both feet to the ankles, in a distribution almost as if she were wearing socks. Using the side of his tuning fork, he found that Mrs. Cohen had poor recognition of cold. Using a disposable needle, Dr. Dierdorf found that Mrs. Cohen's perception of pain was diminished in a similar distribution to that of the diminished vibratory sensation. Deep tendon reflexes were normal throughout except that they were absent bilaterally at the ankles.

He generated his problem list for Mrs. Cohen: (1) peripheral neuropathy, (2) hypertension; (3) asymptomatic right carotid bruit, and (4) obesity. He started to think about the studies that he needed to order. He wanted to evaluate her peripheral neuropathy as efficiently as possible. He asked the nurse to do a fingerstick blood glucose even though Mrs. Cohen had not been fasting. It was 237 mg/dl.

The diagnosis of diabetes mellitus is straightforward and is not different for older adults compared to younger individuals. The criteria for the diagnosis of diabetes mellitus was established by the National Diabetes Data Group (NDDG) in 1979, which revised earlier criteria. The diagnosis of diabetes mellitus can be made in three ways (ADA 1988):

1. Random blood sugar greater than 200 mg/dl with the typical signs and symptoms of diabetes mellitus (polyuria, polydipsia, polyphagia, weight loss).

2. Fasting plasma glucose greater than 140 mg/dl on two separate occasions.

3. Diagnostic oral glucose tolerance test (OGTT) with fasting plasma glucose less than 140 mg/dl, but 2 hour sample greater than 200 mg/dl and one intervening sample (30, 60, or 90 minute sample) greater than 200 mg/dl following a 75 gram oral glucose load. In order to make the diagnosis of diabetes mellitus using an oral glucose tolerance test, two OGTTs must be performed.

Most older adults do not have the typical symptoms of diabetes mellitus. Polyuria, polydipsia, and polyphagia are more commonly found in individuals who present with insulin-dependent diabetes mellitus. In the older age group, the diagnosis of diabetes is typically made through the use of fasting plasma glucose testing. The NDDG criteria require two separate tests with a fasting glucose greater than 140 mg/dl for two reasons: (1) the inability to ensure the fasting state in many individuals:

and (2) the necessity to be as certain as possible before labeling persons with the diagnosis of diabetes.

Glucose tolerance testing should be reserved for individuals whose random blood sugars, while elevated above normal, are under 200 mg/dl and whose fasting plasma glucose levels, again, while elevated above normal, remain below 140 mg/dl. Glucose tolerance testing is also useful for making the diagnosis of impaired glucose tolerance (IGT), a specific diagnostic category with its own concomitant risks. The diagnosis of IGT is made with an oral glucose tolerance test. Individuals meet diagnostic criteria for impaired glucose tolerance when the fasting plasma glucose is less than 140 mg/dl, the 2-hour sample is less than 200 mg/dl and greater than 140 mg/dl, and one of the intervening samples (either 30, 60, or 90 minute samples) is greater than 200 mg/dl. It is important to identify individuals with impaired glucose tolerance for several reasons. These individuals are at risk for the subsequent development of diabetes mellitus, particularly those with fasting glucoses that are greater than 120 mg/dl. In addition, epidemiological studies have revealed that these individuals are also at greater risk for the development of coronary artery disease and myocardial infarction. There is no evidence, however, that treatment of impaired glucose tolerance prevents either the development of diabetes mellitus or cardiovascular disease.

The fingerstick blood glucose of 237 mg/dl made Dr. Dierdorf feel confident that Mrs. Cohen had diabetes. To meet the rigid criteria for diagnosis he needed to get two fasting blood glucoses to be certain. He added a fasting glucose to the panel of laboratory tests that he had already ordered for the following morning: blood, urea, nitrogen (BUN), creatinine, glycosylated hemoglobin, urinalysis for protein, lipid profile, thyroid-stimulating hormone, cell blood count, serum protein electrophoresis, and vitamin B_{12} level. He also ordered an electrocardiogram (ECG) and a chest X-ray. Another fasting glucose would be done a few days after the first. Dr. Dierdorf asked his office nurse to give Mrs. Cohen a diet history booklet and instructions on how to use it for the following week. He told Mrs. Cohen that he felt it was likely that she might have diabetes and that the diabetes was affecting the nerves in her legs, causing her symptoms. He discussed the need for the studies that he was scheduling—to confirm the diagnosis of diabetes mellitus, to screen for risk factors for complications from diabetes, and to look for other explanations for the nerve damage in her feet. He urged her to complete the diet history and asked her to return in one week to review her laboratory results and discuss treatment options.

The American Diabetes Association has established criteria for the history, physical, and laboratory assessment of individuals with diabetes mellitus (ADA

Table 9.1 American Diabetes Association recommendations for assessment of individuals with diabetes mellitus

History	Symptoms of diabetes mellitus
	Symptoms of complications
	Current treatment
	History of acute complications and prior infections
	Dietary and exercise history
	Risk factors for atherosclerosis
Physical Examination	Height, weight, and blood pressure with orthostatic blood pressure
	Ophthalmologic examination
	Thyroid palpation
	Cardiac and peripheral pulses examination
	Foot and skin examination
	Neurologic examination
	Dental and periodontal examination
Laboratory Evaluation	Fasting plasma glucose
	Glycosylated hemoglobin
	Fasting lipid profile
	Serum creatinine
	Urinalysis
	Thyroid function tests
	Electrocardiogram

1989). These are shown in Table 9.1. There are three general goals to be accomplished. First is to ascertain symptoms relevant to acute glucose control and suggestive of complications of diabetes. Second is to screen by physical exam and laboratory studies for the presence of complications of diabetes mellitus, including microvascular, macrovascular, and neuropathic. The third goal is to screen for risk factors for cardiovascular disease. *It is important to note that all individuals with NIDDM need a referral for an ophthalmologic exam at the time of diagnosis.* This strong recommendation reflects the current consensus that NIDDM is present for many years prior to diagnosis (a recent study estimated seven years) and diabetic eye disease has had adequate time to develop. Similarly, newly diagnosed individuals must be screened for the presence of other complications. These are most commonly separated into three groups: microvascular disease (which includes retinopathy and nephropathy), macrovascular disease (which includes cardiovascular disease, peripheral vascular disease, and cerebral vascular disease), and neuropathy. Individuals with NIDDM are at risk for the same complications as those with IDDM although attack rates vary somewhat from site to site and among various populations. Currently, it is not possible to predict which diabetic individuals will develop which complications, so the health care provider must be vigilant with all patients with diabetes.

Cardiovascular disease is prominent among older adults with diabetes mellitus and is the main cause of death in this group. Elderly individuals with diabetes are at twice the risk for coronary artery disease compared to others in their age group without diabetes. Investigation for treatable risk factors of coronary artery disease is a critical aspect part of the initial evaluation of older adults with coronary disease. The risk factors (in addition to diabetes) include hypertension, obesity, tobacco smoking, and family history. While the presence of diabetes and a family history are not modifiable, the other risk factors can be treated and should be sought.

Dr. Dierdorf thought about his patient—she clearly had diabetes mellitus, most likely NIDDM. Her diabetes was complicated by peripheral neuropathy and a suggestion of diabetic nephropathy as evidenced by proteinuria. She was at significant risk for cardiovascular disease because of her weight, hypertension, and lipid abnormalities. He began to consider his treatment options. How tightly should he try to control her diabetes? After all, she was already 75 years old—how much longer could she expect to live?

Before deciding on what to give an older diabetic patient, one must decide how to treat the patient. Specifically, one of the most critical decisions to be made concerns the degree of glycemic control warranted. While this decision is made for each individual patient, patients generally can be separated into two groups. The first group is composed of patients for whom the prevention of symptoms and the acute complications of hyperglycemia is the primary goal of treatment. The symptoms related to hyperglycemia include visual blurriness and symptoms resulting from glycosuria: frequent urination, nocturia, and urinary incontinence. Nonspecific symptoms, such as fatigue and malaise, have also been attributed to chronic hyperglycemia. The acute complications of hyperglycemia include dehydration and hyperosmolar hyperglycemic non-ketotic coma. These complications are the result of glycosuria. Glycemic control for this group aims to keep blood glucose below the renal threshold for glucose and thus preventing glycosuria. The renal threshold for glucose rises with age from about 170 mg/dl in young adults to around 200 mg/dl for older adults, although there is substantial variability from one individual to another. In general, prevention of symptoms requires fasting blood glucose to be kept below 160 mg/dl and random blood glucose below 200 mg/dl. The type of individuals for whom this type of control is appropriate usually have multiple medical problems, a reduced life expectancy, cognitive difficulties that interfere with their ability to follow a treatment regimen, or a combination of psychosocial factors that make them unlikely candidates for more rigid glucose control.

The second group includes essentially all other patients—those for whom the goal of treatment is both the prevention of symptoms and complications of acute hyperglycemia as well as the prevention of chronic complications of diabetes mellitus. While in the past the evidence for rigid control of glycemia in preventing complications was inferential, in 1993 the ten-year Diabetes Control and Complications Trial (DCCT) was concluded six months prior to its scheduled end because of the overwhelming evidence that tight control did prevent both microvascular and neuropathic complications (TDCCTRG 1993). This study used a randomized design where individuals with diabetes were assigned to either an intensively treated group which aimed to normalize glycosylated hemoglobin level or to a group where treatment was aimed at preventing symptoms of hyperglycemia. The intensively treated group monitored their blood glucose several times a day and received either multiple injections of insulin a day or used an insulin pump in an effort to normalize glycosylated hemoglobin. The control group injected insulin one or two times a day and modified their insulin dose to prevent symptoms, but not to normalize glycosylated hemoglobin. The intensively treated group had significantly reduced incidence of both the development and progression of microvascular complications. This study is a major watershed for the treatment of diabetes as it appears to finally resolve the question of whether rigid glycemic control can prevent diabetic complications. Although the trial was conducted only in younger individuals with IDDM, most experts feel that these data can be extrapolated to the population with NIDDM as the mechanism for complications is felt to be the same in both IDDM and NIDDM.

Three points that come from the DCCT are important to keep in mind regarding attempts at tight control of diabetes. First, the risk for significant episodes of hypoglycemia is increased. Surprisingly, there were no hypoglycemia-related deaths in the DCCT, but the overall risk for hypoglycemia was increased about threefold for the intensively treated group. Second, in spite of multiple injections of insulin or the use of insulin pumps, the intensively treated group was not able to maintain normal glycosylated hemoglobin levels. They were able to lower substantially their mean glycosylated hemoglobin level, but it remained elevated above normal levels. Third, the intensively treated group in the DCCT received nutritional counseling and extensive emotional and psychological support. Such services, while valuable in the care of persons with diabetes, are very time consuming, costly, and not generally available in most health care settings. While the DCCT provides a goal at which to aim, patient factors, intrinsic features of diabetes itself, and resource limitations are likely to modify the application of the DCCT to the diabetic population at large.

A number of these considerations are directly relevant to the geriatric diabetic population. There is some evidence to suggest that elderly individuals with diabetes mellitus, especially thin older persons, have more episodes of hypoglycemia than younger individuals when they are treated with either oral hypoglycemic agents or insulin. Some studies also point to less vigorous physiologic responses to hypoglycemia among older diabetics compared to younger diabetics. These two factors combined imply that tight control of glycemia may pose additional risk for older diabetic adults. In addition, the large number of older adults with diabetes, up to 20% of the population over age 65, would require large numbers of health care workers to provide the support necessary for tight control. Given limited health care resources, the selection of patients that are most likely to benefit from intensive therapy is important. While age alone should not be a significant factor for the young old (those less than age 75), age begins to play a role for those 75 and older. If diabetes is diagnosed after age 75, life expectancy may be less than the amount of time needed for the complications to develop and intensive therapy would not be indicated. It is important, however, not to underestimate life expectancy: life expectancy for a 75-year-old woman is 11.5 years and for a 75-year-old man is 8.8 years (Wylie 1984). Clinically apparent complications of diabetes develop, on average, about ten years after the onset of diabetes and, therefore, most older individuals, based on age alone, would be eligible for intensive management.

Dr. Dierdorf considered Mrs. Cohen's case carefully. She already had peripheral neuropathy, so it was likely that she had had diabetes for some time. Her life expectancy was good, about 12 years. Her general health was fairly good, too, and she was cognitively intact. Dr. Dierdorf decided to discuss intensive management with her.

He told Mrs. Cohen that the progression of her peripheral neuropathy might be slowed if they aimed for tight control of her blood sugar. This would mean aiming for normal fasting blood glucoses and normal glycosylated hemoglobin levels. It would mean frequent blood sugar checks and home blood glucose monitoring with a glucometer. Mrs. Cohen wanted to know if this meant that she needed to be on insulin. Dr. Dierdorf told her that it meant that they would use whatever was necessary to get her blood glucose to the level that they desired. If that could be accomplished with diet alone, that would be great, but he thought it likely that Mrs. Cohen would require a combination of dietary therapy and an oral hypoglycemic agent or dietary therapy and insulin.

Mrs. Cohen inquired about alternatives to tight control of her blood sugar. Dr. Dierdorf explained that a less rigorous alternative was to control her blood sugars more loosely, trying to prevent the side-effects of hyperglycemia. He advised her that he did not think that this was in her best interest because he was concerned that the peripheral neuropathy would worsen. In addition, her triglycerides were elevated, a problem that could be improved by better diabetes management. Mrs. Cohen asked to think it over. Dr. Dierdorf recommended that she meet with a dietitian to learn a low fat, high fiber, weight reduction diet and also that she contact the diabetes nurse educator for diabetes classes. He suggested a return visit in two weeks. Dr. Dierdorf completed Mrs. Cohen's health care maintenance by giving her an influenza vaccination, a pneumococcal pneumonia vaccination, and placing a skin test for tuberculosis.

Diabetes management is difficult for both the patient and the physician. It can be frequently frustrating and the difficulties can seem insurmountable. The issues in management will truly be insurmountable if the doctor and the patient are working at cross purposes. The use of a health care team may help smooth the difficulties of diabetes management. Important team members include a diabetes nurse educator and a dietitian. A social worker or psychologist may be helpful if the doctor and patient cannot reach agreement on the type of care, or if the patient seems unwilling to care for himself or herself adequately.

It is critical for the health care provider to appreciate the extent of the impact of diabetes mellitus on a person's life, regardless of the age of the person. Most individuals with diabetes will be forced to change the way they eat. Meals must be carefully planned, frequently with restriction of calories and fat. Meals must also be timed to coincide with medication or insulin injections. No more skipping lunch or a quick mid afternoon ice-cream cone to take the edge off presupper hunger! Many persons with diabetes will have to take new medications, some will have to administer injections to themselves and monitor their blood glucose frequently, perhaps several times a day. Exercise, too, must be added to what may have been a former sedentary lifestyle, or planned around meals and medications so as to prevent hypoglycemia. Few chronic medical illnesses are as invasive as diabetes mellitus.

Patients frequently feel overwhelmed when initially presented with a diagnosis of diabetes, particularly if the physician is a conscientious provider who is determined to educate the patient, prevent diabetes complications, and safely control blood glucose. While the 'You have a touch of diabetes, here's a pill to lower your blood sugar' approach may seem the simplest, this should be avoided in the geriatric population as well as in younger age groups. Geriatric patients are capable of learning the new information and skills necessary to manage their diabetes appropriately. The use of a health care team allows the responsibility for patient

Table 9.2 Commonly used sulfonylurea oral hypoglycemic agents

Drug	Trade name	Duration of action (h)	Starting Dose for older patients (mg)	Maximum daily dose (mg)
Tolbutamide	Orinase	6–12	500	3000
Acetohexamide	Dymelor	12–18	250	1500
Tolazamide	Tolinase	12–16	100	1000
Chlorpropamide	Diabinese	40–72	100–125	500
Glipizide	Glucotrol	8–12	2.5	40
Glyburide	DiaBeta Micronase	24	1.25	20

education and management to be shared across disciplines. It also gives the patient the opportunity to learn about the disease from different vantage points.

Mrs. Cohen returned to her appointment in two weeks. The nurse had helped her to understand why control of her blood sugar was so important. The dietitian had helped her to plan meals with foods she liked and had always eaten. Mrs. Cohen also started a walking program and was walking one mile three times a week with a group of senior (elderly) walkers at the local shopping mall. Physical exam revealed that her weight was down three pounds and her blood pressure was 160/88 mmHg. Fingerstick blood glucose (fasting) was 194 mg/dl.

Dr. Dierdorf praised Mrs. Cohen's weight loss and discussed her diabetes management. Mrs. Cohen felt willing to try more aggressive management of her diabetes with the goals of euglycemia and normalization of her glycosylated hemoglobin levels. Dr. Dierdorf explained that he was going to continue to follow her on diet alone to see if her blood glucose would respond further. He asked her to monitor her blood glucose twice a day, a fasting morning glucose and again at 4 p.m., three times a week. Dr. Dierdorf also advised Mrs. Cohen that she needed treatment for her blood pressure and began her on enalapril, an angiotensin-converting enzyme inhibitor. Dr. Dierdorf asked Mrs. Cohen to make appointments to see a podiatrist and an ophthalmologist. He also asked her to return in two more weeks for a fasting blood glucose.

At the next visit, Mrs. Cohen had lost four more pounds, her blood pressure was 142/82 mmHg and her fasting blood glucose was 166. One month later, she had lost an additional two pounds and her fasting blood glucose was 157. Dr. Dierdorf decided to add an oral hypoglycemic agent.

The decision to add an oral hypoglycemic agent to an exercise and dietary regimen is dependent on whether adequate glycemic control has been achieved with diet and exercise alone. If the goals for glycemic control have not been achieved, either insulin or oral hypoglycemic agents may be used. Most patients and practitioners choose, at this point in management, to add oral hypoglycemic agents for patients with NIDDM because of the ease of use. Currently, six oral hypoglycemic agents are available for use in the United States. All are first or second generation sufonylurea drugs. Another oral hypoglycemic agent, metformin, recently received approval from the Food and Drug Administration.

Characteristics of some of the more commonly used sulfonylureas are shown in Table 9.2. These drugs act primarily by increasing endogenous insulin secretion. Additionally, through the lowering of blood glucose, they further improve glycemic control by enhancing insulin sensitivity. The decision about which of these agents to use is based on the properties of the drug and the individual for whom the drug is to be prescribed. One of the most important features of these medications is drug half-life. Longer-acting oral hypoglycemic agents, such as chlorpropamide and glyburide, have been reported to produce more hypoglycemia than shorter-acting agents. Elderly patients, who may be more predisposed to hypoglycemia as noted above, could potentially benefit from shorter-acting agents. Second generation sulfonylurea agents additionally have reported benefits over first generation agents. Some of these benefits include activity to improve insulin sensitivity directly (although this is a controversial point) and reduced side effects more commonly found in the first generation sulfonylureas, such as hyponatremia and alcohol-induced flushing. When beginning therapy with any of the sulfonylureas, concomitant medications should be reviewed closely because of the large numbers of drugs that interact with these agents. Common drugs that interact with sulfonylureas include aspirin, trimethoprim, alcohol, H2 blockers, anticoagulants, probenecid, allopurinol, and rifampin.

The rule of thumb, 'start low and go slow', when prescribing medications for older adults also applies to the use of oral hypoglycemic agents. These medications should be started at one-half the recommended starting dose for younger adults and should be increased slowly.

Patients must be instructed to eat on a regular basis and warned not to skip meals. Hypoglycemia resulting from oral hypoglycemic agents may be prolonged and can result in coma and death. Patients and their families should be carefully instructed in the recognition of the symptoms of hypoglycemia (sweating, shaking, extreme hunger, confusion, disorientation, and coma) and its treatment (oral sugar-containing solutions like orange juice with sugar dissolved in the juice, or cola beverages, or, for severe hypoglycemia, one milligram of glucagon injected subcutaneously or intramuscularly). The capability to perform home blood glucose monitoring is invaluable for situations where the level of glycemia is unclear.

Although oral hypoglycemic agents are very useful for many patients, 10–20% of individuals who start oral agents will not be adequately controlled, even at maximum doses (Multi-centre Study, 1983). In addition, a substantial number of individuals (perhaps 3–5% per year) will eventually fail treatment with sulfonylurea agents even though they had been previously well controlled. Patients should be warned that they may require insulin therapy to adequately achieve glycemic goals if treatment with a sulfonylurea fails. There is no recognized benefit in switching to a second sulfonylurea following failure with a first.

Mrs. Cohen returned to see Dr. Dierdorf two weeks after beginning glipizide. Her fasting blood glucose was 127 mg/dl. Review of her home blood glucose monitoring revealed that on most mornings her fasting blood glucoses ranged between 105 and 130 mg/dl. Afternoon blood glucoses were between 95 and 125 mg/dl. Dr. Dierdorf was pleased with her response. Mrs. Cohen, however, complained that in spite of all the exercising, dieting, and monitoring she was doing, her feet were still painful. Dr. Dierdorf paused for a moment, remembering that the physician's goals were often not the patient's goals. He then asked Mrs. Cohen about the fatigue and frequent urination that had been bothering her on her first visit. Mrs. Cohen recognized that these problems were largely resolved. He advised her that improvement of the neuropathic symptoms might take months and reassured her that good glycemic control could prevent worsening of her neuropathy. He recommended that she try acetaminophen, 650 mg after dinner, and another 650 mg at bedtime. Dr. Dierdorf also promised that there were other medications to try if her symptoms did not improve with the acetaminophen and better glycemic control.

Peripheral neuropathy is but one of the many chronic complications of diabetes mellitus. All chronic complications develop with both IDDM and NIDDM and can occur in both younger and older diabetic patients. As shown in Table 9.3, diabetes increases the risk of complications for individuals over the age

Table 9.3 Relative risk of complications of diabetes mellitus in the elderly compared to elderly without diabetes mellitus

Complication	Relative risk
Macrovascular disease	
Coronary heart disease	2.0
Stroke	2.0
Amputation	10.0
Capillary microangiopathy	
Retinopathy-macular edema	1.4*
Renal Disease	
Neuropathy	?

* Relative risk of blindness.

of 65 compared to those without diabetes (Herman et al. 1985). Complications are generally divided into three groups: microvascular, macrovascular, and neuropathic. The microvascular complications of retinopathy and nephropathy are believed to be the result of small vessel disease in the retina and the kidney, respectively, and were shown by the DCCT to be related directly to glycemic control.

The risk of blindness is increased by almost 50% for those older adults with diabetes. Visual loss, while devastating at any age, may be the death knell for the independence of an older adult. While retinopathy is most commonly thought of as the eye complication associated with diabetes, glaucoma and cataract are more important causes of blindness in the diabetic geriatric population.

Diabetic nephropathy has been shown in some populations to occur as frequently in NIDDM as it does in IDDM. The appearance of albuminuria is a key sign of the onset of diabetic nephropathy. Preventive measures focus upon the use of angiotensin-converting enzyme inhibitors to reduce proteinuria and slow the rate of decline of renal function. Blood pressure control remains a critical factor in slowing the progression of diabetic renal disease. These interventions, however, have not been studied in older adults with NIDDM.

Macrovascular complications, which include cardiovascular disease, cerebrovascular disease, and peripheral vascular disease, were not definitively shown in the DCCT to be reduced by tight glycemic control, although the data were suggestive. Because the DCCT was performed using young, healthy individuals, very few macrovascular complications occurred during the ten-year study period, limiting the statistical power. However, because of the high prevalence of coronary artery disease, cerebrovascular disease, and peripheral vascular disease in the geriatric population in general, the addition of diabetes as a risk factor for vascular disease magnifies the extent of these problems for older adults. Heart disease is the leading cause of death in the geriatric age group. This poses, once again, the intriguing and unanswered question regarding the effect of glycemic control on the prevention of macrovascular complications for the geriatric diabetic population. While this question may not be definitively answered, other interventions should be made to reduce the risk of macrovascular complications. One critical risk for older adults with diabetes mellitus is that of amputation. Older adults with diabetes have a tenfold higher risk for amputation than older persons without diabetes (Herman *et al.* 1985). Prevention of foot disease among older diabetics is of paramount importance. Interventions that may be helpful include education regarding foot care with daily inspection of the feet, podiatry consultation, appropriate and well-fitting footwear, avoidance of walking in bare feet, and immediate medical attention for even the smallest sign of infection or breakdown.

Neuropathic complications include distal symmetric polyneuropathy, mononeuropathy including cranial nerve palsies, and autonomic neuropathy with orthostatic hypotension, gastroparesis, diarrhea, or atonic bladder. One unusual neuropathic complication, diabetic neuropathic cachexia, tends to occur most commonly in older men and is associated with severe pain in the extremities and weight loss. While poorly understood, this syndrome is usually self-limited and resolves in several months.

Dr. Dierdorf continued to see Mrs. Cohen on a regular basis, every two to three months, over the next year. While her sugars were initially well controlled and Mrs. Cohen was faithful to her diet and home glucose monitoring, about eighteen months after her initial diagnosis she began to slip. Her weight drifted up and she began questioning the need for frequent monitoring. She canceled appointments, stating she felt fine. Dr. Dierdorf heard nothing more for several months until he received a call from the emergency room of the community hospital. Mrs. Cohen had been brought to the emergency room by her daughter who had found her semi-comatose at home. The emergency room physician reported that she was unresponsive with a blood pressure of 80/50, pulse of 128, respirations 24 and shallow, and temperature of 101.8. The rest of Mrs. Cohen's physical exam showed that her neck veins were flat even in the recumbent position and that there were a few crackles at the right base. Her laboratory studies were as follows: white blood count 12.8 with 62% polymorphonuclear leukocytes, 18% bands, 15% lymphocytes, and 5% monocytes; hemoglobin 15.2, hematocrit 47.3; sodium 142, potassium 5.3, chloride 110, bicarbonate 21, blood urea nitrogen 90, and creatinine 3.7. Mrs. Cohen's blood sugar was 780. Blood gases showed that the pH was 7.32, po_2 was 94, pco_2 was 32, HCO_3 was 20 and oxygen saturation 95%. Urine ketones were 1+ and serum ketones were negative. She was then transferred to the intensive care unit.

The acute complications of diabetes mellitus include problems of metabolic control: hypoglycemia, diabetic ketoacidosis, and the hyperglycemic hyperosmolar nonketotic state (also called hyperglycemic hyperosmolar nonketotic coma or the hyperglycemic hyperosmolar syndrome). While hypoglycemia and diabetic ketoacidosis occur in all age groups with diabetes, the hyperglycemic hyperosmolar nonketotic state occurs almost exclusively in individuals over the age of 50 (Wachtel *et al.* 1987). The mortality rate is high for this syndrome, reportedly over 50% in some studies. More recent surveys show mortality rates to be in the range of 20%. Death results from the primary illness which precipitates the hyperglycemic hyperosmolar state and from complications that occur subsequent to the initial insult.

The hyperglycemic hyperosmolar non-ketotic state results from a cycle of events. The major features include a precipitating event, decreased oral intake, rising blood glucose, osmotic diuresis with worsening dehydration and hyperglycemia, and rising stress hormones (including epinephrine and cortisol) which increase insulin resistance and further elevate blood glucose. The precipitating event for this syndrome is frequently an infectious illness such as influenza or pneumonia, which limits a patient's access to fluids. Acute stressful medical events such as burns, myocardial infarction, and stroke can also be precipitants as can drugs which interfere with carbohydrate metabolism (e.g., phenytoin) or drugs, such as diuretics, which increase the risk of dehydration.

The typical patient with the hyperglycemic hyperosmolar nonketotic state has a diagnosis of diabetes mellitus but is 'diet-controlled', although one-third of patients with the syndrome are not previously known to be diabetic. Most patients are ill for four to seven days before developing the acute syndrome and usually present with profound dehydration. Mental status ranges from alertness to obtundation, however most patients are not comatose. The rest of the physical exam is consistent with marked dehydration and patients may demonstrate profound orthostatic hypotension, decreased skin turgor, dry mouth, and flat neck veins in the supine position. The physical exam may also point to the precipitating event. Signs of this should be sought by a careful, complete examination. Initial laboratory investigation confirms the diagnosis and should include a complete blood count, electrolytes, blood urea nitrogen, creatinine, plasma or serum glucose, arterial blood gases, urinalysis, serum ketones, serum osmolality, an electrocardiogram (to evaluate for myocardial infarction) and a chest X-ray (to evaluate for pneumonia).

The results of Mrs. Cohen's initial laboratory investigations confirm severe dehydration, manifested

Table 9.4 Hyperglycemic hyperosmolar non-ketotic syndrome

	Diagnostic Criteria	Normal values*
Plasma glucose (mg/dl)	>600	998
Serum osmolarity (mmol/kg)	>330	363
Arterial pH	>7.3	
Serum bicarbonate	>20 mEq/L	21
Urinary ketones	Negative or small	

* From Wachtel et al. (1987) and Rifkin and Porte (1990).

by elevated blood urea nitrogen, creatinine, and hematocrit. Table 9.4 shows laboratory diagnostic criteria and the usual findings. Plasma glucose is usually around 1000 mg/dl, but there is little or no metabolic acidosis. Any metabolic acidosis present is usually secondary to lactic acidosis. Arterial blood gases confirm that profound metabolic acidosis is not present. Serum osmolality is usually greater than 330 mmol/kg. The differential diagnosis for these patients is usually diabetic ketoacidosis versus the hyperglycemic hyperosmolar nonketotic state. Key features aiding in differentiating these two conditions include arterial blood pH (usually lower in diabetic ketoacidosis than in the hyperglycemic hyperosmolar state), serum bicarbonate levels (usually lower in diabetic ketoacidosis), and the presence of serum ketones (present in high levels in diabetic ketoacidosis but either absent or only present in small amounts in hyperglycemic hyperosmolar nonketotic state).

The primary treatment for the hyperglycemic hyperosmolar nonketotic state is volume replacement and reversal of the fluid deficit. One-half the deficit should be replaced in the first 24 hours and the second half in the next 24–48 hours. Because of the massive volume deficits, patients may require four to six liters of fluid in the first 24 hours. Underlying cardiovascular disease in an older diabetic population can make rapid volume replacement troublesome; central pressure monitoring with a Swan–Ganz catheter may be necessary for some individuals. For those patients who are hemodynamically unstable (i.e., hypotensive and tachycardic), the fluid of choice for initial volume replacement is normal saline. Half-normal saline can by used after hemodynamics are restored. Small doses of insulin should be administered during the initial treatment phase. Insulin is probably best given as an intravenous infusion, in doses similar to those used to treat diabetic ketoacidosis, at 6–10 units per hour. Intramuscular or subcutaneous insulin is not recommended because of poor perfusion and irregular absorption in the severely dehydrated state. Serum potassium levels should be closely monitored and potassium replacement begun when the potassium level drops below 5.0 mEq/l. Potassium replacement

must be cautious when renal function is impaired. Phosphate may be replaced as potassium phosphate although the need for phosphate replacement in the hyperglycemic hyperosmolar non-ketotic state has not been well documented. Treatment of the precipitating event is also important to aid in resolution of the cycle which produced the syndrome.

In the intensive care unit, Dr. Dierdorf examined Mrs. Cohen quickly and reviewed her initial laboratory studies. It was evident on physical examination that she was severely dehydrated. Her laboratory studies were compatible with the hyperglycemic hyperosmolar nonketotic syndrome and he was worried that the precipitating insult was pneumonia. Dr. Dierdorf first turned his attention to stabilizing Mrs. Cohen's hemodynamic status. The nurses told him that her weight was 70.6 kg. He remembered that her usual weight was around 85 kg. Dr. Dierdorf, recalling that total body water is approximately 60% of weight, quickly calculated her total body water deficit at about 8 liters.

$$(85 \text{ kg} \times 0.6 \text{ liter/kg}) - (70.6 \text{ kg} \times 0.6 \text{ liter/kg}) = 51.1 \text{ liter} - 42.3 \text{ liter}$$
$$= 8.8 \text{ liter}$$

He ordered that two intravenous lines be placed and that 0.9% saline be run wide open until one liter had been given. The fluid was then ordered to run at 250 ml/h in each intravenous line (500 ml/h total) until another liter was given. Vital signs were to be monitored every 15 minutes for the first hour and then every 30 minutes for the next two hours. Dr. Dierdorf also ordered an intravenous bolus of 10 units of regular insulin followed by an insulin drip of 6 units per hours. Blood glucose was to be checked hourly and electrolytes checked every two hours for the first six hours.

In a search for a precipitating factor, Dr. Dierdorf ordered cultures of Mrs. Cohen's blood, urine, and sputum. He obtained an emergency chest x-ray, which confirmed a right lower lobe pneumonia. Dr. Dierdorf ordered intravenous broad spectrum antibiotic coverage. He ordered that she have nothing by mouth until she was alert and asked the nurses to place a catheter in her bladder so that urine output could be monitored accurately. Then Dr. Dierdorf waited.

Recovery from the hyperglycemic hyperosmolar nonketotic state is dependent on the severity of the precipitating event, the condition of the person at the time he or she presents for medical care, and the development of complications. Resulting complications include deep vein thrombosis and pulmonary embolism, pneumonia (frequently from aspiration), acute renal failure, and myocardial infarction. Aggressive support and early volume replacement are key elements of care in the prevention of complications and improving survival.

At the time of discharge from the hospital, some

patients will require insulin, some oral agents, and others will require diet alone to control glycemia. Patients should be cautioned to monitor blood glucoses closely during times of illness, to continue insulin or oral agents, to contact their physicians early in the course of illness, and to avoid dehydration by drinking liberal amounts of non-sugar-containing beverages during illnesses. Equally important are daily social contacts for all older diabetic individuals, either with friends, neighbors, or family members to prevent prolonged, undetected episodes of hyperglycemia or hypoglycemia.

Mrs. Cohen went home on a small dose of insulin following her ten-day hospitalization. Dr. Dierdorf thought about how much more frail she seemed now and wondered what he might have done differently to prevent her hospitalization. Perhaps more education, or more reinforcement about home blood glucose monitoring. He arranged for the diabetes nurse to spend more time with her and he asked her daughter to check on her daily from now on. Dr. Dierdorf thought of her, ill for three days in her apartment before her neighbor noticed the mail and the newspapers piling up. He hoped to prevent a recurrence.

After a few weeks, Dr. Dierdorf was able to stop the insulin and put her back on her oral hypoglycemic agent. Mrs. Cohen was more conscientious about her appointments and monitoring but still had occasional lapses with her diet. At her most recent visit she had asked Dr. Dierdorf to help in the care of her elderly aunt, a 92-year-old nursing home resident, for whom Mrs. Cohen was guardian.

Dr. Dierdorf went to see Mrs. Cohen's aunt, Ethel Weiss, the next day. A frail, slightly demented woman, she had moved to the nursing home a few months ago when she could no longer manage her own home, even with heavy support services. Her diabetes had been diagnosed several years earlier and she was now taking glipizide 20 mg twice a day. Her fasting blood glucoses were in the 230–250 mg/dl range and the nurses reported that she was frequently incontinent, both at night and during the day. Her appetite was good, although she was thin.

Care of the frail elderly individual with diabetes is a common problem which many health care providers find perplexing. Many believe that a laissez-faire attitude to the treatment of diabetes is most appropriate for the very old and those with multiple other medical problems. In particular, the decision to use insulin in this group of patients is guided by the same criterion used for other patients. Insulin therapy should be used when blood glucose cannot be adequately controlled with other means. For the frail elderly this translates into using insulin to keep fasting blood glucoses less than 180 mg/dl and random glucoses less than 200 mg/dl. The goal is to prevent glycosuria and the acute complications of diabetes. Prevention of glycosuria

will reduce the risk of dehydration and hyperglycemic hyperosmolar states; it may also reduce episodes of incontinence. Monitoring is best done with blood glucose monitoring, as urine monitoring is relatively inaccurate and does not provide information about hypoglycemia.

Patients will, on occasion, refuse to take insulin. A patient's wishes must be followed provided the individual is capable of informed decision making, understands the risks and benefits of the proposed treatment, and the potential consequences of choosing another course, including non-compliance. Physicians should, however, make the effort to discuss patient fears and resistance, especially regarding insulin therapy, and educate the patient. Physicians should also avoid rejecting treatment options for elderly individuals because the physician perceives them as too difficult or too uncomfortable. Studies have shown that patients adapt well to insulin therapy. Even for very frail individuals, where someone else is available to administer insulin and perform glucose monitoring, insulin therapy may be the most correct decision. Improved glucose control may lessen urinary frequency, incontinence, visual blurring, and improve fatigue and other more generalized symptoms.

Diabetes management of frail nursing home patients, such as Mrs. Cohen's aunt, involves more than glucose control. Preventive medicine remains essential. Tuberculosis screening should be done annually and immunization, including annual influenza and one-time pneumococcal vaccine, given. Careful attention to the prevention of infections, primarily urinary tract and skin infections, can likely lessen morbidity and mortality among nursing home patients with diabetes.

Diabetes mellitus is an illness that challenges both the patient and physician. Management of this condition is, however, possible for even the very old. Management hinges upon treatment goals and treatment should then be directed to achieving these goals. In most older adults with diabetes, treatment goals will be met with diet, exercise, and oral hypoglycemic agents. For the remainder, insulin will be necessary. Appropriate therapy for older adults with diabetes should improve quality of life, and may prevent complications over time.

Questions for further reflection

1. What are the criteria for the diagnosis of diabetes mellitus?
2. Outline a plan of therapy for an overweight 75-year-old man with fasting blood sugars of 180–200 mg/dl. What referrals to other health care professionals are important?
3. Why do individuals with Type I (ketosis prone, juvenile onset) diabetes develop ketoacidosis while individuals

with Type II (adult onset) diabetes develop a non-ketotic hyperosmolar state?

References

ADA (American Diabetes Association) (1988). *Physician's guide to non-insulin dependent (Type II) diabetes*, (2nd edn). American Diabetes Association, Alexandria, VA.

ADA (American Diabetes Association) (1989). Standards of medical care for patients with diabetes mellitus. *Diabetes Care*, **12**, 365–8.

DCCTRG (Diabetes Control and Complications Trial Research Group) (1993). The effect of intensive treatment of diabetes on the development and progression of long-term complications in insulin-dependent diabetes mellitus. *New England Journal of Medicine*, **329**, 977–86.

Harris, M.I., Hadden, W.C., Knowler, W.C., and Bennett P.H. (1987). Prevalence of diabetes and impaired glucose tolerance and plasma glucose levels in the U.S. population aged 30–74 years. *Diabetes*, **36**, 523–34.

Herman, W.H., Teutsch, S.M., and Geiss, L.S. (1985). Closing the gap: the problem of diabetes mellitus in the United States. *Diabetologia*, **24**, 404–11.

Kilvert, A., Fitzgerald, M.G., Wright, A.D., and Nattrass, M. (1986). Clinical characteristics and aetiological classification of insulin-dependent diabetes in the elderly. *Quarterly Journal of Medicine*, **60**, 865–73.

Multi-centre Study (1983). UK prospective study of therapies of maturity-onset diabetes. I. Effect of diet, sulfonylurea, insulin or biguanide therapy on fasting plasma glucose and body weight over one year. *Diabetologia*, **24**, 404–11.

Rifkin, H. and Porte, D. (ed.) (1990). *Diabetes mellitus: theory and practice*, (4th edn), pp. 604–16. Elsevier, Amsterdam.

Wachtel, T.J., Silliman, R.A., and Lamberton, P. (1987). Predisposing factors for the diabetic hypersomolar state. *Archives of Internal Medicine*, **147**, 499–501.

Wylie, C.M. (1984). Contrasts in the health of elderly men and women: an analysis of recent data for whites in the United States. *Journal of the American Geriatrics Society*, **32**, 670.

10

Arthritis in the elderly

K. Lea Sewell

Irene Eakins is 84 and has seen you regularly for annual check-ups. Her health is good, although she has had a peptic ulcer in the past and has always struggled with her weight. She has mild glucose intolerance which, to date, has only necessitated dietary instructions and avoidance of concentrated sweets. She tells you that her knees are bothering her after walking—particularly after climbing stairs—and that this has been gradually developing over two to three years. She never thought to mention it previously, as the discomfort had been minor and she thought that joint pain was normal at her age. But now she has trouble taking the ten stairs at the front of her church and has to park at the rear where the service entrance is located.

The approach to musculoskeletal pain in older patients

Mrs. Eakins' physician, Dr. Susan Bergman, knows that musculoskeletal pain is a very common complaint in elderly patients and that it may range in a full spectrum from a minor problem to a major disabling condition. Nearly 50% of community-dwelling elderly over the age of 70 report arthritis. Of these individuals, physical limitation is present in 60–70% and limitations in activities of daily living (ADLs) in 20–39%. These limitations predict excess risk for both institutionalization and morbidity, after correction for age and co-pathology (Yelin 1992). As musculoskeletal problems worsen, they may lead to significant morbidity, disability, and a host of problems which stem from chronic pain and the social isolation that accompanies functional limitation. Arthritis surveys in the non-institutionalized population over 60 years of age show high medication use (23%), appliance dependency (15%), and home health aide requirements (10%) (Taylor and Ford 1984). In nursing homes, 25% of all residents have arthritis and this condition is the basis for placement in 15% of all non-demented elderly (Guccione *et al.* 1989).

Dr. Bergman also realizes that 'ageism' often prevents patients and their families from seeking attention for musculoskeletal complaints, as the high prevalence of 'rheumatism' in the elderly is often seen as inevitable. She is suspicious that Mrs. Eakins has a significant problem merely on the basis that Mrs. Eakins has decided to mention her problem when it became functionally significant. A presumption of no available therapy may also prevent patients from mentioning even functionally significant musculoskeletal complaints.

A vague complaint of musculoskeletal pain can certainly be overwhelming for the physician unless a structured approach is used. Dr. Bergman's first task is to establish the clinical significance of Mrs. Eakins' arthritis by determining whether there has been a functional impact. A transient ache which caused no other problem would not warrant extensive evaluation. Dr. Bergman asks about additional dependence in dressing, housework, shopping, and ability to arise from bed or a deep chair. She discovers that Mrs. Eakins is no longer able to shop due to pain and must rely on help from a neighbor.

As there does appear to be both pain and functional limitation, Dr. Bergman then asks for further history clarifying the arthritis pattern. Her questions must establish the time course (acute, subacute, insidious onset), the presence or absence of symmetrical joint involvement, the location of involved joints (distal small joints, large proximal joints, or an axial spine pattern), any exacerbating factors, and any response to analgesic or antiinflammatory medications. A family history may also be helpful.

Mrs. Eakins describes, after a day when she would be on her feet, some swelling about the knees by day's end. If she can put her feet up for an hour each morning and afternoon, the knee swelling is much better, and she always gets pain relief with rest. She does not notice any stiffness. In the morning her first few steps are painful and unsteady, but they loosen up after she has moved around for a few minutes. She has used acetaminophen with some relief, but otherwise has been too worried about medicines to try ibuprofen or other agents.

Assessing inflammatory versus non-inflammatory arthritis

Mrs. Eakins' complaints suggest a symmetric oligo-arthritis limited to the knees with an insidious onset. The final historical concern is to differentiate inflammatory versus non-inflammatory types of arthritis with this pattern of joint involvement. Inflammatory arthritis sufferers will typically describe morning stiffness lasting over 15 minutes in involved joints, and 'gelling' or recurrent stiffness with prolonged immobility. Unlike Mrs. Eakins, they do not necessarily obtain pain relief with absolute rest but prefer 'moving a little bit' most of the time.

A challenging differential diagnosis can be polymyalgia rheumatica, an inflammatory condition with morning stiffness in the hips and/or shoulder areas which improves after hot showers, antiinflammatory medications, or mobility for 30 minutes or more. An examination in the afternoon may reveal nothing other than pre-existing degenerative joint changes which are unrelated to the stiffness and discomfort. A second inflammatory possibility may be elderly onset rheumatoid arthritis. In older persons, rheumatoid arthritis involves fewer joints than in younger patients. The joints involved tend to differ, with inflammation in large proximal joints including the knees being common in the elderly, rather than the small joints of the hand and feet as seen in the classic rheumatoid arthritis that afflicts younger individuals. In Mrs. Eakins' case, her symptoms are limited to her knees and are most compatible with a non-inflammatory arthritis.

In evaluating any person with joint complaints, infectious, traumatic, and crystalline arthritis must be considered. Monoarticular complaints are particularly problematic. Features of Mrs. Eakins' history, however, make these concerns unlikely. A bilateral knee problem is unlikely to reflect septic arthritis in the absence of any evidence for septicemia. Additionally, a multi-year, insidious onset is not typical for crystalline arthritis. Should Dr. Bergman have any question about the diagnosis—and any historical clue or finding on examination which raises the possibility of inflammatory or infectious arthritis would be sufficient—then a synovial fluid analysis should be performed.

Age-related changes in the musculo-skeletal system

Dr. Bergman thinks that Mrs. Eakins' complaints are unlikely to be due to 'old age'. The cartilage matrix is modulated by aging, although cartilage strength and flexibility are only moderately reduced. Shorter proteoglycans with altered ratios of adherent glycosoaminoglycans (GAGs) may reflect altered synthetic or post-synthetic processes in senescent chondrocytes (Hamerman 1993). Additionally, bone density decreases as osteoblast activity fails to completely replete osteoclast-remodeled cavities. This decreased bone density may be accentuated in persons with inadequate 25-hydroxy-vitamin D due to vitamin D malabsorption or diminished renal parenchymal activation. Lean muscle mass decreases with aging, although resistance muscle training is still possible at advanced age (Nichols *et al.* 1993).

On examination, Mrs. Eakins weighs 185 pounds (82.6 kg) and is 5 feet 3 inches (97 cm) tall. Her general examination is otherwise unremarkable, including muscle strength and reflexes in her legs, with the exception of her musculoskeletal examination. She has obvious hard 'double bump' knobs on her distal interphalangeal joints (Heberden's nodes) which are slightly painful but do not limit finger flexion or extension. Her hips are not painful nor limited in the three planes of motion (flexion, abduction, rotation). Her toes show bilateral bunion deformities.

Her knees reveal medial joint line tenderness and bony prominence, with crepitus bilaterally. The right knee has mild soft tissue swelling compared to the left, with a small fluid wave and a ballotable patella. Extension is complete; flexion is 100° on the right and 120° on the left. Her stance reveals some valgus deformity.

The examination in osteoarthritis (osteoarthrosis)

Dr. Bergman turns to the examination for evidence of either an inflammatory or degenerative arthritis. Signs of an inflammatory arthritis would include swelling, warmth, and erythema. A degenerative arthritis may reveal crepitance about the involved joint. In addition, Dr. Bergman wishes to determine the extent of joint damage by investigation of the range of motion of the joint, alignment abnormalities, or ligamentous laxity. She evaluates soft tissue structures near the knee for possible bursitis or tendonitis and to exclude the possibility of referred pain from the hip. Mrs. Eakins' limited flexion and the presence of a valgus deformity suggest chronic joint damage.

Another important component of the examination will be to determine Mrs. Eakins' functional grade or status. Dr. Bergman has already questioned Mrs. Eakins about her ability to perform ADLs and instrumental activities of daily living (IADLs) and discovered that she needs assistance in shopping. Dr. Bergman observes Mrs. Eakins' ability to perform difficult lower extremity tasks such as climbing steps and tying shoelaces. She asks Mrs. Eakins to make an X on a 10 cm linear scale measuring overall pain with 'no pain' on the left and 'maximum pain' on the right, which she will compare on future visits.

Dr. Bergman is concerned that the right knee effusion might reflect a superimposed traumatic, inflammatory, or infectious problem in addition to Mrs. Eakins' generalized osteoarthritis (osteoarthrosis). She elected to evaluate the right knee discomfort further.

Right knee standing AP and lateral X-rays revealed medial joint space narrowing with a small tibial osteophyte and no fracture. Synovial fluid analysis of the right knee yielded clear fluid with WBC (white blood count) 1200/mm³ (80% lymphocytes, 20% polymorphonuclear leukocytes), RBC (red blood count) 5400/mm³ and no crystals by birefringent analysis. Laboratory exam of peripheral blood reveals hemoglobin 11.1 gm/l, Westergren sedimentation rate (WESR) 41 mm/h, and a positive rheumatoid factor at 1:80.

Laboratory findings in osteoarthritis (osteroarthrosis)

A non-inflammatory fluid, which may be increased in quantity, is the *sine qua non* of osteoarthritis (osteoarthrosis) (Schumacher and Reginato 1991). The fluid should be clear and contain less than 2000 WBC/mm³ (totally normal joints will have less than 200 WBC/mm³) with fewer than 25% polymorphonuclear leukocytes (PMN) in the differential. Rare cases will show a WBC count of 2000–5000/mm³ and are termed 'inflammatory osteoarthritis'. Individuals with this picture have painful, aggressive cyst formation at their Heberden's (distal interphalangeal joints) or Bouchard's (proximal interphalangeal joints) nodes. Inflammatory osteoarthritis may represent a flare up of primary osteoarthritis or be an overlap with an inflammatory condition, such as gout.

The Westergren sedimentation rate (WESR) varies in an age-dependent manner. It is normal in Mrs. Eakins' case. Dr. Bergman interprets the WESR using the following age correction formula for the upper limit of normal (Sox and Liang 1986):

WESR in Men: Age/2
WESR in Women: (Age + 10)/2

Mrs. Eakins' rheumatoid factor was positive at a low titer, and Dr. Bergman considers this test in view of the known increase in autoantibody production

Table 10.1 The Kellgren grading system for osteoarthritis (osteoarthrosis)

Grade I	Normal joint space, possible osteophyte
Grade II	Possible joint space narrowing, osteophytes present
Grade III	Clear joint space narrowing, moderate multiple osteophytes, and sclerosis
Grade IV	Severe joint space narrowing, osteophytes, sclerosis, and cysts

by normal elderly persons. Rheumatoid factor is a specific autoantibody developed against a person's own immunoglobulins and is often positive at high titer in patients with rheumatoid arthritis. A positive IgM rheumatoid factor is the most common autoantibody finding with normal aging, particularly at low titer. A 40% incidence for rheumatoid factor positive in a titer greater than or equal to 1:100 is reported for healthy persons over age 65 (mean age 70), compared to a 3% incidence in healthy individuals aged 20–45 (mean age 39) (Ruffati *et al.* 1990). Accordingly, Dr. Bergman is reassured that this low titer rheumatoid factor is not strongly supportive of inflammatory arthritis in Mrs. Eakins' case.

Mrs. Eakins' knee X-rays show classic grade II osteoarthritis (osteoarthrosis) changes by the Kellgren grading system, which describes the advancement of radiographic osteoarthritis (see Table 10.1). Dr. Bergman recalls that the X-ray stage and the level of symptoms often do not correlate tightly. Knee osteoarthritis is present radiographically in 30% of all women over age 70, as shown in the independently living population of the Framingham study, with an increasing prevalence as age advances. Approximately 40% of these show severe joint narrowing, yet only 40% of those with severe changes are clinically symptomatic (Felson *et al.* 1987).

Dr. Bergman makes the diagnosis of osteoarthritis of the knee using criteria established by the American College of Rheumatology, which have 91% sensitivity and 86% specificity in this case (Altman *et al.* 1986).

Mrs. Eakins is informed that her diagnosis is osteoarthritis (osteoarthrosis). She is instructed to initiate regular use of acetaminophen (paracetamol) 1 gram two to three times a day to limit pain (analgesia) and is referred to a physical therapist for quadriceps strengthening and a supervised walking program. She receives instructions for symptomatic use of heat and cold applications and dietary consultation to help her attain weight loss. She later noted to Dr. Bergman that the acetaminophen was often insufficient to allow her to climb stairs and therefore she received naproxen with misoprostil (Arthrotec) for morning use, in addition to continuing the acetaminophen as needed.

Conservative therapy in osteoarthritis (osteoarthrosis)

Dr. Bergman employs a comprehensive functional approach to osteoarthritis (osteoarthrosis) care, which aims to maintain joint mobility and patient independence. For each case, she tailors her recommendations to the patient's specific disability, using any of several available therapeutic components in her plan in order to assist the patient to remain as fully functional as possible. She also attempts to fully utilize non-drug therapy in achieving these goals.

The cornerstones of conservative management in osteoarthritis (osteoarthrosis) are: (1) patient education; (2) analgesia; and (3) exercise. Patient education should reassure any fears of disability or disfigurement, emphasize joint protection in all activities, explain the multiple modalities which are available, and involve the patient and family in developing and enacting the plan. Analgesia allows participation in normal activities and in supervised isometric (no movement) or isotonic (movement) exercise. Analgesia can be achieved by application of heat or cold, topical applications of menthol-based or capsaicin-supplemented preparations, simple analgesics containing acetaminophen (paracetemol), narcotic analgesics containing codeine, or antiinflammatory medications. In some cases an intra-articular corticosteroid injection will achieve analgesia for several months. The degree of inflammation in any case of osteoarthritis (osteoarthrosis) will vary but the presence of cytokine (interleukin 1) and enzyme (stromelysin, collagenase) release would suggest that there likely is some inflammatory component in most, if not all, cases.

Rehabilitative services are an important component for movement dysfunction (physical therapy) or for specific functional dysfunction (occupational therapy). Immobilization is never recommended due to the rapid development of muscular atrophy, although adequate rest is appropriate. Isometric quadriceps strengthening would improve knee stability in Mrs. Eakins' case, and assistive appliances such as canes, or adaptive aides such as grab bars, can clearly help to maintain complete independence.

The social support structure and resources available to her patient is another potential concern of Dr. Bergman's. She will frequently suggest an arthritis support group or refer her patient to a social service counselor, particularly when limited family support or advanced functional disability makes in-home assistance or travel support warranted. Rheumatology consultation is requested when treatment complications or failures arise.

Dr. Bergman strongly recommends weight loss for treatment of Mrs. Eakins' knee symptoms and prevention of further cartilage loss. From the Framingham study, Dr. Bergman is aware of the association between obesity and both the prevalence of and clinical symptoms in knee osteoarthritis; individuals in the highest weight quintile are at the highest risk for progressive, symptomatic osteoarthritis (osteoarthrosis) (Felson et al. 1992).

Complications of therapy in osteoarthritis (osteoarthrosis)

The potential risks of non-steroidal antiinflammatory drug (NSAID) therapy in the elderly must be considered when conservative non-drug therapies and simple analgesia have been inadequate in meeting the functional goals of the patient. The elderly are at increased risk for NSAID toxicity for a variety of reasons including: (1) slower hepatic metabolism leading to increased NSAID half-life; (2) the frequent presence of other medications that may decrease protein binding or decrease renal blood flow; and (3) increased dependence on local prostaglandin production for the maintenance of adequate renal perfusion. The major risks from NSAID therapy are gastropathy and renal insufficiency. Gastric erosions which subsequently bleed are more likely to remain asymptomatic in the elderly on NSAIDs. Elderly patients (over age 60) who are admitted to the hospital for upper gastrointestinal hemorrhage are 4.7 times more likely (confidence interval of 3.1–7.2) to have recently used a NSAID (Griffin et al. 1988). The development of edema and worsening hypertension are also potential complications of NSAID use.

When she elects to recommend NSAIDs, Dr. Bergman often initially prescribes non-acetylated salicylates, knowing that their weaker inhibition of prostaglandin synthetase limits risks at the cost of a milder clinical effect. (In the case of Mrs. Eakins, Dr. Bergman went directly to naproxen because her knee pain was so severe.) If these agents are ineffective, she recommends moderate doses of NSAIDs with a short to intermediate half-life (ibuprofen, naproxen, sulindac), avoiding long half-life NSAIDs which, despite their once-daily dosing convenience, run the risk of accumulation to toxic levels. In addition, Dr. Bergman avoids those NSAIDs with very high potency—particularly indomethacin—due to increased risk of gastropathy and reports of central nervous system toxicity.

Dr. Bergman strictly limits NSAID use in high risk elderly patients. Such persons include those with prior ulcer disease, previous NSAID-related symptoms, and those with 'pre-renal' hemodynamics on the basis of diuretic use, congestive heart failure, or pre-existing renovascular disease from hypertension or diabetes (Barrier and Hirschowitz 1989). These patients receive NSAIDs with concomitant prostaglandin E_2 prophylaxis, and additionally should have close monitoring of their hemoglobin, creatinine, and fecal occult blood status every three months. Because of Mrs. Eakins' past history of an ulcer, she was placed on misoprostil along with naproxen.

Over the next two years, Mrs. Eakins manages to lose 10 pounds (4.5 kg) and does well. She returns today to tell Dr. Bergman that her left knee has been particularly painful for three days with night-time throbbing and some morning swelling and stiffness which improved over an hour. Mrs. Eakins had twisted her leg four days ago, but did not fall, when she slipped on a rug edge at home. She has no fever nor has she seen any redness around the knee.

On examination, Mrs. Eakins' left knee was tender but seemed unchanged with the exception of increased ballotable fluid. Her left calf was non-tender and equal in size to the right side. A synovial fluid analysis was performed.

Superimposed inflammatory arthritis flares in osteoarthritis (osteoarthrosis)

Mrs. Eakins' report of a single joint acute flare-up could represent a complicating condition superimposed on her osteoarthritis (osteoarthrosis). Dr. Bergman recalls the broad differential for a monoarthritis and recognizes that the morning stiffness and painful gel with inactivity in bed suggest an inflammatory arthritis. A small fracture would be less likely because there is no point tenderness, but a stress fracture or pseudofracture (osteomalacia) could be possible. In addition, crystal arthritis could present over four days, as could a septic arthritis. Finally, a mechanical problem may have developed at the time she twisted her knee, leading to a ligamentous tear and hemarthrosis, or a meniscal tear with a non-inflammatory effusion. There is no evidence of sufficient fluid or inflammation to result in a Baker's cyst (a rupture of the joint capsule of the knee with subsequent drainage of joint fluid into the calf). All of the diagnostic possibilities require an arthrocentesis. Other imaging studies (X-rays, radionuclide bone scan, magnetic resonance imaging) would be additionally required if mechanical and traumatic problems were still considered following the fluid analysis.

Dr. Bergman removes 20 cc of opaque fluid which is non-hemorrhaghic and has a WBC count of 15 000/mm³ with a differential that includes 88% PMN. Microscopical analysis of the fluid in a polarizing microscope yields positively birefringent intracellular crystals, consistent with calcium pyrophosphate crystals and the diagnosis of pseudogout. Dr. Bergman, recognizing that crystals may coexist with infection, sends a sample of fluid for culture. Mrs. Eakins' attack of knee pain involves one of the two most commonly involved joints in pseudogout, the other commonly involved joint being the wrist. Her subacute onset of pain is entirely consistent with the diagnosis of pseudogout. The positively birefringent crystals differ from the uric acid crystals of a gouty flare, which are negatively birefringent. Osteoarthritis (osteoarthrosis), due to biochemical cartilage changes, is a risk factor for calcium pyrophosphate deposition. Other common risk factors include hyperparathyroidism and a past history of joint trauma. Calcium pyrophosphate deposition disease, in turn, may promote a more fulminant form of cystic osteoarthritis (osteoarthrosis).

As Mrs. Eakins developed her pseudogout while receiving naproxen, Dr. Bergman considers other therapeutic options to treat her acute flare of knee pain. Colchicine and corticosteroids are two possibilities. Colchicine has the disadvantage of significant gastrointestinal toxicity following oral loading but it can be used chronically on a once daily basis to prevent additional attacks. Corticosteroids can be prescribed orally and rapidly tapered, usually avoiding any risk of hyperglycemia or gastritis during their use. Mrs. Eakins, however, has a history of both ulcer disease and glucose intolerance. After establishing the diagnosis of crystalline arthritis and reviewing the gram stain of the fluid to exclude the presence of organisms and a septic arthritis, Dr. Bergman, therefore, elects to avoid any possible systemic complications by injecting the knee with a slow acting corticosteroid.

Mrs. Eakins gained back 5 pounds (2 kg) over the next three years. Her activity had greatly decreased and she had to limit her walking exercise significantly. She noted continuous right knee pain which awakened her in bed whenever she rolled over. She was unable to go downstairs into her basement laundry room. Mrs. Eakins could only take sponge baths as she was unable to get out of her tub. She had switched from naproxen to ibuprofen 600 mg with incomplete pain relief and was now using oxycodone with acetaminophen. Minor activity was extremely difficult and Mrs. Eakins admitted that she felt discouraged. She underwent a total right knee replacement without complication and, following post-operative rehabilitation, she was able to rest comfortably and walk 50 yards (meters) without limitation.

The functional impact of arthritis in the elderly

Dr. Bergman muses that the loss of independence that Mrs. Eakins is experiencing with her arthritis may now be contributing to a depression. This loss clearly stems from her overall functional decline due to immobility and chronic pain. Elderly patients measure their overall satisfaction with life in terms of comfort and ability to participate in desired activities. Dr. Bergman feels that conservative measures and medical treatment are no longer sufficient for Mrs. Eakins's goals of joint mobility and independence.

Surgical therapy of arthritis

Age *per se* is not the determining factor in recommending or withholding surgery, as a patient's cardiovascular status and proper pre-operative preparation factor into peri-operative mortality more profoundly than does absolute age. Procedural options for surgical management now include surgical debridement arthroscopy, a low-morbidity procedure which may offer some pain relief, in addition to joint arthroplasty (or rarely, joint fusion). The traditional indications for surgical management of osteoarthritis (osteoarthrosis) are severe pain at rest and marked dysfunction, with the marginal cost–benefit for total joint replacement being maximal in those patients whose quality of life is most severely affected by their osteoarthritis (osteoarthrosis) pre-operatively (Liang *et al.* 1986).

Dr. Bergman felt that Mrs. Eakins' current severe functional limitation and her night-time pain necessitating narcotic use certainly warranted a discussion of surgery. She was additionally concerned that Mrs. Eakins' forced sedentary lifestyle could lead to quadriceps atrophy and resultant inability to participate in post-operative rehabilitation exercises. After receiving information concerning the period of hospitalization and rehabilitation, as well as the social services that have been arranged to assist her during this period, Mrs. Eakins elected to undergo the arthroplasty.

Mrs. Eakins was very pleased with the result and her increased functional capacity. Her left knee osteoarthritis (osteoarthrosis) was more symptomatic now with her increased post-operative activity level, but her replaced right knee took most of the weight whenever she climbed stairs. She was able to reduce her weight by 5 pounds (2 kg) and to participate in her desired activities. Mrs. Eakins's mood improved as she felt that she was more in control of her problem through active participation in rehabilitation.

Questions for further reflection

1. An older man presents with an acutely inflamed and swollen right knee. What is your differential diagnosis and how would you proceed to establish a diagnosis?
2. What sort of recommendations would you make to an older woman with osteoarthritis in both knees regarding behavioral changes, diet, medications, and possible surgery?
3. How would one differentiate between rheumatoid arthritis and osteoarthritis in an older patient?

References

Altman, R., Asch E., Bloch D., Bole G., Borenstein, D., Brandt, K., *et al.* (1986). Development of criteria for the classification and reporting of osteoarthritis. Classification of osteoarthritis of the knee. *Arthritis and Rheumatism*, **29**, 1039–49.

Barrier, C.H. and Hirschowitz, B.I. (1989). Controversies in the detection and management of nonsteroidal anti-inflammatory drug-induced side effects of the upper gastrointestinal tract. *Arthritis and Rheumatism*, **32**, 926–32.

Felson, D.T., Naimark A., Anderson J., Kazis, L., Castelli, W., and Meenan, R.F. (1987). The prevalence of knee osteoarthritis in the elderly. The Framingham osteoarthritis study. *Arthritis and Rheumatism*, **30**, 914–18.

Felson, D.T., Zhang Y., Anthony J.M., Naimark A., and Anderson, J.J. (1992). Weight loss reduces the risk for symptomatic knee osteoarthritis in women. The Framingham study. *Annals of Internal Medicine*, **116**, 535–9.

Griffin, M.R., Ray, W.A., and Schaffner, W. (1988). Nonsteroidal anti-inflammatory drug use and death from peptic ulcer in elderly persons. *Annals of Internal Medicine*, **109**, 359–63.

Guccione, A.A., Meenan, R.F., and Andersen J.J. (1989). Arthritis in nursing home residents. A validation of its prevalence and examination of its impact on institutionalization and functional status. *Arthritis and Rheumatism*, **32**, 1546–53.

Hamerman, D. (1993). Aging and osteoarthritis: Basic mechanisms. *Journal of the American Geriatrics Society*, **41**, 760–70.

Liang, M.H., Cullen, K.E., Larson, M.G., Thompson, M.S., Schwartz, J.A., Fossel, A.H., *et al.* (1986). Cost-effectiveness of total joint arthroplasty in osteoarthritis. *Arthritis and Rheumatism*, **29**, 937–41.

Nichols, J.F., Omizo, D.K., Peterson, K.K., and Nelson, K.P. (1993). Efficacy of heavy-resistance training for active women over sixty: Muscular strength, body composition, and program adherence. *Journal of the American Geriatrics Society*, **41**, 205–10.

Ruffati, A., Rossi, L., Calligaro, A., Del Ross, T., Lagni, M., Marson, P., *et al.* (1990). Autoantibodies of systemic rheumatic diseases in the healthy elderly. *Gerontology*, **36**, 104–11.

Schumacher, H.R., Jr., and Reginato, A.J. (1991). *Atlas of synovial fluid analysis and crystal identification*. Lea & Febiger, Philadelphia, PA.

Sox, C.S. and Liang, M.H. (1986). The erythrocyte sedimentation rate: guidelines for rational use. *Annals of Internal Medicine*, **104**, 515–23.

Taylor, R. and Ford, G. (1984). Arthritis/rheumatism in an elderly population: prevalence and service use. *Health Bulletin*, **42**, 274–81.

Yelin, E. (1992). The cumulative impact of a common chronic condition. *Arthritis and Rheumatism*, **35**, 489–97.

11

Osteoporosis and hip fractures in the elderly

Rivka Dresner Pollak, Myles N. Sheehan, and Tobin N. Gerhart

Like many January days in Boston, the weather was dreadful this morning with an icy wind off the water bringing with it snow and sleet. Ann Jones left her apartment in the Beacon Hill section of the city. She loved living here. The streets were narrow and made of cobblestones, gas lamps at the corners cast a warm glow on the evenings, and the traditional architecture gave one a sense of what Boston had been like in the nineteenth century. For all the charm, however, Ann was cursing the cobblestones on this wintry morning. Previous snowfall and ice had filled in the crevices between the stones, making the surface treacherous. Two years before, when she was 78, on a similarly wintry day, Ann had slipped and broken her wrist, suffering a Colle's fracture that had taken a long time to knit together. Ann was now on her way to Charles Street, at the base of Beacon Hill, to do some early morning shopping. As she rounded the corner from her side street onto Charles Street she was jostled by a passerby. At that moment, she felt her feet slip on a patch of ice. Ann let out a cry as she realized she was falling. With a loud thump, she came down on her right side. She had tried to break her fall but could not react quickly enough. Ann felt pain in her thigh but tried to pull herself up. A sudden jolt of pain came from the region of her right hip, taking her breath away for a moment and making her lie back on the ground. Several people on the street gathered to help her. A young woman stepped forward as Ann made another attempt to get up.

'Ma'am, I am a doctor. I think you should lie still. I am afraid you might have hurt your hip.' The young woman turned to a passerby and asked him to go into the nearby store and call for an ambulance. 'I will stay with you until the ambulance comes. How are you feeling?'

'Thank you for your kindness, but I feel cold, foolish, and irritated, lying here in the snow. I want to get up and go on about my business, but the way my hip felt when I tried to get up convinces me of the wisdom of your plan. Oh, this is bad news. Here I am, eighty years old, with a broken hip.'

Hip fractures are a major cause of morbidity in the elderly. In the United States, 250 000 people fracture their hip each year (Melton *et al.* 1992). Over 90% of these fractures occur in individuals over age 70. By the age of 90, one in three women and one in six men will have sustained a hip fracture: an event associated with a 12–20% increase in mortality, a 25% chance of long-term institutionalization, and a less than 50% chance that the patient will ever walk independently again (Riggs and Melton 1986). The current annual cost of hip fractures to the American health system is over \$5 billion (Pierron *et al.* 1990). Since hip fracture incidence rates increase exponentially with age, continued expansion of the elderly population will likely be associated with a doubling or tripling of the number of hip fractures by the middle of the next century. As measured by fracture frequency, influence on the quality of life and cost of medical care for those who suffer fractures, hip fractures are an enormous and growing public health problem.

After a brief wait on the snowy sidewalk, the ambulance came and took Mrs. Jones to a nearby hospital. In the emergency ward of the hospital, Mrs. Jones was taken quickly to one of the examining rooms. Dr. Jack McLain, a young orthopedic surgeon, began to take her history. He found out how she had been jostled and fell on the ice covering the cobblestoned street.

'Mrs. Jones, I have a good sense of how your fall occurred. I want to ask you a few questions just to be certain that I understand your situation thoroughly. Did you have any pain in your leg or hip before your fall?'

'No, Doctor. I felt fine.'

'Have you fallen before?'

'Yes, two years ago I fell and broke my wrist under very similar circumstances. I slipped on some ice.'

'In the interim, have you had any problems with falls?'

'No, thank God, I have done well.'

'Mrs. Jones, do you have any problem with your heart, any episodes of fainting, difficulties with dizziness, or maintaining your balance?'

'No, I am in good shape. I saw my physician about three months ago and she told me I was doing well: no evidence for heart disease, good blood pressure, everything checked out fine.'

Good, I am glad to hear that. Do you take any medications? Do you have any allergies to medications?

'I do not have any prescriptions. I will sometimes take some acetaminophen for a headache but that is unusual. I have never had to take much medication. As far as I know I have no allergies.'

'Before this fall, Mrs. Jones, how have you been at home? Are you able to walk around without problems, do your errands, that sort of thing?'

'Oh, I am very independent. My husband died a long time ago. I have a daughter who lives in the area and she comes by now and then, but I take care of myself.'

What are the risk factors for hip fractures? The commonly cited factor for hip fracture in the elderly is falling. Over 90% of hip fractures occur following a fall (Grisso *et al.* 1991). A hip fracture that occurs without a fall should alert physicians to the possibility of a pathologic fracture caused by a malignancy involving the bone, such as multiple myeloma or metastatic cancer. Less than 5% of falls, however, result in hip fractures (Tinetti *et al.* 1988). What protects those who suffer the remaining 95% of falls from a broken hip is still unknown. Fracture risk appears to be related to both the characteristics of the fall and the vulnerability of the individual.

There are three characteristics of falls that are important for the risk of fracture: (1) direction; (2) presence of protective responses, and (3) energy of the fall. Falls can occur four ways: (1) forward; (2) backward; (3) to the side; or (4) straight down. The risk of hip fracture has been found to be 21-fold higher if the direction of the fall is to the side (Greenspan *et al.* 1994). Protective responses in a fall include the ability of the individual to break the force of the fall with an outstretched arm. Young individuals who suffer a bad fall usually stretch their arm out and may suffer a fractured wrist. Older individuals may lack the ability to respond quickly enough to protect themselves and land on their hip with subsequent fracture. An older person who suffers a Colle's fracture rather than a hip fracture likely has good protective responses! The energy of a fall is determined by the weight of the person who falls and the distance they fall. Falls can occur, for example, from a horizontal position (such as a bed), from a seated position, from the standing position, or from walking up or down stairs. The greater the height, the more energy a fall will have for the individual. In Mrs. Jones' case, she fell to the side, was unable to get her arm up in time to protect herself, and fell from the standing position.

Dr. McLain's questions for Mrs. Jones are an appropriate part of careful history taking. After establishing the details of the fall, he also investigated any prior problems she might have had with her hip and leg, previous medical history, and history of falls. Although a fall on snow and ice appears a straightforward cause for injury, it is important to ascertain if there is a history of previous orthopedic problems, medical conditions that could interfere with the operation and post-operative recovery, and the person's functional status prior to the fall. Mrs. Jones' independence and ability to ambulate without problems are important favorable prognostic indicators of a good post-operative course and rehabilitation.

Dr. McLain examined Mrs. Jones. She was obviously in pain. Her physical exam revealed a blood pressure of 140/60 and a heart rate of 90. Her lungs were clear, heart tones were good without a murmur or gallop, abdomen was benign, and stool was guaiac negative. A cursory neurologic exam showed Mrs. Jones to be alert, oriented, and appropriate and with no gross motor and sensory deficits.

Exam of her right leg revealed good distal pulses and intact sensation. The leg was slightly externally rotated and shortened. Dr. McLain tried to gently flex the hip and internally rotate the hip but Mrs. Jones grimaced with pain.

Dr. McLain ordered an anteroposterior radiograph of the pelvis. He examined the film and it confirmed his impression: a completely displaced intracapsular femoral neck fracture. He told Mrs. Jones the news.

'Mrs. Jones, I'm afraid you have broken your hip and we need to repair it with an operation.'

'That's not a surprise, Doctor . . . but will I be able to walk again?'

'Yes, I certainly expect that you will, given how healthy you appear to be and how independent you have been. If you were a very sedentary woman with a lot of physical problems I would not be so optimistic. The operation I plan to do, with your permission, is to give you a new prosthetic hip. Because of the type of fracture you have, simply pinning together the broken bones would probably not work. The hip replacement I will give you is called a bipolar prosthesis. This is a metal hip replacement that I put in your thigh bone, remove the head of your hip, and the ball of the prosthesis fits into the socket where your hip would naturally rest. We will have you walking shortly after the surgery, but you will have a period of rehabilitation after the operation. Getting back to where you will be independent will take a lot of work on your part, but I expect you will co-operate with therapy and do well.'

'What are the risks, what are the complications?'

'Well, Mrs. Jones, there are the risks associated with any anesthetic and surgical procedure. These include death, bleeding, and infection. I do not think any of these are likely in your case and we will monitor you carefully. There are some problems with prosthetic replacements. For the first six weeks after the operation, there is a small chance that the prosthesis may dislocate. We will give you instructions and teaching on how to avoid this problem.

After you recover, a more long term problem with hip prostheses is that you may develop hip pain because the prosthetic hip wears against the acetabulum, the natural hip socket. In the unlikely event that the pain is disabling, then a total hip replacement could be performed. That is not today's problem, however.'

'Why not just do the total hip replacement now, Doctor?'

'That is a longer and more difficult procedure than a relatively simple bipolar prosthesis. If you had had a lot of arthritic problems with your hip prior to today's fracture then we might go ahead and do a total hip replacement. But I expect that your natural hip socket, the acetabulum, is in good shape and not all scarred from arthritis. The portion of the hip replacement that goes into the socket should fit smoothly. I think that you will do well with the bipolar prosthesis and probably not need the total hip procedure. If you were much younger then the wear and tear over years could cause problems.'

'Fine, thank you for answering my questions. When do we do the surgery?'

'Well, we can do the surgery this afternoon. I will have you seen by the anesthesiologist and have some lab tests done and then see you later. Is there anything else I can answer or anything you would like me to do?'

'I would like to speak with my daughter, but other than that I would just like some pain medication.'

Of course. I will take care of that immediately.'

Dr. McLain spoke with the nurse, asked her to help Mrs. Jones contact her daughter, and ordered a dose of morphine intramuscularly with a repeat in three or four hours if needed. He also ordered a complete blood count, electrolytes, urinalysis, a chest X-ray, and an electrocardiogram to prepare for the surgery. Dr. McLain called the anesthesiologist and asked her to come and review Mrs. Jones' case. He ordered that Mrs. Jones receive an intravenous cephalosporin when she was called to the operating room and also ordered her to receive a dose of coumadin. He then took another look at her hip film. As he already knew, there was the hip fracture, but there was also evidence of diffuse osteopenia and osteoporosis.

Hip fractures are usually classified as intracapsular fractures that involve the femoral neck or extracapsular fractures that can involve the intertrochanteric or subtrochanteric portions of the femur. Intracapsular fractures may disrupt the blood supply to the femoral head and can lead to osteonecrosis of the femoral head or non-union of the fracture. The more displaced the intracapsular fracture, the greater the risk of disruption of the blood supply. The blood supply to the femoral head comes through the circumflex artery and branches which run through the capsule. There is an artery in the ligamentum teres that runs to the femoral head, but in adults this artery is usually not functioning. Thus, a severely displaced intracapsular fracture, as is the case with Mrs. Jones, leaves only the sclerotic artery within the ligamentum teres. Although undisplaced intracapsular fractures can be surgically repaired with pins, displaced intracapsular fractures require replacement of the femoral head, usually with a bipolar prosthesis. A bipolar prosthesis consists of a femoral component that fits into the femoral shaft and a metallic sphere that fits into the acetabulum. Extracapsular fractures usually do not involve the blood supply to the femoral head and do not require hip replacement. These fractures, however, may be accompanied by greater initial blood loss and fragmentation of the femoral bone. Although most extracapsular fractures do well with open reduction and internal fixation with pins or orthopedic screws, there is often some leg shortening due to fracture collapse (Gerhart 1987).

Careful pre-operative assessment is essential in caring for older patients with hip fracture. Their medical condition must be stabilized as needed, particularly in the case of a patient who may have suffered extensive blood loss from a comminuted extracapsular fracture. Screening laboratory exams and an electrocardiogram need to be reviewed. Pain control, pre-operative antibiotic prophylaxis, and prophylaxis for venous thromboembolism are all essential elements in preparing the patient for surgery. Pulmonary embolus is a great threat after hip fracture. Beginning coumadin prior to surgery and continuing it post-operatively, with careful monitoring of the prothrombin time, helps to decrease the risk of thromboembolism.

The frail older person with hip fracture may present greater challenges than Mrs. Jones to the operative team. First, the timing of surgery should be delayed until any active medical problems are stabilized. Second, medications should be carefully chosen that will not increase the risk of new medical problems or inducing a delirium. Pain control is an important part of managing persons with hip fractures. Narcotic analgesics used in doses that are adequate to control pain may cause confusion. This may be unavoidable. Patients require frequent reassessment as to the minimal dose that controls pain. Third, decisions regarding the type of surgery to be performed require more than consideration of whether the fracture is extracapsular or intracapsular, displaced or non-displaced. Occasionally, a very frail and bedbound older person may not have hip surgery, as it is unlikely they will walk again. A severely demented nursing home patient with limited mobility may have only a limited procedure because of the inability to co-operate with rehabilitation and the limited potential for ambulation. In considering the type of surgery, the person's pre-fracture status and medical condition are important concerns. People who were only marginally functional prior to a fracture are unlikely to do well.

Although Mrs. Jones fell on an ice covered street, not every person with a similar fall would suffer a hip fracture. A number of factors determine the individual's susceptibility to fracture. These include a family history of osteoporosis, gender, body habitus, diet, alcohol consumption, smoking history, and bone mineral density (BMD). While it is known that a family history of osteoporotic fractures increases the individual's risk for fractures, the genes encoding for this susceptibility have not been defined. Recent studies suggest, at least in some populations, that mutations in the vitamin D receptor allele are associated with decreased bone mass (Morrison *et al.* 1994). Whether these alleles are associated with increased fracture risk is still unknown. Fracture risk is higher in women compared with men and increases with age. The measures of body habitus include height, weight, and body mass index (BMI). (The BMI is the weight in kilograms divided by the height in meters squared.) Fracture rates decrease with increasing BMI, meaning that heavier individuals are more protected from hip fracture than thinner persons! Although a heavier person hits the ground with more force than a thin person falling from the same height, natural padding over the hips of the heavier individual reduces the risk of fracturing when a fall occurs. Low bone mass, as measured by bone mineral density (bone mass/area), is associated with an increased risk of hip fracture (Cummings *et al.* 1990). Considerable overlap exists, however, in bone mineral density value between hip fracture patients and age and gender matched controls (Melton *et al.* 1990). A single measurement of a low bone mineral density does not clearly identify those individuals who will fracture their hip.

Bone mass decreases with aging. Why? Bones undergo continuous turnover and remodeling throughout life. The process of remodeling involves the resorption of bone in some areas while new bone is laid down in other areas. When bone resorption and bone formation match each other there is no net change in bone mass. But when there is uncoupling of these two processes, either because bone resorption increases or bone formation decreases, there is loss in bone mass and increase in bone fragility.

Estrogen deficiency in women is a major risk factor for osteoporosis. The mechanism is not completely understood. It is thought that estrogen suppresses the release of cytokines IL-1 and IL-6, potent stimulators of bone resorption. There is an increase in IL-1 secretion induced by menopause that is limited in time. Interleukin levels return to normal some eight to fifteen years after the menopause. Thus, in the elderly, other factors appear to contribute to ongoing slow bone loss (Resnick and Greenspan 1989).

Among the factors that contribute to bone loss with aging are changes in calcium homeostasis and vitamin D metabolism. Dietary calcium deficiency is not uncommon in the elderly. Although the recommended daily allowance for elderly women is 1500 mg of calcium a day, most community-dwelling elderly women consume less than half that amount. Women who are in institutional settings likely receive even less calcium than those who live in the community. Vitamin D metabolism is also altered with aging. As with calcium, vitamin D intake in the elderly is not uncommonly lower than the recommended daily allowance. In addition, the skin production of vitamin D is decreased in many elderly because of increased amounts of time spent indoors compared to younger individuals. Even when exposed to sunlight, skin production of vitamin D is decreased among older persons. There is also decreased renal activation of vitamin D to calcitriol. In order to maintain normal levels of calcium, older persons have increased secretion of parathyroid hormone. This mild hyperparathyroidism is responsible for calcium mobilization from bone and can result in excessive bone resorption, fragile bones, and increased risk of fractures.

Although not relevant in Mrs. Jones' case, some drugs and diseases can contribute to age-related bone loss. Steroids, anticonvulsant therapy, supplemental thyroid hormone, and alcohol use can cause bone loss by different mechanisms. Older persons who require long courses of steroids are at particular risk of bone loss. Many older women are on supplemental thyroid hormone therapy with thyroxine. Over-replacement with thyroxine may result in enhanced bone resorption. Conditions such as hyperparathyroidism, multiple myeloma, and solid tumors are not uncommon in the elderly and can cause bone loss. In caring for the patient with osteoporosis with a fracture, a review should be done to see if there are any medications or underlying illness that could have contributed to the fracture.

There are some laboratory tests that are useful in evaluating individuals with osteoporosis. A complete blood count, serum calcium, thyroid function tests, and serum and urinary immunoelectrophoresis are indicated in ruling out secondary causes of osteoporosis. A parathyroid hormone level should be measured if calcium levels are elevated. Vitamin D levels are helpful to rule out osteomalacia as contributing to osteoporosis. In ostemalacia, a low 25–hydroxy-vitamin D and a normal dihydrocholecalciferol are commonly found.

Markers of bone turnover have recently been identified. Their clinical utility in osteoporosis in different age groups is currently under investigation. Markers of bone formation include bone specific alkaline phosphatase and the vitamin K-dependent bone matrix protein osteocalcin. Markers of bone resorption are collagen breakdown products and consist of collagen cross linking amino acids hydroxylysyl and lysyl pyridinolines. Currently there is no one laboratory test that is diagnostic of osteoporosis and identifies patients at risk for fractures. A recent study has demonstrated that increased levels of noncarboxylated osteocalcin is associated with increased hip fractures in elderly institutionalized women. The level of the gamma carboxylation of osteocalcin appears to reflect the poor nutritional status of elderly patients with hip fractures through unclear mechanisms (Szulc et al. 1993).

Radiographs are useful in the evaluation of persons with osteoporosis. They are, however, an insensitive tool for the detection of osteopenia. Osteopenia appears once 25% of the bone mineral density is lost. While radiographs can detect existing fractures, they are not a useful measure to predict fracture risk. Radiographic changes suggestive of vertebral osteoporosis include anterior wedging involving more than one vertebrae below the sixth thoracic level.

Bone densitometry is a non-invasive test that has utility in identifying persons at risk for osteoporosis. Low bone mass is associated with increased fracture risk even when adjusted for age. Bone mass can be measured by different techniques at the spine, hip, wrist, and calcaneus. The composition of bone tissue at each of these sites is not identical. Bone tissue is of two types: cortical and trabecular. In the wrist, bone is mostly cortical. In the spine, bone is mostly trabecular. There is a mix of cortical and trabecular bone in the hip. Cortical and trabecular bone are constituted of the same cells and the same matrix elements but there are structural and functional differences. Structurally, cortical bone is 80–90% calcified. Trabecular bone is only 20% calcified. Trabecular bone is considered more metabolically active. Osteoporosis mainly affects the trabecular bone of the spine and hip. Bone mineral density at one site does not correlate with bone density or fracture risk at other sites. Therefore, bone density should be measured at the site of clinical interest. In the elderly, this usually is the hip and spine. The technique of dual energy X-ray absorptiometry is commonly used, involves low amounts of radiation, takes only a brief period of time, and is not excessively expensive.

Bone density is not indicated as a screening test for all women. The test is indicated only when the results will affect decisions about therapeutic interventions. In post-menopausal women who consider estrogen therapy, a low bone mass will provide another reason for replacement therapy, in addition to the beneficial effects of estrogen on lipids and prevention of heart disease. A fracture threshold has been defined in younger subjects as a bone mineral density lower than two standard deviations for age-matched controls. With older subjects, however, there is already significant bone loss even in those without fractures. A considerable overlap exists in bone density between those who fracture and those who do not. A low bone density does not differentiate between those who are eventually going to suffer a fracture from those who are unlikely to fracture. With older women especially at risk for fracture, however, a low bone density would provide added information.

Dr. McLain called Mrs. Jones' daughter early in the evening to let her know that surgery had been successful. He told her rehabilitation would begin the next day with the physical therapists helping her mother. Mrs. Jones' daughter asked Dr. McLain why her mother had had the fracture. He explained that it appeared that she had had a bad fall, but that her mother had frail bones as a consequence of osteoporosis. Most likely, Mrs. Jones' suffered from osteoporosis as a consequence of her age, sex, and post-menopausal status.

The next day, Mrs. Jones, with the help of two physical therapists, stood and supported her weight. They checked her blood pressure and pulse for significant orthostatic changes. Mrs. Jones did well. Unlike an extracapsular fracture with pin fixation, where a prolonged period of non-weight bearing or protected weight bearing may be required after the operation, weight bearing on a hip replacement can begin almost immediately. Dr. McLain visited Mrs. Jones and found that she had some pain but was fairly comfortable. He ordered her narcotic pain medication discontinued and wrote an order for acetaminophen 650 mg every four to six hours, around the clock. Supplemental morphine was also ordered for pain that was not relieved by the acetaminophen. Dr. McLain questioned Mrs. Jones about how she was feeling, her recollection of events, and whether she had any questions. His conversation had two purposes: first, genuine concern as to how his patient felt and second, to ascertain whether there was any evidence of delirium post-operatively. Mrs. Jones appeared attentive and alert, if a bit sleepy. Dr. McLain spoke with Mrs. Jones' nurse prior to finishing his visit. He explained that with a bipolar hip replacement acetaminophen given around the clock usually would control the pain and avoid some of the risks like confusion, constipation, and urinary retention associated with prolonged narcotic use in an elderly person. If Mrs. Jones had pain, he wanted to know about it and wanted her to receive the morphine, but he also knew that there should not be excessive amounts of discomfort with this procedure. Patients who have had an open reduction and internal fixation, however, may have more pain and require narcotics for a longer period. Dr. McLain also asked the nurse to let him know if Mrs. Jones had any problems with constipation or difficulties with urination. Constipation and/or urinary tract infections can slow recovery.

The physical therapists carefully instructed Mrs. Jones on techniques to prevent dislocation of her hip prosthesis. They explained that for at least the first six weeks after surgery that she could not flex her hip beyond 90 degrees, nor move her right leg inward beyond the midline, and not turn her leg inward. To prevent excessive hip flexion, a raised toilet seat was placed in her hospital room and no low chairs were allowed [Stein and Felsenthal 1994].

Mrs. Jones continued to make a good recovery. While he was visiting her three days after surgery, Mrs. Jones asked Dr. McLain if there was anything that could have been done to prevent her hip fracture and what she should do to prevent further problems from osteoporosis?

What could have been done to prevent Mrs. Jones' hip fracture? Strategies are aimed at reducing the risk of fall and prevention of bone loss. In Mrs. Jones' case, there is little that could be done to prevent falls on snowy Boston streets except to caution her in the future to ask for assistance! Bone loss, however, is more amenable to intervention. Bone loss is an asymptomatic process and may not be detected until a woman presents with a fracture. Unfortunately, by the time Mrs. Jones presented with her broken hip, she had already lost a large amount of bone. The key in treating women and avoiding fractures from osteoporosis is early prevention. Prevention of bone loss in asymptomatic women involves two complementary approaches: behavior and pharmacologic intervention. Increased physical activity, alterations in nutrition, reduction in alcohol consumption, and elimination of tobacco use are the main lifestyle factors. Physical inactivity is associated with increased bone loss. Patients at complete bed rest may lose 10–15% of their bone mass in as little as three to six months. Elderly control individuals lose about 1% of bone mass per year (Greenspan et al. 1994). Weight-bearing exercises, such as walking, have been shown to increase lumbar bone mineral density in post-menopausal women. Although some elderly individuals may be at a very low level of fitness and require formal cardiovascular evaluation prior to beginning an exercise program, a reasonable recommendation for many older women is to walk for thirty minutes three times a week. Currently studies are under way to investigate whether exercise alone can reduce fracture risk.

Nutritional changes for women hoping to prevent progressive osteoporosis and decrease their risk of fracture includes calcium and vitamin D supplementation. Although adequate calcium intake is best obtained from nutritional sources, in practice it is difficult for many people to have an intake of more than 800 mg of elemental calcium per day. The major source of dietary calcium in the Western world is dairy products. Self-imposed caloric restriction and the avoidance of cholesterol limits the intake of milk and cheese. There are good sources of calcium other than dairy products, including nuts, green vegetables such as broccoli, and certain fish such as canned salmon. The goal for women is to maintain a positive calcium balance. In post-menopausal women who are not on estrogen, 1200–1500 mg/day of elemental calcium is recommended. Women on estrogen require 1000 mg/day. There are many forms of calcium available as dietary supplements. Calcium absorption is better in an acidic environment. Calcium supplements are, therefore, best taken with food. A schedule of calcium supplementation three times a day with meals provides an easy regimen for most patients. It is important to remember that the calcium content per unit tablet weight is only 30% (for calcium citrate) to 40% (for calcium carbonate tablets). Thus, a 600 mg tablet of calcium carbonate contains only 250 mg of elemental calcium. Absorption of calcium as the citrate is slightly more efficient and not dependent on gastric acidity. This difference may not be biologically important for most patients. Calcium citrate is, however, generally more expensive than calcium carbonate. At the recommended dietary intakes, calcium supplementation is generally free of side effects. In some patients, calcium carbonate may lead to constipation. Citrate provides a useful alternative for those persons. Care should be taken, however, in prescribing calcium supplements to patients with a history of renal stones. Vitamin D supplementation is usually recommended with calcium supplements. The recommended dose of vitamin D is 400–800 IU/day. Multivitamin tablets commonly contain 400 mg of vitamin D. Supplementation with calcium and vitamin D reduces fracture rates significantly in elderly individuals (Chapuy et al. 1992).

The most important pharmacologic intervention to prevent bone loss is estrogen replacement therapy. Numerous studies have shown that estrogen administration to post-menopausal women reduces the rate of bone loss and is associated with a 50% reduction in the rate of hip and wrist fractures (Ettinger et al. 1985). Estrogen replacement therapy is indicated for all recently post-menopausal women with a low bone density, (i.e., two standard deviations below age-matched controls).

There are several general guidelines that can be given for the use of estrogen in recently post-menopausal women. Estrogen primarily reduces the rate of bone loss and, thus, the earlier therapy is started the more likely that the bone mass and structure will be preserved. Recent data, however, suggests that estrogen therapy reduces the rate of bone loss in estrogen-deficient women independent of age and is efficacious at least up to the eighth decade (Lindsay and Tohme

1990). The minimum effective dose is 0.625 mg of conjugated equine estrogen or its equivalent. Efficacy has also been demonstrated for several other estrogen preparations, including estradiol and estrone sulfate. Estrogen can be given effectively via the transdermal route. The effects of estrogen in maintaining bone loss continue for as long as treatment is provided. Bone loss ensues when treatment is discontinued. For the prevention of fractures it appears that long term therapy over at least an 8–10 year period is appropriate. Estrogen's beneficial effects for post-menopausal women are not limited to the bone. Estrogen alleviates menopausal symptoms, improves urogenital atrophy, and reduces the risk of ischemic heart disease by as much as 50% (Stampfer et al. 1991).

There remains concern about the risks of estrogen, especially the risk of endometrial and breast cancer. Estrogens given unopposed by progestin increase the risk of endometrial hyperplasia and carcinoma. Long term estrogen therapy, greater than 15 years, increases the risk of breast cancer by 10–30% (Hulka 1987). Estrogens, however, are associated with improved overall mortality, including a reduction in mortality from breast cancer (Bergkvist et al. 1989). This reduction in breast cancer mortality may be a consequence of the close follow-up of patients receiving estrogen therapy, enabling early detection and treatment of those who develop breast cancer. Because of the concern of increased risk of breast cancer with estrogen use, however, any history of estrogen-dependent tumor and any previous breast malignancy is a contraindication to estrogen therapy. A mammogram should be obtained prior to beginning estrogen therapy. A number of different protocols exist for estrogen use. In the older woman who has previously had a hysterectomy, estrogen alone can be prescribed. (Unless, of course, the woman had the hysterectomy for endometrial cancer, in which case estrogen therapy is contraindicated.) For other older women, estrogen and progesterone are prescribed. This may mean a return to monthly bleeding. A combined continuous regimen of estrogen and progesterone is recommended for those women who wish to avoid monthly vaginal bleeding. Side effects of estrogen therapy include, occasionally weight gain and, rarely, increase in blood pressure. Blood pressure should be measured in all patients after three months of estrogen therapy. Annual follow-up includes semi-annual breast and pelvic exams and, at least, annual mammograms.

What about an older woman like Mrs. Jones who had not previously been on estrogen therapy? Is it still worth starting estrogen therapy? Although the data regarding the efficacy of estrogen in the elderly is sparse, it suggests that the rate of bone loss can be reduced in elderly women who are started on estrogen

even fifteen to twenty years after the menopause. There is no data yet on the reduction of fracture risk. It would be reasonable to recommend to Mrs. Jones that she take estrogen and progesterone along with supplemental calcium and vitamin D.

There is a great need for medications other than estrogen in the effort to prevent osteoporosis and reduce the risk of bony fractures. Although estrogen is a potent inhibitor of bone loss and shows promise of reducing fracture risk, only a minority of post-menopausal women are willing to make the long-term commitment to estrogen therapy. Currently, there are two other classes of medication for treatment of established osteoporosis: calcitonin and bisphosphonates. Salmon calcitonin has been shown to preserve bone mass, alleviate bone pain due to osteoporotic fractures, and reduce vertebral fracture rate (Mazzuoli et al. 1986; Lyritis et al. 1991; Rico et al. 1992). Unfortunately, calcitonin must be given by subcutaneous or intramuscular injection and is expensive. An intranasal spray preparation of calcitonin has been shown in a European study to be efficacious in the prevention and treatment of osteoporosis (Meunier et al. 1990). Bisphosphonates are derivatives of pyrophosphate and are potent inhibitors of bone remodeling. Bisphosphonates can be given orally and preserve bone mass in early post-menopausal and elderly women. A few recent studies suggest that these drugs also reduce fracture risk (Harris et al. 1993). Various bisphosphonate agents are under clinical investigation. There are other classes of agents on the horizon for osteoporosis treatment and prevention. These include analogues of estrogen and parathyroid hormone as stimulators of bone formation.

Dr. McLain made arrangements for Mrs. Jones to be discharged to a rehabilitation facility for several weeks of therapy prior to returning home. He discussed with her a regimen for reducing her risk of fracture. First, she would need to be cautious and avoid hazards that might make her likely to fall. He explained about making her apartment a safer place by eliminating throw rugs and other hazards. Second, he encouraged her to continue a program of regular walking. Although the cobblestoned streets of Beacon Hill are hazardous, he suggested that she walk daily through the Public Gardens or the Boston Common. Third, he recommended that she take two 600 mg tablets of calcium carbonate three times a day with meals to provide her with 1500 mg of elemental calcium daily, and that she take one multi-vitamin tablet a day with 400 IU of vitamin D. Finally, he suggested that she see her primary care physician once she was discharged from the rehabilitation center and strongly consider estrogen replacement therapy to lessen the rate of bone loss and, he hoped, the risk of another fracture. The recommendation on estrogen would depend on several factors: the lack of history of any estrogen-dependent tumors, no previous breast malignancy, and a negative mammogram. If all was well, then he would suggest that Mrs. Jones take 0.625 mg daily of conjugated estrogens along with a daily dose of 2.5 mg of medroxyprogesterone acetate.

Questions for further reflection

1. An elderly woman falls. What factors influence whether she will suffer a hip fracture?
2. Why do orthopedic surgeons generally perform place a bipolar prosthesis rather than pin fractures that involve the femoral neck?
3. A 75-year-old woman comes to you concerned about her risk for hip fracture. What would you discuss with her? What further testing do you think would be appropriate? What would you recommend with regard to Vitamin D and calcium supplementation? Estrogen replacement? Before considering hormonal therapy, what further history and/or testing must be obtained?

References

Bergkvist, L., Adami, H.O., Persson, I., Bergstrom, R., and Krusemo, U.B. (1989). Prognosis after breast cancer diagnosis in women exposed to estrogen and estrogen-progestogen replacement therapy. *American Journal of Epidemiology*, 130, 221–8.

Chapuy, M.C., Arlot, M.E., Duboeuf, F., Brun, J. Crouzet, B. Arnaud, et al. (1992). Vitamin D$_3$ and calcium to prevent hip fractures in elderly women. *New England Journal of Medicine*, 327, 1637–42.

Cummings, S.R., Black, D.M., Nevitt, M.C., Browner, W.S., Cauley, J.A., Genant, H.K., et al. (1990). Appendicular bone density and age predict hip fracture in women: The Study of the Osteoporotic Fractures Research Group. *Journal of the American Medical Association*, 263, 665–8.

Ettinger, B., Genant, H.K., and Cann, C.E. (1985). Long-term estrogen replacement therapy prevents bone loss and fractures. *Annals of Internal Medicine*, 102, 319–24.

Gerhart, T.N. (1987). Managing and preventing hip fractures in the elderly. *Journal of Musculoskeletal Medicine*, 4, 60–8.

Greenspan, S.L., Maitland, L.A., Myers, E.R., Krasnow, M.B., and Kido, T.H. (1994a). Femoral bone loss progresses with age: A longitudinal study in women over age 65. *Journal of Bone and Mineral Research*, 9, 1959–65.

Greenspan, S.L., Myers, E.T., Maitland, L., Resnick, N.M., and Hayes, W.C. (1994b). Fall severity and bone mineral density as risk factors for hip fracture in ambulatory elderly. *Journal of the American Medical Association*, 271, 128–33.

Grisso, J.A., Kelsey, J.L., Strom, B.L., Chiu, G.Y., Maislin, G., O'Brien, L.A., et al. (1991). Risk factors for falls as a cause of hip fracture in women. *New England Journal of Medicine*, 324, 1326–31.

Harris, S.T., Watts, N.B., Jackson, R.D., Genant, H.K., Wasnich, R.D., Ross, P., et al. (1993). Four-year study of intermittent cyclic etidronate treatment of postmenopausal osteoporosis: three years of blinded therapy followed by one year of open therapy. *American Journal of Medicine*, 95, 557–67.

Hulka, B.S. (1987). Replacement estrogens and risk of gynecologic cancers and breast cancer. *Cancer*, 60, 1960–4.

Lindsay R. and Tohme, J.F. (1990). Estrogen treatment of patients with established postmenopausal osteoporosis. *Obstetrics and Gynecology*, 76, 290–5.

Lyritis, G.P., Tsakalakos, N., Magiasis, B., Karachalios, T., Yiatzides, A., and Tsekoura, M. (1991). Analgesic effect of salmon calcitonin in osteoporotic vertebral fractures: a double-blind placebo-controlled clinical study. *Calcified Tissue International*, 49, 369–72.

Mazzuoli, G.F., Passeri, M., Gennari, C., Minisola, S., Antonelli, R., Valtorta, C., et al. (1986). Effects of salmon calcitonin in postmenopausal osteoporosis: a controlled double-blind clinical study. *Calcified Tissue International*, 38, 3–8.

Melton, L.J. III, Eddy, D.M., and Johnston, C.C. Jr (1990). Screening for osteoporosis. *Annals of Internal Medicine*, 112, 516–28.

Melton, L.J. III, Chrisschilles, E.A., Cooper, C., Lane, A.W., and Riggs, (1992). How many women have osteoporosis? *Journal of Bone and Mineral Research*, 7, 1005–10.

Meunier, P.J., Delmas, P.D., Chaumet-Riffand, P.D., Gozzo, I., Duboeuf, F., Chapuy, M.L., et al. (1990). Intranasal salmon calcitonin for prevention of postmenopausal bone loss. Placebo controlled study in 109 women. In *Osteoporosis*, (ed. C. Christiansen and K. Overgaard), pp. 1861–67. Handelstagkkeviet Aalberg, Aalberg, Denmark.

Morrison, N.A., Qi, J.C., Tokita, A., Kelly, P.J., Crofts, L., Nguyen, T.V., et al. (1994). Prediction of bone density from vitamin D receptor alleles. *Nature*, 367, 284–7.

Pierron, R.L., Perry, H.M., Grossberg, G., Morley, J.E., Mahon, G., and Stewart, T. (1990). The aging hip. *Journal of the American Geriatrics Society*, 38, 1339–52.

Resnick, N.M. and Greenspan, S.L. (1989). 'Senile' osteoporosis reconsidered. *Journal of the American Medical Association*, 261, 1025–9.

Rico, H., Hernandez, E.R., Revilla, M., and Gomez-Castresana, F. (1992). Salmon calcitonin reduces vertebral fracture rate in postmenopausal crush fracture syndrome. *Bone and Mineral*, 16, 131–8.

Riggs, B.L. and Melton, L.J. III (1986). Involutional osteoporosis. *New England Journal of Medicine*, 314, 1676–86.

Riggs, B.L. and Melton, L.J. III (1992). The prevention and treatment of osteoporosis. *New England Journal of Medicine*, 327, 620–7.

Stampfer, M.J., Colditz, G.A., Willett, W.C., Manson, J.E., Rosner, B., Speizer, F.E., et al. (1991). Postmenopausal estrogen therapy and cardiovascular disease. Ten-year follow-up from the nurses' health study. *New England Journal of Medicine*, 325, 756–62.

Stein, B.D. and Felsenthal, G. (1994). Rehabilitation of fractures in the geriatric population. In *Rehabilitation of the aging and elderly patient*, (ed. G. Felsenthal, S.J. Garrison, and F.U. Steinberg), pp. 123–39.

Szulc, P., Chapuy, M.C., Meunier, P.J., and Delmas, P.D. (1993). Serum undercarboxylated osteocalcin is a marker of the risk of hip fracture in elderly women. *Journal of Clinical Investigation*, **91**, 1769–74.

Tinetti, M.E., Speechey, M., and Ginter, S.F. (1988). Risk fractures for falls among elderly persons living in the community. *New England Journal of Medicine*, **319**, 1701–7.

12

Falls in the elderly

Juergen H. Bludau and Lewis A. Lipsitz

Dr. Marie Alemian sipped a cup of coffee as she reviewed her clinic schedule for the coming day. She had already done the ward round of her hospitalized patients and now, at 9 a.m., she was ready to begin in the Geriatric Assessment Clinic. The first patient, Steven Rosenblatt, was due any minute. His daughter had called last week and requested the first available appointment. Dr. Alemian stepped out into the reception area and saw an elderly man enter the clinic, accompanied by a young woman and young man. The older man held on tightly to the arm of the woman. The younger man carried a cane and handed it to Mr. Rosenblatt as they came to the reception desk. The trio approached the receptionist and the woman announced that Mr. Rosenblatt had come for his appointment. Dr. Alemian noticed a bruise on Mr. Rosenblatt's forehead. Dr. Alemian came out from behind the desk and introduced herself:

'Mr. Rosenblatt, I'm Dr. Alemian, it's nice to meet you.'

'Nice to meet you, Doctor. This is my daughter Barbara and her husband Jeff Steinsaltz.'

'Pleased to meet you. Mr. Rosenblatt, would it be all right if your daughter and son-in-law came with us to the office so we could talk together?'

'Why do you want Barbara and Jeff there?'

'Well, it helps me get more of the story as to what's going on and allows everyone to speak up about their concerns. Is that OK.'

'I guess, Doctor. Sure, why not? Is that OK. with you, Barbara? Jeff?'

'I think that's a good idea, Dad, let's go.' Mr. Rosenblatt's daughter took his arm again and they walked to the office where Dr. Alemian ushered them in. After all were settled in the chairs of the examining room, Dr. Alemian began: 'What can I do for you, Mr. Rosenblatt? How can I be most helpful to you?'

Mr. Rosenblatt hesitated and shrugged his shoulders, 'I'm not quite sure myself. Barbara, you tell the doctor why you and Jeff made me come here.'

'Doctor, my father can no longer live by himself,' Barbara began. 'Jeff and I are both working and we can't be with him all the time.'

Her husband interrupted: 'We thought we could manage but over the past few months things have not

been right. His age seems to have caught up with him, he's 78.'

Dr. Alemian listened for a moment. The interview had begun in a manner that was common in the assessment of older patients. The patient and the family did not have one specific medical problem or illness that concerned them. Instead, the concern was a practical one: a parent is no longer able to live alone. Dr. Alemian thought of how she would unravel the story.

'Mr. Rosenblatt, let me ask Barbara and Jeff a few questions. Can you tell me what you mean when you say your father is not able to live by himself?'

'Well,' Barbara answered, looking at her father, 'Mom died about two and a half years ago. Initially, dad seemed to cope pretty well by himself in the big house. Actually, he had looked after mom during the last few months of her life—he had done the shopping, cooking, and cleaning. But things aren't the same anymore.' She turned to her husband. 'My father-in-law lives only a few houses down the road from us,' Jeff said, 'and after my mother-in-law passed away he would walk over to our house and have a cup of coffee a few times during the week and then, every Friday night, he would come over for dinner. But in the last few months, he has not been able to drop by for coffee and I have to pick him up and drive him from his house to ours and then back again after dinner.'

'Dad stopped driving himself because he said he could not see well after dark. But you know,' Barbara continued, 'it is not just the driving. He doesn't get out and walk anymore. He won't go down the street to the store to buy groceries, he doesn't go out into the garden, Jeff has to mow the lawn now. The neighbors called me to ask if Dad is OK, since they miss seeing him on the porch. This is not my dad. He has always loved gardening. In fact,' Barbara chuckled a bit, 'mom would get angry about all the time that he would spend out in the garden or talking with friends.'

Dr. Alemian turned to Mr. Rosenblatt: 'So, Mr. Rosenblatt, what do you think about what your daughter and son-in-law are telling me?'

Mr. Rosenblatt looked at Barbara and Jeff and gave them a smile. He answered quietly: 'Dr. Alemian, I just do not have the energy to do all the things I used to. I am finding it hard to walk up and down the stairs or get around the house. I don't even want to go into the garden after that fall I had on the backstairs a few weeks ago. You see this?' Mr. Rosenblatt pointed to the bruise on his temple that Dr. Alemian had noted when he came into the office. 'I got that when I went down the stairs head first. Luckily, I didn't break my neck!'

Barbara interrupted her father: 'Oh yes, Doctor, I'm glad Dad mentioned it. We forgot to tell you that he is falling a lot. Jeff found him on the kitchen floor two months ago. We got scared when he didn't answer the phone. He had slipped or something and was lying on the floor for hours. He couldn't get up by himself.'

Dr. Alemian is appropriately alarmed by the history of repeated falls. Falling is a symptom, not a diagnosis. When an older person tells a physician about falls, it is not unlike when a person confides that he or she is having bouts of chest pain: it is an alarming symptom that may have a variety of causes, some of which are life-threatening. Falls are one of the most common geriatric syndromes. Studies of community-dwelling elderly have shown that up to one-third of individuals over the age of 65 fall each year. Approximately one-half of those who have fallen report multiple episodes of falling. Although the likelihood of falling increases with age, both frail and healthy community-dwelling elderly fall, indicating that falling is not simply a marker of frailty and old age. Falls are associated with major morbidity, functional decline, and increased health care expenditures (Tinetti 1994).

About 5% of falls result in fractures, the most serious being a hip fracture, with about 10% per cent suffering soft tissue injuries, the majority of which are minor abrasions or contusions (Tinetti *et al.* 1987). Often worse than the fall is the morbidity that results from an older person's subsequent fear of falling. Many older persons, after a serious fall, will start to restrict their activities, trying to avoid situations in which they may be more likely to fall. The fear and limitation on activity can lead to impaired self-care, depression, and functional decline.

Dr. Alemian, having learned that Mr. Rosenblatt has been falling, has the task of sorting out the relationship between his falls and apparent decline. Has he undergone a functional decline for an as yet undiscovered reason and, as a consequence, begun to fall? Or have the falls, for whatever reason, caused Mr. Rosenblatt to stay in his house and limit his activities? In getting a good history about a fall, information often must be obtained not only from the patient but from family, friends, or caretakers. Falls in the older person may be the atypical presentation of an underlying acute illness that would not usually be associated with falling. Pneumonia, a urinary tract infection, another infectious process, or congestive heart failure may be heralded in an older person by falls. A person who has made a good recovery from a stroke, for example, may suffer repeated falls in the setting of a urinary tract infection.

There are a variety of points that need to be covered. *The exact circumstances of the falls, associated symptoms, and the onset and nature of recovery provide valuable information.* Patients whose falls are associated with fainting may have a cardiovascular cause such as cardiac arrhythmias or heart block. A fall caused by vasovagal syncope, as can occur, for example, after blood is drawn, is often preceded by nausea, vomiting, dizziness, a feeling of fatigue, and sweating. A slow recovery from a fall, marked by lethargy and confusion, may be a symptom of a seizure disorder. *The caregiver needs to think about possible precipitants of the fall.* What was going on around the time of the fall? Had the patient been taking any medications, eating a meal, or straining in the toilet? Other portions of a complete history deserve special attention. *Past historical information about falls and fractures can help to identify how chronic or acute the current problem may be. A thorough review of all medications, those prescribed as well as over-the-counter medicines, may give important clues about why the person is falling.* In older persons, the sedative, hypotensive, neuropsychiatric, and anticholinergic side-effects of medications can cause substantial problems. *Social history should not be neglected.* Alcohol use can impair postural control mechanisms as well as cause sedation or confusion. A single glass of wine or a lifelong habit of an evening cocktail may be the culprit in causing falls because of interactions with medications or because of age-related changes in hepatic metabolism or the central nervous system. *A careful review of systems will help identify other factors that may be contributing to the falls.* Questions should evaluate cardiovascular, musculoskeletal, and neurological symptoms as well as the past history. Inquiry about eyesight and hearing may reveal sensory loss as a possible cause of falling.

In trying to plan a line of questioning with a person who has falls, it is useful to consider falls as either episodic events or chronic locomotor problems (Table 12.1). Episodic falls are the result of syncope, orthostatic hypotension, acute cerebrovascular accidents, or transient ischemic events. Chronic locomotor problems, however, often present as a gait and/or balance disorder, due to, for example, Parkinson's disease, sensory deficits, Alzheimer's disease, other neurodegenerative diseases or musculoskeletal problems, such as myopathies and arthritis. It is important to recognize that acute or episodic falls may evolve over time into chronic locomotor problems as a result of worsening balance and gait. Likewise, a person with Parkinson's disease who has been doing well may

Table 12.1 Key points in history taking for elderly patients with falls

Obtain history from patient and family and caregivers, with patient's permission
Ask about associated activities and symptoms for one hour prior to the fall (e.g., meals, rushing to bathroom, incontinence, changes in position)
Neck turning prior to fall, tight collars (carotid sinus hypersensitivity)
Detailed medication history (prescription and non-prescription)
Alcohol use
Previous falls
Inability to get up after a fall
Effect of falls on the person's life
Presence of environmental hazards
Sensory deficits (vision, hearing)
Symptoms of cardiovascular, neurologic, or musculoskeletal disease
Differentiate acute, episodic falls from chronic locomotor problems

suddenly experience episodic falls, particularly in the presence of another factor such as orthostatic hypotension or a complication from a new medication.

Dr. Alemian has much ground to cover in her history taking with Mr. Rosenblatt and his family.

'Mr. Rosenblatt,' Dr. Alemian asked, 'tell me more about these falls of yours. How often have you been falling?'

'Well, Doctor, I really don't know, I fall a lot!'

Barbara interrupted, 'Dad, tell the doctor about your fall last weekend in the bathroom.'

'Oh, Barbara, that was not a big deal. I was trying to reach for a towel when I was getting out of the shower and I just lost my balance. I hit the back of my head against the wall of the shower. Still hurts here a bit when I touch it. But I just got myself up and felt fine.'

'Mr. Rosenblatt,' Dr. Alemian asked, 'with the fall in the bathroom, or any of your falls, have you ever been knocked out or lost consciousness?'

'Not yet, Doctor! So far I have been lucky: no broken bones and just a bunch of bruises.'

'Can you tell me more about the fall your daughter mentioned, when Jeff found you on the kitchen floor?'

'I can't really remember too much about that right now. I just couldn't get up by myself. Usually, when I fall I can pull myself up by holding on to some furniture. But that time on the kitchen floor I felt lousy and weak. I just stayed on the floor and hoped that somebody would come by. Jeff and Barbara are so good to me I figured that they would worry about me. But that bothers me, too. I don't want them to be worrying. I've become a burden. I was very frightened by that fall.'

Barbara and Jeff shifted a bit in their chairs and looked down at the floor.

Dr. Alemian continued: 'Mr. Rosenblatt, we should talk more about how you and your children are doing with some of the problems you are having. It concerns me that you feel like you might be a burden. For now, though, I want to see if I can figure out what's going on with all these falls. Have you had any other episodes like the time you were on the kitchen floor, when you felt too weak to get up?'

'No, that was the only time.'

'Good. This may be a bit harder to answer, but try and remember the first time that you fell.'

Mr. Rosenblatt concentrated and then shook his head: 'Oh, Doctor, I really can't recall, it was several years ago.'

Barbara added: 'Dad, remember when mom was still living. You had that fall after you came home from the hospital when you had the stroke. That's when they gave you the cane that you never use!'

'Barbara ... Mr. Rosenblatt ... can either of you tell me more about the stroke?'

Mr. Rosenblatt looked at his daughter and shook his head. She began: 'Actually, dad had his stroke about three years ago. He was in the garden and mom found him lying on his side, yelling for help. We got him to the hospital. He had a lot of trouble moving his left side. They did many tests there. One of them was some kind of brain X-ray, a CAT scan I think, and the doctor told us that it showed that he had a stroke on the right side of his brain. The doctor also told us that he had had some other small strokes that showed up on the scan. Dad was weak for a while but has gotten a lot better. When he first came home from the hospital, he took a big fall in the living room when he was walking to the bedroom. I remember Jeff and I went over to help him up since mom had called all upset, but he had gotten up by himself and was OK.'

'Mr. Rosenblatt,' Dr. Alemian asked, 'have you felt at all like the time you had the stroke? Have any of your falls been like the time you were out in the garden?'

'No, Doctor, the stroke was different. I really couldn't move my left arm and my leg was very clumsy. Sometimes I feel weak and when I am tired my left arm is not as good as the right, but it is not the same as when I had the stroke.'

'What other medical problems do you have?'

'Barbara knows this stuff better than I do. Tell the doctor.'

Barbara smiled at Dr. Alemian. 'Dad had a small heart attack a few years ago. He also has some mild diabetes.

His regular doctor said he did not need any medication but we should keep an eye on his weight. And then he has some arthritis, especially in his knees. The doctor called it degenerative joint disease.'

'What medications do you take, Mr. Rosenblatt?'

Barbara reached into her purse and pulled out a brown bag. 'When we made the appointment your secretary said to bring everything that dad takes. It's all in here.'

'That's right, Doctor. I take an aspirin, a vitamin my daughter buys for me, a heart tablet, and a sleeping pill. I don't know their names, though.'

Dr. Alemian reached into the bag and brought out four bottles. She looked over the labels. There was a bottle of low-dose aspirin at 81 mg a pill, a generic multi-vitamin, and another bottle labeled 'isosorbide dinitrate, 40 mg. Take one three times a day.' The fourth bottle contained diazepam, 2 mg.

'Tell me Mr. Rosenblatt, how often do you take the diazepam?'

'Oh, most every night. It helps me to sleep. My doctor gave it to me after my wife died. I just call for a refill each month.'

Dr. Alemian wondered about how the prolonged half-life of diazepam in older people might be affecting Mr. Rosenblatt. 'Do you ever feel sleepy during the day.'

'No, not really.'

'After you take your heart pill, the isosorbide dinitrate, do you ever feel weak or dizzy?'

'No, Doctor Alemian. But now that you mention it, when I first started taking them I would get light-headed when I stood up. That is better now, though.'

'Mr. Rosenblatt, Jeff mentioned earlier that you had stopped driving because your eyesight is bad?'

'Yes, the eye doctor told me that I have cataracts and he is thinking of operating on them.'

'How is your hearing, Mr. Rosenblatt.'

'Oh, it's not too bad.'

Barbara nearly jumped from her seat: 'Not too bad! You leave the TV on so loud no one can talk. And you don't seem to hear half the things I tell you!'

'Well, it does seem, Mr. Rosenblatt, there is some concern about your hearing!' Dr. Alemian continued: 'I would like to pick up on something you mentioned earlier that concerns me. You said you were very scared after your fall in the kitchen, when you couldn't get up by yourself. What are you frightened about? How have you been dealing with all these falls?'

'Oh, Doctor, I'm afraid of many things. I am afraid that I might fall and break something.' Mr. Rosenblatt began to hesitate. Tears came to his eyes and his voice became thick. 'And I'm afraid of falling and being alone on the floor again. Mostly, I'm afraid of what's happening. It's getting hard for me to get around.' Mr. Rosenblatt paused and seemed to regain his composure. 'This is what I do. I am very, very careful when I walk. I hardly ever go upstairs during the day because I felt unsafe on the stairs. I stay inside the house most of the time since I don't like to walk around

outside by myself and there is no one who can walk with me. That's why I don't go into the garden anymore. If I bend down or turn around I tend to lose my balance and fall.'

'So how do you spend the day, Mr. Rosenblatt?'

'Not doing much! I sit in the chair and watch TV. I'm lucky with all the help that my daughter and Jeff give me. She does my shopping and cooking and Jeff helps around the house and checks up on me.'

'Mr. Rosenblatt, let me ask you some questions about personal habits. Do you drink alcohol?

'No, Doctor, very rarely. Occasionally on a holiday I will have a glass of wine with dinner. But that's the exception.'

Dr. Alemian has learned a great deal from her interview with Mr. Rosenblatt and his family. She needs to consider in more detail some of the specific risk factors for falls.

As has been noted, falling is not a diagnosis but a symptom. It often occurs as the end result of an interaction between chronic age-related changes, acute disease processes, and situational or environmental factors. Sensory deficits, such as visual impairment from cataracts and an age-related decline in hearing (presbycusis), both of which are present in Mr. Rosenblatt, can predispose to falls. Likewise, common musculoskeletal problems, such as arthritis, can exacerbate any difficulty with walking and increase the chance of a fall. Abnormalities in gait can have a number of causes, among them diseases affecting the central nervous system. The multiple strokes that Mr. Rosenblatt suffered over the years may well have resulted in the gait impairment that Dr. Alemian noted when he came into the office. As a consequence of his problems walking, Mr. Rosenblatt may have restricted his activities, leading to muscle atrophy and further weakness that exacerbated the initial gait and balance problem. There are a number of other concerns that Dr. Alemian will need to explore. Mr. Rosenblatt's reliance on his daughter and lack of historical information may be a sign of some cognitive impairment. The history of multiple strokes lends added weight to this possibility. In addition, Mr. Rosenblatt's loss of interest in activities, his withdrawal from the outside world, and his increasing need to be cared for by others, could be signs of depression.

There are other risk factors that make older people especially vulnerable to falls. Degeneration of receptors in the large joints (ankles, knees, hips, and cervical spine), which signal joint position to the brain, may result in postural instability and impaired compensation for sudden changes in posture. Similarly, age-related changes in baroreflex control of blood pressure may result in falls by predisposing patients to orthostatic hypotension and syncope. As in the case of Mr. Rosenblatt, medications that lower blood pressure (e.g., isosorbide dinitrate), ischemic heart disease, and autonomic dysfunction related to cerebral vascular disease can result in substantial decreases in blood pressure during postural change, especially from supine to standing, and predispose to fainting. Orthostatic hypotension and syncope are important considerations in evaluating older persons with falls.

One should not overlook the obvious in considering factors that predispose a person to falls. Foot problems may cause a person to walk abnormally. Likewise, poorly fitting shoes may play a role in gait impairment. Household clutter and throw rugs can turn a house into an obstacle course and increase the chance of falling. Rugs are easy to trip over or they may cause a slip on to the floor. Wires from lamps and electrical appliances are particularly hazardous.

Medications, not only those that affect blood pressure, are potential causes or contributors to falls. Dr. Alemian is particularly concerned about Mr. Rosenblatt's regular use of diazepam. Long-acting benzodiazepines have been known to be associated with falls and hip fractures in the elderly (Ray *et al.* 1989).

Dr. Alemian recognizes that she probably will not find a single cause that explains Mr. Rosenblatt's falls. Most likely, there are multiple coexisting conditions that contribute to the falls. Although she is aware of his risk factors for falls, more historical information as well as a careful physical examination are needed in order for Dr. Alemian to integrate the information gathered and construct a plan to decrease or eliminate Mr. Rosenblatt's falls. Dr. Alemian already feels quite certain that Mr. Rosenblatt's fear of falling is contributing to his restricted life style and the increasing difficulties with independent living.

'Mr. Rosenblatt,' Dr. Alemian continues, 'I have been asking many questions this morning, but I have a few more for you. What I am particularly interested in learning from you is what you remember happening at the time of the fall or right afterwards. Now you told me that you have never lost consciousness or passed out, right?'

'That's correct, Doctor. I would remember that.'

'Good. Have you ever lost control of your bowel or bladder when you have fallen?'

'No.'

'How about chest discomfort or pressure? Any palpitations or racing heart beat?

'No, nothing like that.'

'Has your daughter or son-in-law, or anyone else, for that matter, seen you fall?'

'Well,' Jeff interrupted, turning to his father-in-law, 'you almost fell last Friday night when you were getting up from the table after dinner. Remember? You were swaying and staggering? I was afraid that you were going to fall so I helped you sit down in the chair?'

'Jeff, I really don't remember it.' Mr. Rosenblatt turned to Dr. Alemian. 'But what it does help me to recall is that I do get light-headed, especially at night, when I get up to go the bathroom. I have to hold on to the furniture. I haven't fallen yet, so I did not mention it.'

'Mr. Rosenblatt, you mentioned earlier that you were light-headed when you first took the isosorbide dinitrate but that you thought it got better over time. Do you have any other episodes of light-headedness?'

'Yes, I'm sorry Doctor, it seems I am remembering more. As I think about it, I can be dizzy during the day, especially when I get up fast from my chair.'

'What do you do when this happens?'

'Well, I just sit back down slowly and then in a minute or two I try and get up again, but I take it real slow.'

'Mr. Rosenblatt, I know this is a hard question, but when you say "dizzy", what do you mean? Do you mean that you feel light-headed or do you have the sensation that the room is spinning around you.'

'Oh Doctor, I don't know. Dizzy is dizzy. I feel light-headed I guess. The room does not spin around.'

'OK Mr. Rosenblatt. How often do you have the dizzy feeling?'

'I'd say several times a week.'

'Mr. Rosenblatt, I want to ask you a series of questions that are related to some types of falls. First, do you ever feel faint or fall after you pass water or empty the bowels?'

'No, Doctor.'

'Any problems with falling or passing out after a cough?'

'No, Doctor.'

'What about if you turn your neck to one side?'

'No, Doctor, what are you getting at?'

'Well, I did not think that any of these problems were related to your fall but it always pays to ask. Sometimes people will pass out after emptying their bladder or bowel, or coughing. The fancy names are elimination (micturition or defecation), and tussive syncope, respectively. Finally, in some individuals, a sharp turn of the head, especially if the person is wearing a tight collar, can lead to collapse. There is an area in the carotid artery that, with pressure, can slow the heart rate. Again, your history is not so much one of loss of consciousness as it is of falls, but I want to be thorough.'

'Well, Doctor, I guess that's why my daughter brought me here.'

'Just one more question. Can you tell me what activity most likely causes you to fall?'

'I don't know, Doctor, I have never thought of it that way. Barbara, can you help me out?'

'I know that you are most frightened of the stairs. Although you can fall doing a lot of things, when you go up or down the stairs, especially if the lighting is not right, you are prone to a fall.'

'Thank you. What I would like to do now is to examine you, Mr. Rosenblatt. The nurse will come in and help you undress. I will ask your daughter and son-in-law to excuse us and go out into the waiting room for a few minutes. When we're through, then we can all get together and I will tell you what I have found, think about ways to help improve your situation, and you and your family can ask me any questions that you may have.'

Dr. Alemian stepped out of the room and asked her nurse to come in and help Mr. Rosenblatt. She thought a bit about Mr. Rosenblatt while he was changing for the exam. She was especially concerned by the light-headedness or 'dizziness' as well as his gait impairment. Also worrisome was the question of memory impairment. Dr. Alemian made a mental note to check for orthostatic hypotension, do a mental status exam, and perform a careful neurologic examination. She also reminded herself not to forget to look at his feet and see how his shoes fit!

The physical exam of the person with falls must, like the history, be comprehensive (Table 12.2). This is because falls may be the atypical presentation of almost any pathological process. As Dr. Alemian notes, checking for orthostatic hypotension is an important task. Changes in blood pressure and the presence or absence of changes in pulse rate can give clues as to why a person is falling. The absence or blunting of cardioacceleration during postural change could suggest the presence of underlying autonomic insufficiency or diminished baroreflex function. Paying attention to the peripheral pulse and central heart rate for changes in rate and rhythm can help to diagnose brady- or tachyarrhythmias as well as irregularities suggestive of ectopic ventricular beats or atrial fibrillation. The eyes and ears should be inspected, looking for cataracts and cerumen impaction. A quick check of visual acuity may reveal a profound deficit that could be easily correctable with appropriate eyeglasses. Listening to the lungs and heart may reveal evidence of congestive heart failure or murmurs suggestive of aortic stenosis, dynamic outflow obstruction as occurs in hypertrophic cardiomyopathy, or other valvular lesions. Examining the extremities and joints will give information about muscle bulk or wasting as well as

signs of arthritis. A complete neurologic examination, including mental status testing and an evaluation of balance and gait, are crucial.

Dr. Alemian re-entered the examining room and saw Mr. Rosenblatt sitting on the examining table in a gown. She asked Mr. Rosenblatt to lie back on the examining table and explained that she would check his pulse and blood pressure while he was lying, sitting, and standing to see whether there was any significant change.

Dr. Alemian checked Mr. Rosenblatt's blood pressure and pulse rate while he was lying comfortably. She then asked him to sit up with his legs dangling over the edge of the table. She waited for one minute and measured his blood pressure and pulse rate again. Dr. Alemian asked Mr. Rosenblatt to get off the examining table carefully and stand next to it. She and the nurse stood close to Mr. Rosenblatt and were ready to lend a hand should he feel dizzy or show signs of falling. Mr. Rosenblatt began to complain of some dizziness after standing for a few minutes. Dr. Alemian and the nurse helped him back onto the examining table. Dr. Alemian reviewed the blood pressure and pulse rate readings: the blood pressure was 150/80 mmHg with a regular pulse rate of 76/min when lying down, the readings were the same when sitting, but when Mr. Rosenblatt had stood up for three minutes, his pressure had dropped to 110/60 mmHg with a pulse rate of 80/min. The trivial change in pulse rate with the marked drop in blood pressure upon standing gave the doctor important information, suggesting the presence of an impairment in baroreflex response to postural change.

Dr. Alemian continued her exam, looking at the old bruise on Mr. Rosenblatt's forehead and palpating the occipital area where he had fallen several days previous. She was concerned about the possibility of a subdural hematoma given the falls. Dr. Alemian was impressed by the marked temporal wasting. Ophthalmoscopic exam showed, as Mr. Rosenblatt had mentioned, bilateral cataracts. A quick glance in the ears revealed cerumen impactions. Dr. Alemian took the opportunity to clean out Mr. Rosenblatt's ears. That task completed, Dr. Alemian asked Mr. Rosenblatt to open his mouth. His dentures fit poorly. Mr. Rosenblatt's lungs were clear to percussion and auscultation, without a hint of rales that

Table 12.2 Key points in the physical examination of the older person with falls

Check blood pressure and pulse rate in the supine, sitting, and standing positions
Vision and hearing: cataracts, need for glasses that properly correct for visual impairments; hearing deficit, hearing aid
Detailed cardiovascular exam: carotid bruits, murmurs, signs of congestive heart failure
Joint and muscle examination: muscle strength, signs of arthritis, myopathy
Detailed neurological examination (including sensation, balance, and motor function with careful attention to tone, tremor, and focal weakness)
Mental status examination, including memory and visual/spatial orientation
'Get up and go' test

could indicate congestive heart failure. Cardiac exam was remarkable for a 2/6 holosystolic murmur heard from the apex into the left axilla. The abdomen was soft, without masses or organomegaly. Rectal exam revealed a large but non-tender prostate gland without nodules. Stool was brown and tested negative for occult blood. Dr. Alemian was struck by the muscle wasting that was evident in Mr. Rosenblatt's thigh muscles. His knees appeared arthritic, with crepitance when she passively moved the left knee. There was no evidence of an effusion. There was no peripheral edema. Dr. Alemian spent considerable time trying to find distal pulses in both legs. The femoral pulses were strong bilaterally but only weak pulses could be palpated in the popliteal fossae. She could find neither a dorsalis pedis pulse nor a tibialis anterior pulse. Dr. Alemian carefully examined Mr. Rosenblatt's feet. His nails were overgrown and could use a trim but she found no evidence of calluses, corns, bunions, or toe deformities that could increase his risk of a fall. She picked up his shoes and saw no signs of excessive wear or breakdown.

Dr. Alemian continued her examination with a detailed neurologic examination. Cranial nerves showed a slight central VIIth nerve weakness with a minor left facial droop. Motor exam was remarkable for the thigh wasting bilaterally, the absence of a tremor, and some increase in tone on the left side, especially the left arm. Strength was decreased on the left, a bit more in the arm than the leg. Sensory exam revealed bilateral decrease to pin and vibration in the lower extremities, from the feet up to the knees. Cerebellar exam showed Mr. Rosenblatt had some difficulty running his heel down the shin of the opposite leg but it seemed to be due as much to his arthritic knees as anything else. He was able to perform regular alternating movements, touching the tip of his nose with his index finger and then touching Dr. Alemian's finger. Reflexes were increased on the left as compared to the right. When stroking the plantar surfaces of Mr. Rosenblatt's feet, his toe was upgoing on the left and downgoing on the right.

Dr. Alemian asked Mr. Rosenblatt to get off slowly from the examining table and then walk about the room. She noted that he had a slight tendency to circumabduct his left leg, that is, to swing it out a bit when taking a step. He also carried his left arm in a mildly flexed posture. Dr. Alemian asked Mr. Rosenblatt to attempt to walk a straight line, placing one foot in front of the other like a circus performer on the high wire. Mr. Rosenblatt made two tentative steps and then began to lose his balance, causing Dr. Alemian to reach out to steady him. 'I would have been in big trouble if that really was a high wire, Doctor.' Dr. Alemian laughed with Mr. Rosenblatt and then asked him to stand still and shut his eyes. He began to sway and, again, she had to reach out to steady him. She then asked that Mr. Rosenblatt try and stand on the tips of his toes and then the back of his heels. He could do neither.

Dr. Alemian asked Mr. Rosenblatt to sit down in the chair in the examining room. 'Now we're going to try what's called the "Get up and go test." I want you to get up from the chair, if possible without using your arms, and then walk across the room and back' [Mathias et al. 1986]. Mr. Rosenblatt could not get up from the chair without using the arms of the chairs like parallel bars. Once again, Dr. Alemian noted that Mr. Rosenblatt circumabducted his left leg. She also saw that he had a wide based gait. Dr. Alemian, reflecting for a moment, felt comfortable in deciding that it was unlikely that Mr. Rosenblatt had Parkinson's disease. He walked slowly but without the difficulty initiating movement nor the tiny steps of Parkinson's disease. Neither had she detected the characteristic tremor nor any masking of facial expression.

Dr. Alemian asked Mr. Rosenblatt to be seated while she sat down next to him at the desk in the examining room. 'Mr. Rosenblatt, what I would like to do now is to test your memory formally with a series of questions. Some of the questions and tasks may seem a bit silly, but bear with me and do your best. Don't worry if you cannot answer all the questions.' She proceeded with a Folstein mini-mental state exam. Mr. Rosenblatt scored well overall, showing intact orientation, attention, language, and visuospatial abilities. He did, however, have a problem with short-term memory, remembering only one out of three items when asked to recall them after three minutes.

Dr. Alemian also asked Mr. Rosenblatt if there was anything he wanted to bring up without his family in the room. He appreciated the opportunity but said that he had little to add. She asked him some questions about depression and received equivocal responses, not clearly suggesting depression but making her aware that this was a possible diagnosis that would need to be monitored and reconsidered.

Dr. Alemian thanked Mr. Rosenblatt for his co-operation, asked him to get dressed, and thanked the nurse for her help during the exam. She told Mr. Rosenblatt she would be back in a few minutes.

See Table 12.3 for risk factors associated with falls in the elderly.

Pathophysiology of falls

Mr. Rosenblatt, by history and from the physical examination, clearly has difficulties with gait and balance. As Mr. Rosenblatt's case demonstrates, falls can occur as a combination of the effects of chronic conditions along with more acute changes. A major contributing cause of his falls is the previous stroke, in the distribution of the right middle cerebral artery, with its attendant left hemiparesis. At least four other coexisting problems, however, make Mr. Rosenblatt have more trouble with falls than would be the case

Table 12.3 Risk factors for falls in the elderly

Intrinsic factors	
Sensory impairment	Visual deficits
	Proprioceptive deficits
	Vestibular dysfunction
Cardiovascular impairment	Decreased cardiac output
	Arrhythmias
	Hypotensive reflexes (vasovagal, cough or elimination (micturition or defecation) syncope, carotid hypersensitivity)
Musculoskeletal diseases	Arthritis
	Previous fractures
	Deformities
	Myopathies
Neurologic Diseases	Central:
	Degenerative diseases, e.g.,
	Alzheimer's disease
	Parkinson's disease
	Vascular diseases, e.g.,
	Strokes
	Transient ischemic attacks
	Multi-infarct dementia
	Myelopathy (spinal cord)
	Peripheral neuropathies
Extrinsic factors	
Medications	Examples:
	Neuroleptics with extrapyramidal side-effects (Haloperidol)
	Tricyclic antidepressants with anticholinergic side-effects (amitriptyline)
	Benzodiazepines with active metabolites and prolonged half life (flurazepam, diazepam)
	Cardiovascular medications with hypotensive side-effects (diuretics, nitrates, other vasodilators)
Environmental hazards	Examples:
	Poor lighting, uneven floors, throw rugs, lamp cords

if the stroke had been the only difficulty. These four problems are commonly found in older patients with falls: (1) sensory deficits; (2) orthostatic hypotension; (3) musculoskeletal problems; and (4) neurological and cognitive deficits.

First, the presence of sensory deficits increases the chances for falling. Mr. Rosenblatt has significant visual impairment because of bilateral cataracts. He has difficulty seeing where he is going! Many older people also have difficulty with visual acuity as a consequence of macular degeneration, with a resultant loss of central vision. Other common ophthalmologic disorders in the elderly are glaucoma and diabetic retinopathy.

Although probably not as important for his problems with gait and balance as the decreased vision and sensation, Mr. Rosenblatt's diminished hearing can also increase his sense of isolation and limit his ability to get about. Bilateral symmetric hearing loss that occurs as part of normal aging is known as presbycusis. It is a sensorineural loss that predominantly affects the high frequency sounds. People with presbycusis often complain of difficulty understanding conversation in large groups or in areas with significant background noise. Not only does it often cause embarrassment and social isolation, this hearing loss can also result in serious accidents, especially on busy streets when an older person may not hear the approaching traffic. Simple cerumen impaction, as occurred with Mr. Rosenblatt, may also result in significant hearing impairment and compound the hearing loss of presbycusis.

Second, the presence of orthostatic hypotension is an important physical finding (Lipsitz 1989) (Table 12.4). A major function of the cardiovascular system is to ensure perfusion and oxygenation of the organs, especially the brain. Usually, changes in posture should not result in major changes in blood pressure; a postural decrease in blood pressure is normally compensated for by a reflex increase in the heart rate to maintain cardiac output at a nearly constant level. The absence of a rise in the pulse rate in the presence of a drop in blood pressure is a sign of decreased baroreflex sensitivity, a finding that is common with aging. Autonomic dysfunction, cerebrovascular disease, and diabetes may also impair autonomic control of blood

Table 12.4 Orthostatic hypotension

Definition	A drop of 20 mmHg or greater in systolic blood pressure and/or 10 mmHg or more in diastolic blood pressure when changing position from the supine to upright
Evaluation	Measurement of blood pressure and pulse rate while the patient is in the supine position, followed by a repeat measurement of blood pressure and pulse rate at 1 min and 3 min while the patient is seated, then repeated again while standing, at 1 min and 3 min

pressure and produce orthostatic hypotension. Other changes that are common with aging can also contribute to decreased blood pressure. Normal aging is frequently associated with stiffening of the vasculature as well as the heart. There is diminished ventricular relaxation, resulting in decreased filling of the heart in diastole (Wei 1992). A reduction in preload can cause a marked decrease in cardiac output and subsequent hypotension. Mr. Rosenblatt's orthostatic hypotension could be caused or worsened by his heart medication (Jonsson *et al.* 1990). Isosorbide dinitrate, and other nitrate medications like nitroglycerin, are vasodilators. They affect both arteries and veins but especially the latter, resulting in venous pooling and reduction of the blood returning to the heart (cardiac preload). Nitrates are extremely valuable drugs in the management of patients with coronary artery disease and congestive heart failure. They must, however, be used with caution because of the risk of orthostatic hypotension and falls or syncope. Diuretics, although not used by Mr. Rosenblatt, can also cause problems in older persons through an excessive reduction in ventricular filling.

Post-prandial hypotension is experienced by many older people. It may or may not be accompanied by orthostatic hypotension (Lipsitz *et al.* 1983). The mechanism involves decreased ventricular filling caused by pooling of blood in the splanchnic circulation after a meal. The history given by Mr. Rosenlatt's son-in-law, of staggering and weakness upon getting up from the table is suggestive of a post-prandial blood pressure drop exacerbating his tendency to an orthostatic blood pressure decrease. Although Dr. Alemian did not obtain any clear history of a relationship between falls and meal ingestion, the first one or two hours after meals appears to be a period of increased risk for Mr. Rosenblatt.

A third factor that contributes to Mr. Rosenblatt's problems with falls are his musculoskeletal problems. Fear of falling has caused Mr. Rosenblatt to adopt a more sedentary lifestyle, which, in turn, has caused him to become more deconditioned. The wasting of his thigh

muscles indicates the presence of quadricep weakness and atrophy. Weak legs are unable to resist small changes in balance and the little stumbles that are part of everyday life, predisposing the individual for a major fall. Compounding the muscle weakness is the bilateral degenerative joint disease afflicting Mr. Rosenblatt's knees. Pain and stiffness make it difficult for Mr. Rosenblatt to walk, adapt to an uneven surface, or to react promptly to a change in position, such as turning.

A fourth possible contributor to Mr. Rosenblatt's woes is neurologic and cognitive impairment. The mental status exam done by Dr. Alemian was only a screen, but in a patient who is attentive and alert it can give evidence for a significant problem in short term memory. Mr. Rosenblatt's memory loss, functional decline, and seeming failure to thrive are worrisome. It may be that the memory problems are a consequence of multiple small strokes along with the lower right hemispheric stoke. Of particular concern, however, given Mr. Rosenblatt's many falls, is the possibility of a subdural hematoma. It would be reasonable to order a computerized tomographic (CT) scan of the head to exclude the presence of a subdural hematoma. Another factor that may be clouding the picture is Mr. Rosenblatt's use of diazepam. His memory deficits, and his falls, may be the result of a daytime hangover from the drug.

Neurologic impairment from a variety of causes can create gait abnormalities and postural instability. Atrophy of dopaminergic neurons in the basal ganglia produce Parkinsonian features of slow gait and increased motor tone. Increased tone may also be the result of atrophy of Betz cells in the motor cortex which innervate large proximal antigravity muscles of the arm, trunk, back, and lower extremities. Along with increased tone, Betz cell atrophy results in a stiff, shuffling gait (Lipsitz 1992). Dementing illnesses, such as Alzheimer's disease, increase the risk of falling in the elderly by impairments in both motor activity (apraxia) and interpretation of sensory information (agnosia). Deficits in judgement and impulsive activity also make individuals with Alzheimer's disease or other dementing illnesses prone to falls. Postural instability and balance disorders are also common features of Parkinson's disease, stroke, and normal pressure hydrocephalus.

Mr. Rosenblatt also has a peripheral neuropathy with decreased sensation to pin and vibration in both legs below the knees. Mr. Rosenblatt does not receive the precise information about where his feet are at any one time that another person, with intact sensation, would receive. Mr. Rosenblatt began to fall when Dr. Alemian, in the Romberg maneuver, asked him to close his eyes. Much of his balance was dependent on visual input. Deficits in proprioception are common with aging and impair the body's ability to correct its position when spatially perturbed, with resultant unsteadiness while standing or walking. As a consequence many older patients show an increase in body sway and increased likelihood of a fall.

In the case of Mr. Rosenblatt, it is clear that his falls are the result of multiple interacting factors, complicating the evaluation and treatment. Reviewing his history and physical has covered a large amount of the assessment of a person with a history of falls. Dr. Alemian has reached a preliminary impression that Mr. Rosenblatt is predisposed to fall as a consequence of his stroke-related left-sided weakness. She feels that the orthostatic hypotension, his visual impairment, and diazepam use are the most likely major culprits for exacerbating this tendency to falls. As noted, however, Dr. Alemian remains concerned about the possibility of an underlying process that may also be contributing to the falls. In evaluating a person with new onset of falls, the caregiver must have a high degree of suspicion that the falls are a warning of a new medical illness. Although Mr. Rosenblatt has been falling for some time, Dr. Alemian still wants to make sure that some treatable condition, such as a urinary tract infection, is not increasing his likelihood of falls.

Laboratory testing in the evaluation of falls

In ordering laboratory studies in the evaluation of a person with falls, one is often forced to order a number of tests because of the possibility that an occult process is contributing to the falls. In the case of Mr. Rosenblatt, Dr. Alemian has some very specific concerns that require follow-up. She will order a CT scan of the head to exclude the presence of a subdural hematoma. The head CT scan also will provide information about the presence or absence of hydrocephalus. In addition, given Mr. Rosenblatt's cardiac history, an electrocardiogram will be useful to exclude new ischemic injury and the presence of any new conduction defects. In any person with a cardiac history, abnormal electrocardiogram, or symptoms suggestive of a cardiac origin of the fall, it is wise to exclude the possibility of arrhythmias or heart block. With 24-hour ambulatory electrocardiographic monitoring, one can reveal arrhythmias or intermittent heart block. Although not a concern at the present time in Mr. Rosenblatt, a fall associated with syncope, or syncope without a fall in a person with an abnormal electrocardiogram, usually warrants admission to the hospital for monitoring and exclusion of a myocardial infarction.

Dr. Alemian decides to recommend 24-hour outpatient ambulatory monitoring for Mr. Rosenblatt because of the fall in the kitchen two months previously when he could not get up for several hours. The prolonged period of weakness and inability to get up is worrisome for a possible cardiac cause to the fall. Dr. Alemian also wants to get a urinalysis with culture to check for a urinary tract infection. A broad metabolic screen is also reasonable. A complete blood count with hematocrit will determine if there is an anemia or evidence of infection. Checking serum electrolytes and kidney function can exclude concerns about hyponatremia (and associated baroreflex attenuation) or renal failure. Likewise, a serum calcium will provide evidence that hyperparathyroidism or some other process is not causing hypercalcemia and predisposing to confusion and falls. Liver function tests can determine if there is a hepatic process that may cause a mild encephalopathy and thus interfere with balance and gait. Thyroid function tests are especially useful as both hypothyroidism and hyperthyroidism can have a broad range of effects on muscle strength, mood, and overall ability to function.

Dr. Alemian finishes the laboratory requisition slips and the radiology request form to order a CT scan. She considers her plan, assuming that there are no unpleasant surprises in any of the laboratory results, CT scan, or electrocardiogram. Dr. Alemian invites Mr. Rosenblatt's daughter and son-in-law back into the room with Mr. Rosenblatt.

'With your permission, Mr. Rosenblatt, I would like to write to your doctor about some of the findings and my recommendations regarding your falls.'

'That would be fine, Dr. Alemian. I wish we lived closer to you so that you could become my doctor. What do you think is the problem?'

'Well, there are a few things. I think that your stroke from a couple of years back may give you a tendency to fall. But I also think that there are some things that we could do that may be of help. First, I want you to go back to your ophthalmologist and ask him about having your cataracts operated on and about new eyeglasses. If your regular doctor agrees and it is the recommendation of the ophthalmologist, I think it is a good idea for you to get the cataracts taken care of. Second, I do not like the diazepam that you are taking. One cannot abruptly stop medications like diazepam, as there is a risk of withdrawal. I am going to write to your doctor and suggest that it be slowly tapered and then discontinued. You may have some difficulty sleeping, but, frankly, I think that would be better than falling and getting hurt. Third, I will recommend a change in the timing of your isosorbide dinitrate. I also want to ask your doctor if you can be changed to a lower dose. And, in addition, I think it is a good idea for you to stagger the time that you take the isosorbide at least two hours from mealtime. People can drop their blood pressure after a meal and from taking medications such as isosorbide. It is a good medication, but we need to pay attention to the time that you take it so that we can maximize its positive effects. Let's say you have breakfast at 7 a.m., lunch at noon, and supper at 6 p.m. I would suggest you take the isosorbide at 9 a.m., 3 p.m., and then at about 9 or 10 p.m. Do you understand what I am suggesting?'

'Yes, Doctor, I do.'

'Good, Mr. Rosenblatt. To be sure, repeat it to me!'

'Brother, you can be tough! OK, I think I got it. You want me to go to the eye doctor about the cataracts, see my regular doctor about getting off the diazepam and decreasing the dose of the isosorbide. You also want me to take it at times different from meal times. Can you please write all that out? I won't remember it otherwise.'

'Of course. I also want to go over something else with you. One of the problems I found in examining you is that when you get up from the lying or sitting position to standing, your blood pressure drops. I want you to pay careful attention to how you get up and always go very slowly. OK?'

'Yes, Doctor.'

'I want to ask Barbara and Jeff if they can help with something, too. Please take a look around the house and see if there are things that make it easy for your dad to trip or fall. I am talking about excess furniture, slippery floors, throw rugs, poor lighting, and unsafe stairways. It would also help if you could consider getting a raised toilet seat that fits over the toilet so that it will be easier for your father to get up from the toilet.'

'Thanks, Doctor, we'll give it a look.'

'What I would like to do is see Mr. Rosenblatt again in about three months and see if the falling has improved. I also am going to suggest that the Visiting Nurse association arrange for a therapist to visit your dad and do a home visit, looking for safety issues as well as helping him with a cane and giving him some exercises to improve his muscle strength. The therapist may also consider if another kind of cane, or perhaps a walker, would be more useful. I will ask the nurse to visit weekly and check his blood pressures, to see if there is any improvement in the decrease in pressure when he stands. Does anyone have any questions?'

'Do you think I will get any better or should I be heading for a nursing home?'

'I cannot promise you, Mr. Rosenblatt, that all will be well. But I do think that some simple interventions can make a difference. I know it is worth trying. I would like to see you again in three months to see how things are going.'

Therapeutic approach to falls

Dr. Alemian's approach in attempting to alleviate Mr. Rosenblatt's falls is fourfold. First, she urges that easily correctable risk factors be addressed: cataract removal, medication changes, and a check of his home for environmental hazards. Second, Dr. Alemian screens for potential contributors to the falls that may not be detectable on history and physical alone: lab tests, CT scan, and electrocardiogram. Third, she enlists Mr. Rosenblatt as a partner in his own care by suggesting a new medication schedule and giving him careful instructions on changing from a lying or sitting position to standing. Fourth, Dr. Alemian makes plans for continued evaluation by arranging for a home visit from a nurse and a therapist, as well as suggesting a follow-up appointment in three months to see how things have progressed.

It may not be possible in every setting to arrange for a home visit by a district nurse or visiting nurse association. The involvement of other professionals, however, can be of prime importance in monitoring the situation, relieving family anxieties, supplying valuable information to the physician, and detecting hazards or other problems that would otherwise go unnoticed. Ideally, the physician could make a home visit or house call, although this may prove difficult for a variety of reasons. Arranging for assistance from physical and/or occupational therapists in evaluating and caring for individuals with a history of falls not only helps to identify common precipitants of falls, but can be invaluable in supplying the faller with appropriate devices (canes, walkers, etc.) and providing instruction in the appropriate use of such devices.

After three months, Mr. Rosenblatt returns to see Dr. Alemian, accompanied by his daughter and son-in-law. The screening labs drawn at the time of his last visit revealed little new information. The electrocardiogram revealed sinus rhythm with first degree heart block and evidence of an old inferior myocardial infarction. Twenty-four hour ambulatory electrocardiographic monitoring did not give a clear etiology to the falls. There were periods of sinus bradycardia at a rate in the range of 45 beats per minute as well as several short bursts of a supraventricular tachycardia at a rate of 120 beats per minute. Neither the bradycardia nor the tachycardia lasted for more than a minute or two and was unaccompanied by symptoms. In Dr. Alemian's mind, she wondered if Mr. Rosenblatt was showing signs of a sick sinus syndrome. There was little to do at this time, however, based on the findings except to be ready to repeat the study and consider other measures if symptoms occurred. The head CT scan did show a small area of fluid in the left frontal area, suggestive of a very small subdural hematoma. No treatment was indicated other than watching for any signs of neurological deterioration and a repeat scan in about six months.

In the time since the last visit, Mr. Rosenblatt reported sustaining two falls, one while coming down the stairs in the morning while it was still dark and the second while getting out of the bathtub. Neither resulted in injury. Mr. Rosenblatt only occasionally takes a sleeping pill and he has been switched from diazepam to a low dose of oxazepam, 10 mg. The change in the isosorbide dinitrate dose (from 40 mg to 20 mg) and the time it is taken has reduced feelings of dizziness after meals. The nurse has reported that Mr. Rosenblatt continues to have orthostatic blood pressure changes but they are only rarely symptomatic.

Because of the falls since the last visit, Mr. Rosenblatt's son-in-law put in new lighting around the stairs and is beginning work on converting a first-floor room into a bedroom so that Mr. Rosenblatt can live on one floor. At the suggestion of the visiting nurse, a chair seat and railing were put in the tub to make it safer for Mr. Rosenblatt. In addition, Mr. Rosenblatt has begun to wear a signaling device around his neck, so if he falls and cannot get immediate help, he can push a button on the device to summon assistance. He has also begun to use a quad-cane, with a four-prong base to provide greater stability and support. The visiting therapist has been working with Mr. Rosenblatt in exercises to strengthen his quadriceps as well as instructing him in the appropriate use of the cane.

He had one cataract removed (from the left eye) two months after seeing Dr. Alemian the first time. He is scheduled for the second cataract removal in another four months. When asked how he felt he was doing, Mr. Rosenblatt responded: 'Well, Doctor, I am not as good as new. I wish I did not have to use the cane and I dislike having to be careful all the time. But I would rather be careful then fall and break something. I am better than when you last saw me. I'm looking forward to getting my other eye fixed, since I see much better now even with only one cataract removed. I realize that I may still fall, but I feel safer and am grateful for the help that I got. I think things are easier for my family, too.'

Questions for further reflection

1. What is the appropriate way to check for orthostatic changes in pulse and blood pressure?

2. Why might falls be a particular problem for an older person after a meal? Why might they fall some time after taking nitrates for coronary artery disease?

3. What sort of hazards in a person's home can contribute to falls?

References

Jonsson, P.V., Lipsitz, L.A., Kelley, M., and Koestner, J. (1990). Hypotensive responses to common daily activities in institutionalized elderly. A potential risk for falls. *Archives of Internal Medicine*, **150**, 1518–24.

Lipsitz, L.A. (1989). Orthostatic hypotension in the elderly. *New England Journal of Medicine*, **321**, 952–7.

Lipsitz, L.A. (1992). Falls in the elderly. In *Textbook of internal medicine*, (2nd edn), (ed. W.N. Kelley), pp. 2420–3. W.B. Lipincott Co., Philadelphia.

Lipsitz, L.A., Nyquist, R.P.Jr, Wei, J.Y., and Rowe, J.W. (1983). Postprandial reduction in blood pressure in the elderly. *New England Journal of Medicine*, **309**, 81–3.

Mathias, S. Nayak, U.S., and Isaacs, B. (1986). Balance in elderly patients: the "get up and go" test. *Archives of Physical Medicine and Rehabilitation*, **67**, 387–9.

Ray, W.A., Griffin, M.R., and Downey, W. (1989). Benzodiazepines of long and short elimination half-life and the risk of hip fracture. *Journal of the American Medical Association*, **262**, 3303–7.

Tinetti, M.E. (1994). Falls. In *Principles of geriatric medicine and gerontology*, (3rd edn), (ed. W.R. Hazzard, E.L. Bierman, J.P. Blass, W.H. Ettinger, Jr, and J.B. Haller), pp. 1313–20. McGraw Hill, New York.

Tinetti, M.E., Speechley, M., and Ginler, S.F. (1987). Risk factors for falls among elderly persons living in the community. *New England Journal of Medicine*, **319**, 1701.

Wei, J.Y. (1992). Age and the cardiovascular system. *New England Journal of Medicine*, **327**, 1735–9.

13

Breast cancer and the older woman

Lidia Schapira

While showering, Eleanor Picard, a 75 year old retired bookkeeper, felt a pea size mass in her left breast. Her thoughts immediately went to memories of her neighbor and friend, Rose Burton, who had died two years earlier from cancer. As Eleanor brought her fingers back to the small lump in her breast, she sensed this was a tumor and became very frightened. She remembered Rose's slow and painful decline, and wondered if she had begun to head down the same difficult road.

She finished her shower, dried, and dressed. Eleanor sat on her bed and prepared to call her doctor. She felt a bit calmer once she came up with a plan. She would call her doctor, find out what he recommended, and then decide what to do. After all, she reasoned, not all cancer is alike and she wanted the best care possible. Mrs. Picard reached the physician's office, spoke to the secretary, and explained her problem: 'This is Eleanor Picard. I would like to see Dr. Yamashita as soon as possible. I discovered a lump in my breast this morning. I am very upset and I need to see the doctor.' Mrs. Picard received an appointment for the next morning. She reached for the phone again and called her daughter Kathy, asking her to come with her to her appointment as she was frightened and wanted her support.

That night was a restless one for Mrs. Picard. She tossed and turned, finally getting out of bed and dressed by 7 a.m. Her daughter, anticipating her mother's anxiety, came over early. 'Mom, let's get out of here and get a cup of tea instead of sitting around and worrying.' After a drive, a cup of tea, some tears, and a good conversation together, the two women drove to Dr. Eiji Yamashita's office for their 10 a.m. appointment. The nurse took Mrs. Picard into the examining room and told her the doctor would be in shortly, and that she was glad that she did not delay in seeking attention for the breast mass. Mrs. Picard changed into an examining gown. Dr. Yamashita entered the room shortly after. He asked her how she was feeling, and was not surprised at her answer of 'Nervous and scared.' The doctor confirmed the presence of the mass. In his notes he wrote: 'two centimeter firm mass located in upper outer quadrant of left breast. No axillary nodes, no skin changes, no nipple discharge, no other masses felt in either breast.'

Dr. Yamashita reminded Mrs. Picard that she had had a normal mammogram nine months ago, so it was likely that this was a new finding and that it had been discovered early. He explained that she would need a repeat mammogram and that the mass would have to be biopsied. Dr. Yamashita arranged for an appointment for a mammogram that afternoon and called a surgeon, Dr. Mary Guevara, to see Mrs. Picard next week. Mrs. Picard asked Dr. Yamashita if he thought this breast mass was cancer. 'It may well be,' he replied, 'but we need to take one step at a time and, unfortunately, will not know for sure until we have the results of the biopsy. What do you think?' Mrs. Picard was a bit surprised by the question but glad for the chance to share her fears with her doctor. 'I think it's cancer, and I'm afraid I will die, and I'm terrified I will have a long and painful time like my neighbor Rose.' Dr. Yamashita took Mrs. Picard's hand as she began to cry. 'I cannot promise you that all will be well, but I can promise to care for you the very best I know how, to listen to how you are feeling, and to do what I can to get you the help and support you need. I am going to give you a small prescription for a mild anxiety-reducing drug, if you are getting uncomfortable take one and if you think it would help, please give me a call. I won't be able to talk right away, but I will call you back later. Dr. Yamashita gave Mrs. Picard a prescription for oxazepam 15 mg. He shook her hand and whispered: 'Good luck—I will call you when I have the results.'

Breast cancer is probably the most feared cancer in women, both because of its frequency and its psychological impact. In the United States, there has been a slowly progressive increase in the incidence of breast cancer which has paralleled the aging of the general population. Mortality rates for breast cancer, however, have remained relatively stable, suggesting that the increase in incidence most likely reflects the identification of earlier stage cancers with a fairly good prognosis (Scanlon 1991).

Although the cause of breast cancer remains unknown, there are certain risk factors which are associated with an increased likelihood of developing the disease. It must be remembered, however, that about 70%

of patients with the disease have no apparent risk factors (Seidman *et al.* 1982). Despite this, it is widely accepted that the risk of developing breast cancer increases with advancing age. Early menarche and late menopause, as well as nulliparity or a first pregnancy after age 30 are all associated with a greater risk for the development of breast cancer. A prior history of cancer, either in the contralateral breast or in other organs such as colon, ovary, endometrium, or thyroid, also confers an added risk for developing breast cancer. In younger women, additional risk factors include a strong family history and the presence of certain microscopic findings associated with fibrocystic changes (i.e., moderate or florid hyperplasia, atypical hyperplasia, or papilloma with a vascular core).

There are regional differences around the world in the incidence of breast cancer. The highest incidence world-wide is among women living in northern Europe, Canada, and the United States. In the developing nations and Asia, the incidence is much lower. Some studies have attempted to explain these regional differences on the basis of dietary fat and alcohol intake, noting higher fat and alcohol consumption in the regions with higher incidence of breast cancer. No definitive statement, however, can be made about the potential causal relationships between these factors and the development of breast cancer.

Although there is uncertainty about the causes of breast cancer, the initial approach by a primary care physician to a woman with a new breast mass is well established. As the case of Mrs. Picard illustrates, any 'lump' deserves prompt clinical evaluation followed by biopsy. Mrs. Picard's physician, Dr. Yamashita, understood the sense of urgency a woman experiences when she discovers a mass. Although not immediately life-threatening, any delay in the diagnostic work-up of a new breast mass should be avoided because of the impact on the woman's emotional well-being. A careful physical examination is part of the assessment. The size and location of the mass should be noted as well as indications of attachment to the underlying chest wall or changes in the overlying skin or nipple. Checking for the presence or absence of axillary or supraclavicular nodes is an important part of the examination. The clinical assessment is not definitive, however. Approximately one-quarter of women who have palpable nodes on exam turn out to have no cancer identified at the time of pathological analysis.

In an older woman, such as Mrs. Picard, a new breast mass is most likely to be malignant and, as a general rule, any mass in an older woman is presumed to be malignant until the biopsy proves otherwise. Younger women may repeatedly develop cysts or other benign lesions associated with cyclic hormonal changes. A simple cyst in a young woman can be aspirated in a surgeon's office. If the fluid is clear and the cyst disappears after the aspiration, no further diagnostic tests are indicated. Occasionally, cysts will be seen in an older woman on hormonal therapy but a malignant lesion should still be the presumption.

Any palpable mass in a post-menopausal woman needs to be biopsied. There is no place for watchful waiting with a new mass, unless the woman is critically ill from other causes. Prior to the biopsy, mammography assists in the evaluation of the lesions. It is the most reliable and least expensive imaging technique for screening and evaluation (Blustein 1995). Mammography aids in determining the size and contour of the lesion and also helps in the identification of other areas of disease in either breast. Since breast cancer is often multifocal (i.e., a new focus identified in the vicinity of the primary tumor) and, less often, multicentric (i.e., a second primary tumor), both breasts need to be examined prior to surgery.

In considering the diagnostic work-up of a new breast mass, clinicians often wonder if the patient's chronological age should influence the extent of the evaluation. More helpful than simply considering the patient's age, however, are questions about the degree of independence of the patient, the presence and severity of cognitive impairment, the presence and type of any physical disability, as well as the nature and severity of any comorbid illnesses (Satariano and Ragland 1994; Lindsey *et al.* 1994; Wei 1995). Another crucial consideration is the presence of family and other social support networks. Answers to these questions give more guidance to the physician than the patient's age. Mrs. Picard, for example, is an independent, active older woman. A woman of 75 years of age has a life expectancy of about 12–13 years. Given that Mrs. Picard has no concurrent illness, is alert and intellectually intact, and has a supportive network of family and friends, she is among the most 'fit' elderly patients. Mrs. Picard should receive the same care as a younger woman. Although it is commonly held that breast cancer in older women is a biologically more indolent tumor than when it occurs at a younger age, the outcome is likely to be worse both for those women under 35 or over 75 years of age at the time of presentation (Host and Lund 1986).

In Mrs. Picard's case, her care plan did not consider her age as an appropriate factor in providing less aggressive treatment. Unfortunately, physicians' recommendations for therapy are often influenced by age alone. With increasing age, an older woman may have a number of other illnesses, referred to as comorbid conditions, that may affect the presentation of symptoms and delay the diagnosis of breast cancer. In the presence of major medical illnesses other than breast cancer, there may be good reason to alter the diagnostic approach and therapeutic plan. Studies of older women, however, have revealed that older women tend to get less aggressive therapy than their younger counterparts, often because of the assumption of comorbidity and the fear of the treating physician that they will be unable to tolerate the standard therapy (Allen *et al.* 1986; Samet *et al.* 1986; Chu *et al.* 1987; Greenfield *et al.* 1987; Mor *et al.* 1989).

The workup and treatment of breast cancer may be modified depending on the special situation of the woman. In the case of very elderly women, (i.e., those older than 85 years of age), it may not be unreasonable to consider a more conservative evaluation and treatment. For older woman with widespread disease at the time of evaluation, tissue diagnosis should be obtained in the least invasive fashion possible and treatment should be directed at disease stabilization, palliation of symptoms, and improving the woman's functional status.

Mrs. Picard was pleased with the appointment with the surgeon, Dr. Anna Guevara. She was impressed by Dr. Guevara's kindness, her straightforward manner, and her willingness to share information. After greeting Mrs. Picard, Dr. Guevara had reviewed her mammogram with her. The lesion that Mrs. Picard had felt was seen on the films to be a 1.5 cm solid lesion in the upper outer quadrant of the left breast. There were no associated microcalcifications, frequently seen in breast cancer, nor were there any other suspicious areas seen in either breast.

Dr. Guevara took a history from Mrs. Picard that focused on known risk factors. She had had her first period at age 13 and her first pregnancy at age 26. She had had a second pregnancy a few years later. She had nursed both children. Menopause was at age 45 and she had not taken estrogen replacement therapy. There was no family history of breast cancer.

Dr. Guevara examined Mrs. Picard. Her exam agreed with that of Dr. Yamashita. She explained to Mrs. Picard her recommendations. 'You need to have an excisional biopsy. What that means is that, under local anesthesia, I will take out the lump as well as a surrounding rim of normal tissue. Some people call this a lumpectomy. The tissue from your breast biopsy will be given to the pathologist. He or she will examine the tissue, make slides to study under the microscope, and save a small amount of tissue for hormonal studies that will be very helpful in planning further treatment if the biopsy comes back as cancer. Where we go after the lumpectomy depends on the results we get back from the pathologist. If, as I am afraid, it is cancerous, then we may need to plan a second surgical procedure.'

Mrs. Picard was appreciative that Dr. Guevara, after asking her permission, invited her daughter into her office and reviewed the same information and let them both ask questions. She reassured Mrs. Picard's daughter that the surgery could be safely done under local anesthesia and that her mother should be well enough to go home that same day.

That evening Mrs. Picard got a phone call from Dr. Yamashita. Dr. Guevara had given him a call after her appointment and reviewed her plans with Dr. Yamashita. He told Mrs. Picard that he agreed with Dr. Guevara's recommendations. A surgical date was scheduled for early next week.

The definitive diagnosis of cancer is made by histological examination of tissue obtained by biopsy. There are different methods of biopsying breast lesions. Needle aspiration of a palpable lesion can provide enough material for cytological analysis and may help establish a diagnosis of malignancy. This is easily performed by a surgeon in the surgery but it requires skilled cytopathologic analysis of the sample obtained and, unfortunately, has a fairly high false negative rate. A core biopsy involves the use of a larger instrument than a needle aspiration biopsy. It, too, can be done in the surgery and is relatively easy as well as quick to perform. Like needle aspiration, however, it also requires an expert cytology department for examination of the sample. Incisional biopsy is done when a mass is extremely large or inoperable. The standard method of biopsy, however, is what is recommended to Mrs. Picard: an excisional biopsy or lumpectomy. The lesion and surrounding normal tissue is removed. On receiving the tissue, the pathologist will place indelible blank ink on the outside of the specimen. When the microscopic slides are examined, the ink will be seen as a thin black line. The pathologist can then see if the surgical borders, known as the margins, are free of tumor or whether breast cancer cells are seen adjacent to the inked line. If the margins are involved with tumor, a wider excision is usually required. Frequently, however, the excisional biopsy is both diagnostic and therapeutic: a diagnosis of malignancy can be made or rejected and, if the margins are tumor-free, no further local breast surgery may be required. With the increased utilization of screening mammography, areas suspicious for carcinoma, often marked by characteristic specks of microcalcification, can be identified prior to the appearance of a palpable mass. Excisional biopsy can be performed in these cases by the use of needle localization, where the mammographer, in co-operation with the surgeon, places a thin needle in the woman's breast and, by mammogram, localizes the needle in the center of the microcalcifications. The surgeon then can remove tissue surrounding the needle. Frequently, the pathologist will bring the excised tissue to the mammographer. A mammogram done of the biopsy specimen should reveal the suspicious microcalcifications that were present in the original mammogram. If not, further localization and biopsy may be indicated.

Mrs. Picard's biopsy went smoothly and without complications. She spent the night of the surgery at her daughter's home but returned to her own home the next morning. She did have some soreness in the breast but it was relieved by acetaminophen. Two days after the procedure, Dr. Guevara reviewed the slides with the pathologist. Mrs. Picard did have breast cancer, with the biopsy revealing a 1.8 cm infiltrating ductal carcinoma, grade II/III, extending to the inked resection margin. There was no evidence of vascular or lymphatic invasion.

A sample of the tissue had been processed for hormonal receptor analysis and revealed both positive estrogen receptors (40 femtomoles per nanoliter:fm/nl) and progesterone receptors (140 fm/nl).

A week after the procedure, Mrs. Picard returned to Dr. Guevara's office. In the examining room, she removed the sutures from Mrs. Picard's breast and examined the healing wound. Dr. Guevara then asked Mrs. Picard, and her daughter if she wished, to join her in her consultation room.

Dr. Guevara quickly made Mrs. Picard and her daughter comfortable, then sat down herself, pushed away the other papers on her desk, and picked up Mrs. Picard's chart with the pathologist's report. 'It was cancer,' she said, looking directly at Mrs. Picard. Tears welled up in Mrs. Picard's eyes. Dr. Guevara paused, grabbed some tissues from a box on the side of her desk, and handed them to Mrs. Picard. Mrs. Picard's daughter put her arm around her mother as she cried. After a brief silence, her daughter said, 'Mom, let's listen to what the doctor has to say.' Mrs. Picard looked up at Dr. Guevara and asked her to continue.

Dr. Guevara began to explain the options to Mrs. Picard and her daughter. For the treatment of the local disease in the breast there were two surgical approaches which were equivalent in efficacy. A mastectomy could be performed where the entire breast would be removed along with a sampling of the axillary nodes. The other possibility was a wide excision of the original biopsy site, to remove any residual cancer as the resection margins of the biopsy were positive, along with a sampling of the axillary nodes, followed by radiation therapy to the entire breast and possibly to the surrounding nodal regions.

Mrs. Picard continued to cry as Dr. Guevara explained the two possibilities. She felt overwhelmed and asked her daughter and doctor to help her decide. Her daughter asked many questions about safety and side-effects of the treatments, their efficacy, and cosmetic appearance. Finally, after listening to Dr. Guevara's answers, she asked her for her recommendation.

Dr Guevara did not hesitate in recommending the breast conserving approach over the mastectomy. She explained that she was quite convinced that the outcome, both in terms of disease-free survival and overall survival, was identical between the two procedures. Dr. Guevara felt that the lumpectomy was less disfiguring and was the preferred procedure provided that the entire cancer could be removed. Dr. Guevara explained that the margins of Mrs. Picard's biopsy had shown some malignant cells, so she would need 'to clean out the area a bit more.' She emphasized that this could be done using the same incision and would minimize cosmetic damage. Dr. Guevara also explained that if the tissue from the original biopsy site continued to be involved with tumor, then a mastectomy would be necessary.

Mrs. Picard sat quietly for a moment and turned to her daughter: 'I trust Dr. Guevara's recommendation, what do you think?' Her daughter responded that although it was her mother's choice to make, she, too, thought that the doctor's recommendation made good sense even if it was hard to weigh the different possibilities. Dr. Guevara made arrangements for surgery in two weeks.

Dr. Guevara ordered a series of tests in preparation for the second surgery. The chest roentgenogram, bone scan, and blood tests of liver function were all normal and showed no evidence of tumor spread. Mrs. Picard called her daughter with this good news shortly after Dr. Guevara had let her know these results. Although she was happy that there was no sign of tumor spread, Mrs. Picard confided that she was uncertain about everything Dr. Guevara had told her in the office. She was afraid of the radiation treatments and was worried it would take up several months and interfere with her life. She began to cry as she told her daughter she was afraid she would be a burden to her family and wondered if she had made the right decision. Mrs. Picard's daughter listened to her mother, told her of her love for her, her willingness to help her if she needed help, and assured her that although the choices were complicated ones, she felt her mother had made the right one. They both agreed that they trusted Dr. Guevara and felt that she would give Mrs. Picard the best care possible.

As scheduled, Mrs. Picard underwent the operation under general anesthesia. She was discharged to her home the next morning. The report from the pathologist showed no residual tumor in the sample taken from the original biopsy site. One of the twelve axillary lymph nodes in the axillary sampling was positive for tumor.

There are several types of breast cancer. Despite the variety, the vast majority of breast carcinomas are adenocarcinomas, arising in the glandular epithelium that lines the ducts and lobules of the breast. There are also a number of subtypes of breast adenocarcinomas. The most common type of tumor is referred to as infiltrating ductal carcinoma. Important pathological features in examining breast cancer include the size of the lesion, the histological grade of the tumor, as well as the presence of blood vessel or lymphatic invasion. Analysis for the presence or absence of steroid hormone receptors is a crucial part of the proper examination of breast cancer biopsies. In some centers, other types of tissue analysis including DNA analysis and determination of oncogene translocations (e.g., C-myc, C-erbB-2, and HER 2/neu) are used to help identify more aggressive tumors (Bacus et al. 1994; Camplejohn et al. 1995).

With the widespread use of mammography, carcinoma-in-situ is becoming a more commonly diagnosed condition. The term refers to the proliferation of breast cancer cells within the ducts or lobules of the breast but without cells traversing the basement membrane, as would be the case in infiltrating carcinomas. In-situ breast carcinomas may present as microcalcifications on mammograms and be diagnosed by needle localization biopsy. The natural history and prognosis of carcinoma-in-situ, whether ductal or lobular, is more favorable than the infiltrating varieties.

The staging of breast cancer allows for the determination of prognosis and is essential in formulating recommendations for treatment. Despite the different types of breast carcinoma, it appears that the long-term prognosis for all types of breast cancer is the same, depending on the stage of the tumor at the time of diagnosis. Staging refers to the process by which the size of the tumor, spread to lymph nodes, and presence or absence of metastases to other areas in the body is determined. An internationally recognized system, the TNM system, provides the basis for documenting the results of staging in an individual patient by noting the extent of tumor (T), the involvement of lymph nodes (N), and the presence of metastases (M). Tumors can be staged both clinically and pathologically. With clinical staging, the physical exam provides the information to assess the suspected involvement by tumor. Pathologic staging is the preferred method and involves the gross and microscopic analysis of biopsied tissue. In Mrs. Picard's case, with a 1.8 cm tumor, one positive node, and no evidence of metastases, review of staging criteria allows us to classify Mrs. Picard's tumor as a T1N1M0 lesion, or stage II in the TNM system. (A review of the various stages in the TNM system is beyond the scope of this chapter. Further information can be found in De Vita et al. 1993).

The staging evaluation serves as a guide for treatment. As in Mrs. Picard's case, it involves a careful history and physical exam, a chest roentgenogram, liver function tests, and a complete blood count. Bone scans are also commonly ordered, although the likelihood of identifying distant metastases in early stage disease is extremely unlikely. For women with stage I and stage II breast tumors, the incidence of occult bone metastases is well under 5%. In older women, the scans are frequently falsely positive due to the presence of concurrent degenerative arthritis, and scans in this age group need to be interpreted with caution. The use of radionuclide liver scans, computerized tomographic scans, or magnetic resonance imaging scans is reserved for the evaluation of abnormal liver function tests or palpable hepatomegaly. Although serum tumor markers, such as CEA and CA 15-3, are sometimes present in women with breast cancer, these markers are not helpful in the initial staging procedures and may only add confusion to the evaluation.

The most important prognostic determinant is the number of involved lymph nodes and the size of the original tumor. For those women with in-situ cancer, survival approaches 100% at five years after diagnosis. In the case of women with non-palpable cancers found to be less than 1 cm at biopsy and with negative nodes, survival is approximately 95%. For all women with lymph nodes negative for tumor, the overall survival is about 85%–90%. The prognosis becomes progressively worse with more involvement of regional lymph nodes. The lymph nodes that are most frequently involved in carcinoma of the breast are the axillary, mammary, and supraclavicular lymph nodes. The axillary nodes are the most accessible for biopsy. The presence or absence of lymph node involvement as well as the number of nodes involved influences the choice of chemotherapeutic regimen used as adjuvant therapy (Bonadonna et al. 1995). It is also useful in the identification of women thought to be at very high risk of recurrence who may benefit from investigational therapies. Older women (above 75 years) with estrogen receptor positive nodes are usually treated with hormonal therapy alone, regardless of the total number of involved nodes.

Surgeons commonly perform an 'axillary sampling' after the lumpectomy in women whose original biopsy has been proven cancerous. For those women whose surgical margins are found to be positive at the time of the biopsy, this axillary sampling can be done at the same time as a wider excision of breast tissue is performed at the site of the biopsy. In the past, prior to the widespread use of non-surgical treatments for breast cancer, it was thought that extensive removal of the axillary lymph nodes, usually along with removal of the entire breast and the attached pectoralis major and minor muscles, had a therapeutic value. Results, however, were disappointing with high local recurrence rates. The current trend to minimize surgical procedures has been shown to result in survival rates comparable to the more extensive and disfiguring procedures (Jacobsen et al. 1995).

In older patients it may be possible to forego the

axillary sampling if the decision to proceed with adjuvant endocrine therapy can be made based solely on the characteristics of the primary tumor. This remains a controversial area.

Careful planning and thought must go into decisions about the diagnostic and therapeutic options for a woman with breast cancer, and attention paid to the emotional and psychosocial aspects of the diagnosis. In Mrs. Picard's case, both Dr. Yamashita and Dr. Guevara spoke openly and in a forthright manner about her diagnosis. The discussion with Dr. Guevara allowed Mrs. Picard to participate actively in choosing between two equally effective treatments. Dr. Guevara respected Mrs. Picard's independence and judgement. Unfortunately, some physicians fail to communicate directly with the elderly patient, preferring instead to deal with their adult children, as if the patient were somehow not present or unable to participate in decision making. Aside from treating the woman as if she were an infant, it leaves her alone and vulnerable, out of control and dependent on others. Although anyone with cancer requires support from friends and family, that support should not rob a woman of the right to participate in decisions. Some physicians may be more comfortable with a paternalistic style, arguing that they hope to spare their patient the bad news and distress. This may reflect their own fears and opinions regarding the ability of women to take an active role in their care. In caring for elderly women, some may neglect the impact of breast cancer on a woman's self-image (Patterson 1992). Many physicians and surgeons treating older women may forget that older women may be as interested as younger women in breast conservation.

A diagnosis of breast cancer is invariably frightening. The physician needs to take the time to understand the patient's values, strengths, weaknesses, and fears. There is also need to explore for each woman the meaning for her of the diagnosis of breast cancer. In Mrs. Picard's case, she immediately assumed that she would suffer a painful and lingering death like her neighbor. Any diagnosis of cancer usually triggers a number of negative memories and associations. The physician must be sensitive to this and allow the patient to talk about her experience. In showing concern and taking the time to establish a personal relationship, the doctor helps to create a relationship of confidence and trust. This relationship may be a difficult one when cancer treatments are not successful or the diagnostic workup reveals advanced cancer. In such cases, physicians have an obligation to assure a patient that her wishes will be respected, that pain will be aggressively treated, and suffering will be minimized. Even when all the options for the treatment of the tumor have failed, there remains the physician's personal compassion, concern, and assiduous efforts at symptom relief. Such compassionate care can keep a woman from feeling abandoned in her illness.

Mrs. Picard and her daughter made the follow-up visit to Dr. Guevara's office. She removed Mrs. Picard's sutures and explained a few exercises to regain the full range of motion to her left arm after the surgery. Dr. Guevara also explained that although one node was found to be involved with tumor, the general outlook remained extremely favorable. An appointment was made for Mrs. Picard to consult with Dr. Hari Bharat, a specialist in radiation therapy. Dr. Guevara also suggested that Mrs. Picard arrange for an appointment with a medical oncologist to discuss the value of adjuvant hormonal therapy once the primary radiation was completed.

Mrs. Picard, again accompanied by her daughter, went to Dr. Bharat's office to discuss radiation therapy. On the way to his office, Mrs. Picard wryly commented that she was getting a bit tired of meeting doctors. She also thanked her daughter for her help and support over the past few weeks. The two had been able to talk about the options for treatment and, although there had been plenty of tears, Mrs. Picard felt ready to go on with treatments and get going again with her life. Although there was still some local discomfort in the area of the surgery, she felt well and was cautiously optimistic.

Dr. Bharat reviewed the history and physical exam with Mrs. Picard. He questioned her about any other illnesses she might have, with special attention to any potential problems with her heart or lungs, which could interact with her treatment. Mrs. Picard discussed the details of the radiation treatments with Dr. Bharat. She received a treatment schedule and, after discussion with her daughter, made it clear that she would be able to manage these appointments on her own.

Mrs. Picard began her radiation therapy, visiting the radiotherapy suite located in the basement of the hospital where she had had her surgery. Every morning she met the same group of women in the waiting room and felt a certain camaraderie develop among them. Mrs. Picard often spoke with Elizabeth King, a 54-year-old woman who was receiving chemotherapy along with radiation therapy for her stage II breast cancer. Frances Harrison was 70 years of age, and had breast cancer that was metastatic to her lumbar spine. She was receiving radiation to the metastatic tumor as well as taking tamoxifen. Mrs. Picard felt especially drawn to a young woman, only 39, who wore a scarf over her bald head and brought her two young children along with her each morning. Mrs. Picard offered to watch the young woman's children when she was called into the treatment room; she learned that her name was Joan McMahon and that her cancer had been discovered shortly after the birth of her second child.

Joan had already had a mastectomy, had just finished a course of chemotherapy, and was now undergoing radiation to the chest wall. Mrs. Picard found herself talking easily with Joan, as they discussed their hopes and fears.

The goal of definitive local therapy with radiation for breast cancer is to reduce the incidence of local recurrence. The combination of surgery and radiation needs to be individualized for each patient, both to achieve the best cosmetic results and to minimize the risk of relapse. This combined therapy may not be suitable for women with advanced dementia because they are often unable to co-operate with the treatments. Breast-conserving surgery followed by radiation therapy is also not indicated for women with large primary tumors (i.e., greater than 5 cm), women with small breasts and large tumors where a suboptimal cosmetic result would be likely, or women with extensive intraductal carcinoma. Women in this last category have been found to have an unacceptably high risk of relapse.

Post-operative radiotherapy was first used for women with breast cancer as an adjunct to mastectomy. As with Mrs. Picard, however, it is currently used for women who have opted for breast-conserving therapies. Although radiation has been conclusively shown to decrease the rate of relapse, it has not been shown to affect overall survival. Further studies are needed to determine the optimal timing of radiation, chemotherapy, and hormonal therapy after surgical excision.

Standard radiation therapy consists of a dose to the breast of approximately 4500 to 5000 centigray (cGy), given in fractions of 180 to 200 cGy five days a week for five weeks. Areas of controversy include the need to provide additional radiation to the primary tumor bed, called 'boost radiation', and the need to treat regional lymph nodes. Boost radiation raises the total dose to the primary site to 6000 cGy, is delivered via an electron beam or an interstitial implant, is widely used in the United States, and may be particularly helpful in cases of limited surgery or where surgical margins are positive. The decision to include the supraclavicular, axillary, and mammary nodes in the radiation field is not clear cut. It depends on the experience of the treating radiotherapist as well as the extent of the nodal sampling.

In the elderly, carefully determining how a proposed treatment, such as radiation therapy, will affect a woman's life is a fundamental part of the decision making process. Such quality of life decisions need to be made in concert with the woman. Although radiation is generally well tolerated, it is not without side-effects. Local side-effects of radiation therapy include edema, erythema, hyperpigmentation, and skin telangiectasias in the treated field. Lymphedema

of the arm is a rare but troublesome complication, as is radiation pneumonitis. Late complications include cardiac dysfunction with congestive heart failure and secondary tumors in the radiation field. As crucial as the consideration of possible complications is the need to consider the unsettling effect of a series of radiation therapy treatments in the life of an older woman. Daily radiation treatments for five or six weeks may not be feasible in many cases. The need for transportation may become an insurmountable hurdle for many of the frailest older women. The need to remain immobile and co-operate with the technician excludes many demented patients. Because of such difficulties, there has been a recent re-examination of the need for post-operative radiation therapy in women of advanced age. Although there is controversy about the actual rate of local recurrence in older women treated with breast conservation only, it may be appropriate to consider either breast-conserving surgery alone or in combination with adjuvant hormonal therapy (tamoxifen) as reasonable alternatives for older women who are unable to undergo radiation therapy for medical and/or social reasons. Omitting post-operative radiation therapy, however, should be an exception, not a routine choice, and should always based on considerations of the woman's functional abilities.

Many women find support in sharing their experience of illness with others who have undergone similar situations. For Mrs. Picard, the daily visits and constant exposure to other cancer patients provided her with a stabilizing routine and the opportunity to meet other women. She began to experience her breast cancer as part of her life, albeit an unwelcome part, rather than a devastating event that robbed her of life. Her contact with others gave her a perspective on her own situation and taught her that life goes on, even in the midst of problems. Women may find strength in support groups, in a strong family, and/or in their religious tradition. In caring for a patient with breast cancer, the physician should inquire about the patient's experience, ask how she is coping, and provide counseling and referral when a woman needs more help with anxieties or depression.

Mrs. Picard completed the five weeks of therapy with few side-effects. She developed some minor edema and hyperpigmentation in the treated breast. Dr. Bharat suggested a consultation with a medical oncologist.

Once again, Mrs. Picard found herself in the office of another doctor. Dr. Angela Panzino greeted her and, after the by now familiar ritual of questions and examination, discussed the indications for use of tamoxifen and the possible side-effects. Dr. Panzino gently raised the possibility of a recurrence of cancer, despite careful monitoring and appropriate therapy. She made it clear that

Mrs. Picard should contact either her or the primary care physician, Dr. Yamashita, if new symptoms occurred.

Mrs. Picard had difficulty sleeping the night after the visit with Dr. Panzino. Now that the routine of the radiation treatments was over, she felt more anxious. In the weeks and months that followed, Mrs. Picard often felt herself panic over any new ache or pain, wondering if it could be a recurrence of cancer. Mrs. Picard, on several occasions, called Dr. Yamashita and Dr. Panzino and was reassured that her complaint was unlikely to indicate a recurrence.

The role of the medical oncologist in the care of the woman with breast cancer is to determine if the woman would benefit from adjuvant therapy. Adjuvant therapy is defined as treatment given to women who have no demonstrable evidence of residual breast cancer after their initial treatment. The goals of adjuvant therapy are to reduce the risk of relapse, increase the disease-free survival, and improve overall survival. For a woman with a nearly 100% five-year survival rate with local therapy alone, there is little to be gained by any additional therapy. For a woman with a 50% chance of recurrence within five years, however, it makes sense to intervene early if an effective form of therapy is available. The greater the risk of relapse, the more pressing the need for adjuvant therapy.

Adjuvant therapy has been shown to be beneficial in the treatment of women with positive axillary nodes. Ongoing research efforts are studying the impact of adjuvant therapy in women with node negative breast cancer. It appears that many subgroups of women with node-negative breast cancer may also benefit from adjuvant therapy. This topic, however, is controversial and beyond the scope of this chapter.

Those therapies that are used in the adjuvant setting are those that have been shown to be the most effective in the management of metastatic disease. They must also have a reasonable toxicity profile without delayed toxicity. The differing risks for relapse as well as the differing risks of toxicity explain why the women with breast cancer whom Mrs. Picard met in the radiation therapy treatment suite were receiving different therapies. For women over the age of 70, adjuvant therapy for breast cancer is limited to hormonal treatment, generally with the antiestrogen medication tamoxifen. For women in the 60–70 year age group, tamoxifen is also considered standard adjuvant therapy. There is, however, some evidence that other adjuvant chemotherapy may be beneficial for these women and practice recommendations vary considerably.

Tamoxifen is the most widely prescribed form of adjuvant endocrine therapy. It is a synthetic anti-estrogen with weak estrogenic activity. Like post-menopausal estrogen replacement therapy, tamoxifen has a beneficial effect on the lipid profile and confers protection against age-related loss of bone density. Tamoxifen also decreases the incidence of contralateral breast cancer. Side-effects include 'hot flashes' and a slight increase in the risk of developing endometrial cancer. Although many other side-effects have been reported, most occur infrequently and tamoxifen is generally well tolerated.

Adjuvant tamoxifen reduces the risk of relapse of breast cancer by about 30% in post-menopausal women with positive nodes. A recent meta-analysis of the effect of adjuvant therapy on mortality included approximately 30 000 women enrolled in tamoxifen trials. Only 10% of these women, however, were over 70 years of age. The results of the ten-year analysis showed that adjuvant tamoxifen produced a significant reduction, on the order of 20%, in mortality at five and ten years. The benefit was greater for women with strongly positive estrogen receptor tumors (EBCTCG 1992). The optimum duration of adjuvant endocrine therapy has not been established. Most oncologists treat women at a dose of 20 mg daily for five years. Since it appears that more prolonged therapy is beneficial, many physicians are recommending 'indefinite' courses of therapy.

After some time, visits to her doctors, and conversations with her family and friends, Mrs. Picard learned to live with the idea of cancer, realizing that if she sat around and waited for something ominous to happen she would be wasting her life. She began to accept her fears of illness and death but recognized her greater desire to enjoy her life. Her family noted a significant improvement in her mood.

Mrs. Picard kept in touch with her friend Joan from the group of women receiving radiation. Joan had a recurrence of her tumor with a metastasis to her spine two years after her treatments were completed. She received a course of radiation therapy to the spine. Mrs. Picard volunteered to accompany her to all her treatments and helped Joan with her children. Unfortunately, liver metastases developed shortly thereafter. Despite a resumed trial of chemotherapy, Joan died six months later.

Some three years later, Mrs. Picard celebrated her eightieth birthday. Her family arranged a big celebration, rejoicing at her birthday as well as the milestone of five years free of recurrent breast cancer. Her family was proud of her and the courage that she had shown in her battle against breast cancer. Mrs. Picard still felt afraid when the time came for a follow-up exam, but she understood that there was little that she could do to control the outcome and that, at her age, as much as she might hope otherwise, she could not expect everything to always work perfectly.

The follow-up of women with breast cancer consists of periodic visits and examinations. This is routinely done every three months for the first eighteen months, then every four to six months thereafter. There is no consensus on what constitutes a 'standard' follow-up visit or what is the best interval for these visits. There is no reason for ordering multiple scans or roentgenograms in asymptomatic women, as there is no proven advantage in discovering early metastatic disease. At the time of follow-up, a careful interval history, physical exam, complete blood count, and chemistry profile (liver function tests) are necessary. For patients with a mastectomy, a yearly mammogram of the remaining breast is recommended. Women who have had breast-conserving surgery require bilateral mammograms yearly, with some centers performing mammograms of the treated breast every six months for the first few years. Serum tumor markers (e.g., CEA, CA 15–3) are useful only in women with known metastatic disease and have not been approved for other uses.

Primary care doctors and geriatricians who treat older women with breast cancer need to take a careful review of systems and perform a thorough physical exam at the time of follow-up. Many older patients often have a reluctance to mention to their physicians new symptoms or discomforts. For example, older women may mistakenly assume that urinary incontinence is part of normal aging, that back pain is unavoidable, and that a lack of energy is part of life in old age. While incontinence in women may be found more frequently with advanced age, it is not a normal finding and, in a woman with a history of breast cancer, it may be the presenting symptom of a spinal cord compression from tumor that is metastatic to the spine. Likewise, back pain in older persons can have a variety of benign causes but in a woman with a history of breast cancer, new complaints or worsening of back pain requires further evaluation. Similarly, the 'lack of energy' that an older woman experiences may be a symptom of anemia from bone marrow involvement by tumor. The physician has the difficult task of carefully evaluating minor complaints without causing undue alarm. Recurrence may occur many years after the initial diagnosis, so physicians need to be vigilant for symptoms that herald the presence of metastatic disease. Metastases are usually detected by investigating symptoms rather than as surprise findings on test results. Whenever possible, a tissue diagnosis of metastatic disease should be obtained since non-malignant processes can be mistaken for metastases. The decision to obtain histologic proof of tumor depends on the site involved, the ease or difficulty with which a biopsy can be performed, and the underlying condition of the patient.

The treatment of metastases depends in part on the site of involvement. Vertebral or bony metastases are usually treated with radiotherapy directed to the area of tumor spread. Control of disseminated disease is often achieved with endocrine therapy or chemotherapy. Tamoxifen is considered first line hormonal therapy. If the woman was already receiving adjuvant tamoxifen, this is stopped and second line hormonal therapy with megestrol is usually tried, particularly if the tumor was positive for progesterone receptors. Hormonal therapy is usually continued as long as it can be shown to be effective. Chemotherapy is reserved for women in whom hormonal therapy fails, for reduction of large tumor burdens, or when a prompt response is required.

Chemotherapy regimens use single drugs or combinations of drugs known to be effective in the treatment of breast cancer (such as cyclophosphamide, methotrexate, 5–fluorouracil, adriamycin, etc.) and can be administered safely to older women. These same drugs are also used as adjuvant therapy in the premenopausal group. The treatment plans may appear similar in comparing older women with metastatic disease with younger women undergoing adjuvant chemotherapy. The intention of treatment, however, is different, with palliation being the goal in the former group and cure being the hope in the latter.

Women with metastatic disease are typically considered to have an average life expectancy of approximately eighteen months. Because of the biological heterogeneity of breast cancers and the varied responses to hormonal and chemotherapy, this statistic is quite misleading as many older women can live for a decade or more after the diagnosis. Tumor resistance to drugs is probably the most important predictor of a poor outcome.

Over the past decade, there has not been much progress in the treatment of metastatic disease. Although younger women with advanced disease are often treated with very aggressive high-dose chemotherapy regimens as part of investigational trials, the approach to the older woman with metastatic disease remains palliative. In caring for an older woman with metastatic breast cancer, the goals are to reduce symptoms, control the tumor burden, and, most importantly, preserve the woman's functional ability and quality of life. It is irrational and unethical to propose treatments which are more distressing than the disease itself.

In caring for patients with metastatic disease, it is important to document the response to therapy. Responses may be complete or partial, with progression of disease requiring a change in therapy. Each case needs to be individualized and the treatment methods selected based on the woman's underlying medical condition, her functional status, and her personal wishes. As the disease advances, one can expect a progressive loss of function and general decline, often accompanied by pain. As mentioned earlier, the relief of pain and suffering is an integral part of the treatment plan. This may be achieved using local measures (e.g., nerve blocks and radiotherapy) or by using drugs. Medications include antiinflammatory drugs, narcotics, other analgesics, and antidepressants. These are often used in combination until an acceptable degree of pain relief has been achieved and the side-effects of medications are tolerable. In older women, this may be more difficult because of the increase in adverse reactions to medications, the longer half-lives of many drugs, and the presence of comorbid conditions (Ferrell 1991).

Summary

In this chapter, the approach to a palpable breast mass, the staging work-up, and the treatment options for early (stages I and II) breast cancer have been reviewed. The treatment may vary according to regional differences as well as patient and physician attitudes. The treatment of locally advanced and metastatic disease has been considered, with an emphasis on the management of symptoms and the control of pain. The approach outlined is based on the available published literature. In considering the older woman with breast cancer, this raises some problems as older women have been traditionally under-represented in clinical trials, necessitating extrapolation from studies of younger women. The lack of data for older women is particularly distressing as the majority of women with breast cancer are in the older age groups. Even studies of post-menopausal women have frequently excluded women over the age of 70 years. Because of the lack of data and changing approaches in the clinical care of women with breast cancer, there are a few clinical problems for which there are still no acceptable answers. These problems include: (1) the need for axillary sampling for older women who will receive adjuvant therapy; (2) the need for post-operative radiotherapy for the older and frail woman; and (3) the value of adjuvant chemotherapy either alone or in combination with endocrine therapy for women over 70 years of age.

Questions for further reflection

1. An older woman presents with a 1-cm mass in the upper outer quadrant of her right breast. What are the appropriate steps in diagnosis?
2. How would you explain the various options in surgical treatment for the woman in Question (1) should the breast lesion prove to be cancerous?
3. What are the risk factors for development of breast cancer? Prognostic factors if breast cancer is discovered?

References

Allen, C., Cox, E.B., Manton, K.G., and Cohen, H.J. (1986). Breast cancer in the elderly: Current patterns of care. *Journal of the American Geriatrics Society*, 34, 637–42.

Bacus, S.S., Zelnick, C.R., Plowman, G., and Yarden, Y. (1994). Expression of the erbB-2 family of growth factor receptors and their ligands in breast cancers. Implication for tumor biology and clinical behavior. *American Journal of Clinical Pathology*, 102, S13-24.

Blustein, J. (1995). Medicare coverage, supplemental insurance, and the use of mammography by older women. *New England Journal of Medicine*, 332, 1138–43.

Bonadonna, G., Valagussa, P., Moliterni, A., Zambetti, M., and Brambilla, C. (1995). Adjuvant cyclophosphamide methotrexate and fluorouracil in node-positive breast cancer: The results of 20 years of follow-up. *New England Journal of Medicine*, 332, 901–6.

Camplejohn, R.S., Ash, C.M., Gillett, C.E., Raikundalia, B., Barnes, D.M., Gregory, W.M., et al. (1995). The prognostic significance of DNA flow cytometry in breast cancer: Results from 881 patients treated in a single center. *British Journal of Cancer*, 71, 140–5.

Chu, J., Diehr, P., Freigl, P., Glaefke, G., Begg, C., Glicksman, A., et al. (1987). The effect of age on the care of women with breast cancer in community hospitals. *Journal of Gerontology*, 42, 185–90.

De Vita, V., Hellman, S., and Rosenberg, S.A. (1993). *Cancer, principles and practice of oncology*. Lippincott, Philadelphia, PA.

EBCTCG (Early Breast Cancer Trialists' Collaborative Group) (1992). Systemic treatment of early breast cancer by hormonal, cytotoxic, or immune therapy. 133 randomized trials involving 31,000 recurrences and 24,000 deaths among 75,000 women. *Lancet*, 339, 71–85.

Ferrell, B.A. (1991). Pain management in elderly people. *Journal of the American Geriatrics Society*, 39, 64–73.

Greenfield, S., Blanco, D.M., Elashoff, R.M., and Ganz, P.A. (1987). Patterns of care related to age of breast cancer patients. *Journal of the American Medical Association*, 257, 2766–70.

Host, H. and Lund, E. (1986). Age as a prognostic factor in breast cancer. *Cancer*, 57, 2217-21.

Jacobsen, J.A., Danforth, D.N., Cowan, K.H., d'Angelo, T., Steinberg, S.M., Pierce, L., et al. (1995). Ten-year results of a comparison of conservation with mastectomy in the treatment of stage I and II breast cancer. *New England Journal of Medicine*, 332, 907–11.

Lindsey, A.M., Larson, P.J., Dodd, M.J., Brecht, M.L., and Packer, A. (1994). Comorbidity, nutritional intake, social support, weight, and functional status over time in older cancer patients receiving radiotherapy. *Cancer Nursing*, 17, 113–24.

Mor, V., Guadagnoli, E., Silliman, R., Weitberg, A., Glicksman, A., Goldberg, R., et al. (1989). Influence of old age, performance status, medical and psychosocial status on management of cancer patients. In *Cancer in the elderly, approaches to diagnosis and treatment*, (ed. R. Yancik and J.W. Yates), pp. 127–48. Springer, New York.

Patterson, W.B. (1992). Cancer in older people: An overview. In *Oxford textbook of geriatric medicine* (ed. J.G. Evans and T.F. Williams), pp. 287–94. Oxford University Press.

Samet, J., Hunt, W.C., Key, C., Humble, C.G., and Goodwin, J.S. (1986). *Journal of the American Medical Association*, **255**, 3385–90.

Satariano, W.A. and Ragland, D.R. (1994). The effect of comorbidity on 3–year survival of women with primary breast cancer. *Annals of Internal Medicine*, **120**, 104–10.

Scanlon, E.F. (1991). Breast cancer. In *Textbook of clinical oncology*, (ed. A.I. Holleb, D.J. Fink, and G.P. Murphy), pp.177–93. American Cancer Society, Atlanta, GA.

Seidman, H., Stellman, S.D., and Mushinski, M.H. (1982). A different perspective on breast cancer risk factors: Some implications of the non-attributable risk. *CA: A Cancer Journal for Clinicians*, **32**, 301–13.

Wei, J.Y. (1995). Cardiovascular comorbidity in the older cancer patient. *Seminars in Oncology*, **22**, 9–10.

14

Fluid and electrolyte disorders in the elderly

Catherine L. Kelleher

Dr. Nalu Mobutu was impressed by the new medical student taking the fourth-year geriatric medicine elective. Elizabeth Kim seemed enthusiastic and anxious to participate actively in the care of Dr. Mobutu's older patients. Dr. Mobutu gave Ms. Kim a cup of coffee and reviewed with her the schedule for the coming day. Dr. Mobutu told Ms. Kim about her practice in geriatric primary care. She had a close relationship with many of her patients in the surrounding neighborhood. It was, in many ways, a very special place to practice geriatrics. The elderly neighbors had set up a community network system to evaluate each other when they were ill and to let the doctor know when they thought there was a problem. The first patient this morning, Mrs. Agnes Speen, was a 79-year-old woman who had been Dr. Mobutu's patient for several years. She was accompanied by her neighbor and friend, Doris Miller. Dr. Mobutu noticed that Mrs. Miller looked worried.

'Agnes,' Mrs. Miller said, referring to Mrs. Speen, 'is not right, Doctor. I'm worried about her.'

Dr. Mobutu introduced the medical student, Ms. Kim to Mrs. Speen and Mrs. Miller. She explained that with Mrs. Speen's permission, Ms. Kim would also be seeing her today. Mrs. Speen said she had no objections to being seen by a medical student and they proceeded into the examining room.

Dr. Mobutu noticed that Mrs. Speen seemed a bit lethargic and slow compared to her usual alert and talkative style. In the examining room, Dr. Mobutu explained to the medical student that she was concerned about Mrs. Speen and would quickly evaluate her prior to taking the complete history.

Dr. Mobutu checked Mrs. Speen's vital signs and found a blood pressure of 110/80, pulse of 100, respiratory rate of 24 and an oral temperature of 101°F. She listened to Mrs. Speen's heart and lungs and examined her abdomen. Dr. Mobutu invited Ms. Kim to examine Mrs. Speen with her.

While performing the examination, Dr. Mobutu asked Mrs. Speen how she was feeling.

'I don't feel well, Dr. Mobutu. I am not myself. I feel hot and I have a cough. I don't seem to be thinking clearly. I feel foggy. I have felt bad for a couple of days and my breathing is getting difficult.'

'What do you think might be happening?', Dr. Mobutu asked the student.

'Well,' she replied, 'Mrs. Speen does look like she is sick. She has told us she is not herself. She does seem slow and different from what you described as her usual self. Her respiratory rate is elevated, she has a fever, and her eyes are glazed. Both Mrs. Speen and her friend Mrs. Miller feel she is not herself. I think that is important, as I have read that older persons may have a change in their cognitive ability when they are acutely ill. Also, she mentioned her cough and difficulty breathing. When I examined her I heard crackles in her right lung base. She is tachycardic but her cardiac exam was otherwise unremarkable. We need to review her cardiac status but I am concerned she might have a pneumonia.'

'Excellent. Mrs. Speen, I think you and I are fortunate to have Ms. Kim with us today. Tell me what you think we need to do next and any other concerns.'

Ms. Kim felt more confident. 'We need to make sure she is stable so that we can decide how rapidly to move with diagnostic and treatment efforts. I want to review her record quickly to compare her current blood pressure to previous readings. Let me see,' she said, thumbing through the chart, 'Mrs. Speen's blood pressure is normally about 130/80 mmHg, so today's reading is a bit low for her. We should check orthostatic vital signs on her. Also, given the increased respiratory rate and concern about pneumonia, it is important to check Mrs. Speen's oxygenation.'

Dr. Mobutu suggested that Ms. Kim do the orthostatic vital signs while she went out of the office to get the pulse oximeter. On return, Ms. Kim reported that there were no significant postural changes so she doubted Ms. Speen was significantly volume depleted. The pulse oximeter revealed an oxygen saturation of 88%. As this was less than 90%, Ms. Speen was given supplemental oxygen at a flow rate of 2 liters a minute. Her tachycardia was probably due to her fever and underlying illness.

'Good,' said Dr. Mobutu, 'what are you going to do about her oxygenation? Do you have any concerns about simply giving her some supplemental oxygen?'

'Well, it would be nice to check an arterial blood gas because we do not know Mrs. Speen's arterial carbon dioxide. A potential complication from supplemental oxygen is retention of carbon dioxide. Carbon dioxide retention causes patients to become lethargic and has the potential to cause respiratory arrest. However, given that this is an office and not a hospital, I think for now it would be best to begin with nasal oxygen at a low flow rate and watch Mrs. Speen carefully. We can recheck her pulse oximetry to see if it improves her oxygenation.'

Over the next ten minutes, Mrs. Speen's color improved and she said she felt a bit better. There was no sign of increased lethargy or confusion. As Ms. Speen was now stabilized, Dr. Mobutu asked Ms. Kim to continue with the history and physical examination. She said she would be in the examining room across the hall if the student needed any assistance.

Ms. Kim learned that Mrs. Speen had been sick for the last two days. She had not visited any of her neighbors because she felt so poorly and had no appetite. Her neighbor, Mrs. Miller, had been over to her apartment frequently, urging her to drink fluids. Yesterday, Mrs. Speen began to feel short of breath. It was worse today. She had also begun to cough up yellow sputum. She denied any pleuritic chest pain, hemoptysis, or wheezing. Although she had felt warm, she denied any rigors. Mrs. Speen told Ms. Kim that she had no previous history of lung disease, emphysema, pulmonary emboli, or tuberculosis. She had never had any heart trouble. Ms. Kim did discover that Mrs. Speen had lost ten pounds over the last few months without a clear reason. She had also had a right lower lobe pneumonia six months ago that had been treated in the hospital for two weeks. Otherwise, she had been in good health. Mrs. Speen had lived alone in the community since the death of her husband thirty years ago. Her regular medications included one aspirin a day and a multi-vitamin. She had been a heavy smoker. Beginning at age 25, she smoked one to two packs of cigarettes a day until her episode of pneumonia 6 months ago, when she quit smoking completely.

On physical exam, vital signs were essentially unchanged. The remainder of the exam revealed no abnormal physical findings other than the tachycardia and crackles at the right lung base. Neurologic examination was significant for some difficulty in short term memory.

Ms. Kim asked Dr. Mobutu to return to the examining room so she could summarize her findings. Ms. Kim felt that pneumonia was still the most likely diagnosis. She felt the patient should be admitted to the hospital because she needed intravenous antibiotics, supplemental oxygen, and had difficulty caring for herself. Ms. Kim also remarked that it would be important to get an electrocardiogram (ECG). Based on her history and physical exam, Ms. Kim did not feel Mrs. Speen's symptoms were secondary to cardiac ischemia, but she wanted an ECG to be sure that it had not changed from those done previously. In addition, Ms. Kim ordered a chest X-ray to confirm her diagnosis of pneumonia. She asked that the sputum be sent for gram stain and culture. Ms. Kim also ordered a complete blood count and serum electrolytes, creatinine, blood urea nitrogen, and glucose.

Dr. Mobutu completely agreed with Ms. Kim's assessment and plan. She told Mrs. Speen about the need

for admission and arranged for her transfer to the hospital. She went into the waiting room, told Mrs. Miller about the plans to admit Mrs. Speen to the hospital, and thanked her for accompanying Mrs. Speen to the office.

Later that afternoon, Dr. Mobutu and Ms. Kim made rounds at the hospital. Mrs. Speen looked fairly comfortable. Antibiotics had been started for a community-acquired pneumonia. Her laboratory results revealed an elevated white blood count. Also noted was a serum sodium of 125 mEq/l and a decreased serum chloride.

Ms. Kim wondered why Mrs. Speen's sodium was low.

In considering abnormalities of serum sodium, it is important to review the functional anatomy of body fluids. Approximately 60% of total body weight is water. Total body water (TBW) is divided into two compartments: the intracellular space and the extracellular space. Most of the body's water is in the intracellular space, which makes up 60% of the total body water. The extracellular space is normally 40% of total body water and is divided into three smaller compartments: the interstitial space, plasma (about 5% of body weight), and the transcellular space (space separated by a layer of epithelium, such as the intraocular or the pleural space).

Clinically, problems that develop with alterations in body fluids are often referred to as 'dehydration' or 'overhydration'. However, these terms may be misleading and should be interpreted with caution. They do not accurately distinguish between problems primarily involving the extracellular fluid versus problems involving the TBW, which includes both the intracellular and extracellular fluid. Alterations in the intracellular fluid volume result primarily from changes in plasma osmolality or the serum sodium concentration. They reflect the hydration status of individual cells. For the purposes of the present discussion, these changes will be referred to as alterations in water balance. This will be discussed in detail later. As discussed above, the extracellular fluid (ECF) is divided into three compartments. The portion (5%) in the intravascular space (plasma) determines the effective circulating volume. In general, the intravascular or plasma volume reflects the ECF volume (though this may not be the case in certain patients with edema). Alterations in the ECF volume are also sometimes referred to as changes in the absolute amount of sodium or salt present. The etiology, clinical presentation, and treatments of predominantly intracellular versus extracellular fluid problems are different. In practice, problems with sodium content and water balance often exist together. When they do exist together they should be evaluated separately because they represent different physiologic processes.

Disproportionate retention or loss of water is reflected in the serum sodium concentration. The serum sodium concentration does not give any information on the total amount of sodium or water in the body or indicate whether the patient is hypovolemic, euvolemic, or hypervolemic. It tells us about the concentration of sodium relative to the amount of water present. This determines the hydration status of individual cells. It is important that this concentration remain within defined limits (generally 137–143 mEq/l) because sodium is the primary cation in the extracellular fluid and the major determinant of plasma osmolality. An elevated serum sodium or elevated plasma osmolality means that the patient has a free water deficit relative to the amount of sodium present, and the individual cells are dehydrated. A low serum sodium or low plasma osmolality means that free water is present in excess of sodium and individual cells are overhydrated.

The concentration of serum sodium (not the total body sodium) is primarily controlled by excretion or absorption of water by the kidney and by the sensation of thirst. When the serum sodium concentration rises, either secondary to an increase in total body sodium or due to loss of free water, the increase in plasma osmolality is usually sensed by hypothalamic osmoreceptors. In response, antidiuretic hormone (ADH) is secreted from the posterior portion of the pituitary gland and causes increased absorption of water at the renal tubules. Consequently, water excretion is decreased and plasma osmolality returns to normal. In addition, an elevated serum sodium concentration would cause a patient to be thirsty. The intake of water would also help to lower the serum sodium concentration. When the serum sodium concentration is low or there is water excess, ADH secretion is inhibited and more water is excreted in the urine.

The second clinically important problem seen with alterations in body fluids involves the absolute amount of sodium and water in the ECF. With sodium problems we are talking about the absolute amounts of sodium and water in the body and not about the relative concentrations of these substances. With water problems we are talking about changes in osmolality or the concentration of sodium relative to water. Another way to help separate these two problems is to remember that when we give someone free water, it disperses between the intracellular and extracellular space. For example, when a liter of water is administered, two-thirds of it goes into the intracellular space and only one-third stays in the extracellular space. This produces a fall in plasma osmolality. On the other hand, when we give one liter of isotonic saline (0.9% sodium chloride) and it is retained, it all stays in the extracellular space which is expanded. There is no change in osmolality (assuming it was normal before the fluid was administered). Renal excretion of sodium

is determined by the interplay among several neural hormonal systems, including atrial natriuretic peptide (ANP), the renin–angiotensin–aldosterone system, the sympathetic nervous system, and other factors. These factors respond to changes in the effective circulating volume that is sensed by receptors in the carotid sinus, the kidney, and the atrium of the heart.

In interpreting an abnormal serum sodium, an important question to consider is whether the reported sodium is an accurate reflection of serum osmolality.

Dr. Mobutu asked Ms. Kim what should be done to determine if the sodium of 125 mEq/l was indicative of serum osmolality.

Ms. Kim suggested that it would be important to repeat the serum sodium measurement to make sure that the result was consistent. Assuming that the repeat value remains low, Ms. Kim reasoned that any condition that increased the amount of water in the body might lead to a decrease in the serum sodium concentration without a decrease in the amount of total body sodium. She thought a bit about water and solute transport. Salt is lost or added to the extracellular space while water is lost or gained from both the intracellular and extracellular space. Water moves in and out of cells by osmosis. Any substance that stays in the extravascular space but was unable to move freely into the intracellular space would increase the osmotic gradient for water to move from inside the cell to the extracellular space. Ms. Kim recalled that glucose stays in the extracellular space and causes water to move from inside the cells to the extracellular space. Ms. Kim turned to Dr. Mobutu and said: 'An elevated serum glucose will cause water to move from inside the cells to outside the cells and artificially lower the serum sodium measurement. That is not the case in Mrs. Speen. Her glucose is fine.'

'That is excellent,' Dr. Mobutu replied. 'Now remember, when serum sodium is measured it is often in relation to the plasma water. In general, 93% of the plasma is water, with the remaining portion consisting of lipids and proteins. Any increase in protein or lipids can artificially lower the serum sodium concentration without changing the measured serum osmolality. This is a relatively rare occurrence. I do not think that it is the problem with Mrs. Speen for two reasons. First, she has never had a problem with significant hyperlipidemia or elevated serum proteins. To the best of my knowledge she does not have multiple myeloma. Second, we can measure her serum osmolality and compare it to a calculated serum osmolality. If her measured and calculated osmolality are approximately the same, then there are no unaccounted osmoles to worry about. If, on the other hand, her measured serum osmolality is different from the calculated osmolality, we need to find out why. We can order a serum osmolality and a stat serum sodium to verify our previous result.

After finishing rounds, Dr. Mobutu called the lab and found that the repeat serum sodium was 125 mEq/l.

The measured serum osmolality matched the calculated osmolality. She asked Ms. Kim what she thought.

'Well, Dr. Mobutu, Mrs. Speen's low sodium must reflect a true decrease in her serum sodium concentration or serum osmolality.'

As stated above, osmolality is defined as the number of osmoles of solute per liter of solution. Sodium concentration is the major determinant of plasma osmolality. Under normal physiologic conditions, other ions such as glucose, contribute only minimally to the osmolality. In fact, you can estimate the plasma osmolality by multiplying the serum sodium concentration by 2. The reason you multiply by 2 is because of the presence of dissociated anions like chloride (if we were to be exact, we would multiply by 1.75 because, in fact, NaCl is only 75% dissociated). Plasma osmolality may be calculated as follows:

Plasma osmolality = 2 × plasma sodium (mEq/l) + glucose (mg/dl) /18 + blood urea nitrogen (mg/dl)/2.8

In using this formula, it should be remembered that the blood urea nitrogen does not dehydrate cells because it moves freely through most plasma membranes. It is not an effective osmole. Therefore, the effective plasma osmolality is calculated as follows:

Effective plasma osmolality = 2 × plasma sodium (mEq/l) + glucose(mg/dl)/18

In the process of evaluating hyponatremia it is important to exclude laboratory error as well as pseudo-hyponatremia. Pseudohyponatremia is associated with hyperglycemia, hyperproteinemia, and hyperlipidemia. A convenient method for calculating the effect of hyperglycemia on the serum sodium concentration is to recall that for every 100 mg/dl rise in blood glucose, the serum sodium falls approximately 1.6 mEq/l. (For example, if an individual is admitted with a serum glucose of 520 mg/dl and a serum sodium concentration of 130 mEq/l, then the serum sodium concentration can be predicted to rise to 136.4 mEq/l after insulin is given and blood glucose falls to normal.) As glucose moves into the cells after treatment with insulin, water travels down its osmotic gradient into the cells. This will result in a return of the serum sodium concentration to normal levels. With hyperlipidemia and hyperproteinemia, the measured plasma osmolality is normal. In practice, hyponatremia secondary to hyperlipidemia and hyperproteinuria is rare. There is no acute need to treat the increase in lipids or proteins.

Since the possibility of both laboratory error and pseudohyponatremia are excluded, a diagnosis of true hyponatremia may be confirmed. To fully evaluate a decreased serum sodium, it is essential to determine the patient's volume status.

'Since it is clear that Mrs. Speen has true hyponatremia, how would you proceed to sort it out?', Dr. Mobutu asked the student.

'As we discussed, Dr. Mobutu, the serum sodium is a measure of concentration. It tells us nothing about the absolute amount of sodium present in the intravascular space or plasma. Mrs. Speen could have this serum sodium concentration of 125 mEq/l and be hypervolemic, normovolemic, or hypovolemic. In other words, the total amount of sodium in her intravascular space may be normal, depleted, or expanded.'

'Good, so which of the three is she? You have the information available for the answer.'

Ms. Kim thought for a moment. Dr. Mobutu was certainly putting her through a rigorous teaching session, but she seemed to be holding her own! She reasoned aloud with Dr. Mobutu:

'If Mrs. Speen has hyponatremia from an increase in intravascular volume, then there would be evidence for this in the history and on the physical exam. There was no weight gain, mention of tight belts or clothing, or symptoms of congestive heart failure, ascites, or nephrosis. On physical exam there was no edema, jugular venous distention, ascites, or pulmonary congestion. So it seems unlikely that Mrs. Speen is hyponatremic from a volume expanded state.

'If Ms. Speen has hyponatremia from a decreased intravascular volume, then there would also be evidence for this in the history and on the physical exam. The patient did not complain of increase thirst, decreased weight, or new dizziness on standing to suggest orthostatic hypotension. There is no evidence in the history of loss of a sodium-containing fluid. On physical exam the patient was not orthostatic and her mucous membranes were moist.'

'You are doing very well, Ms. Kim. Based on your careful history and physical exam you have adequately excluded intravascular volume expansion or depletion. It looks like the problem in this woman, along with her pneumonia, is normovolemic hyponatremia.'

As Dr. Mobutu noted, hyponatremia associated with a hypervolemic state may be seen with congestive heart failure, cirrhosis with ascites, and renal disease with nephrotic syndrome. In these conditions, there is an excess of total body sodium.

The diagnosis of hypervolemia can usually be made on the basis of the history and physical examination.

It can be a little more difficult to determine whether a patient is hypovolemic as opposed to euvolemic. As Dr. Mobutu noted, in a person who has a decrease in total body sodium, there should evidence of sodium loss. Possibilities include: (1) extrarenal losses of a sodium-containing fluid, as would occur with vomiting, diarrhea, excessive nasogastric tube drainage, or burns; (2) renal losses of sodium as may be seen with diuretic use or a salt losing nephropathy; or (3) adrenal insufficiency. There are, however, some situations in the elderly where a history of sodium loss is not obvious. For example, elderly patients who are on a sodium-restricted diet may become sodium-depleted. This is especially the case if they are also taking a diuretic.

In differentiating hypovolemia from euvolemia, the urine sodium concentration and fractional excretion of sodium may be helpful. With hypovolemia, if the kidneys are normal, they would increase sodium absorption in an effort to increase intravascular volume. The fraction of excreted sodium is generally low (less than 1%). With euvolumia, the fraction of excreted sodium is usually greater than 1%. These measurements are not useful in patients receiving diuretics. The fraction of excreted sodium can be calculated by the following formula:

$$\text{Fractional excretion of sodium} = 100 \times \frac{(U_{Na} \times P_{Cr})}{(P_{Na} \times U_{Cr})}$$

Dr. Mobutu and the student continued to discuss the case.

'On the basis of her physical exam, history, and decreased serum sodium and chloride with otherwise normal lab results, we think Mrs. Speen has normovolemic hyponatremia. Therefore, she has a normal amount of sodium in her intravascular space and the decrease in the serum sodium concentration is on the basis of a problem with the regulation of water. How do you think this could occur, Ms. Kim?'

'Mrs. Speen has too much free water in the intravascular space producing a decrease in the serum osmolality. Serum osmolality is in part regulated by ADH. When the osmolality or serum sodium increases, this is sensed by the hypothalamic osmoreceptors and ADH is released. ADH then acts on the collecting tubules to increase the absorption of free water and the plasma osmolality decreases. Conversely, when the osmolality decreases, the secretion of ADH decreases and there is an increased excretion of free water in the urine. I know that ADH secretion can be affected by many medications. For example, the oral hypoglycemic agent chlorpropamide can potentiate the effects of ADH.'

'You have done some good work and accomplished quite a bit in medical school, Ms. Kim! Great! Now, tell me, do you know about something that can suppress the effect of ADH?' Dr. Mobutu was smiling at Ms. Kim.

'Well, Dr. Mobutu, I remember from physiology, and I also remember from a personal experience or two, that ethanol can suppress ADH and lead to a diuresis. After a few drinks, more free water is excreted!'

Among hospitalized elderly patients in the acute hospital setting, 11% have hyponatremia (Nunez et al. 1987). Prevalence studies done in the chronic care setting show 22% of patients have chronic hyponatremia (Kleinfeld et al. 1979). As discussed below, a significant number of hospitalized elderly patients have hypernatremia at the time of admission. On a day-to-day basis the aging kidney is able to meet the challenges of electrolyte and water homeostasis. However, when older kidneys are faced with a stress, their response is less flexible and slower compared to younger kidneys. Clinical studies suggest that older individuals may have more difficulty than younger persons in maintaining a normal serum sodium concentration. As an example, clinical studies have shown that renal concentrating ability decreases with age. In individuals studied after 12–24 hours of water deprivation, the maximal urine osmolality in the younger individuals (age 20–39) was 1109 mosmol/kg compared to 882 mosmol/kg in persons aged 60–79 (Rowe et al. 1976). The extent to which this is clinically significant is not entirely clear and is the subject of ongoing research. Clinical studies have also shown that while the response of ADH to a decrease in the intravascular volume is decreased in the elderly (Rowe et al. 1982), the response to an osmolar stimulus is actually increased (Helderman et al. 1978). The urinary diluting ability of the kidney is also thought to decline with age. The pathophysiology behind these changes is currently under investigation.

In assessing a patient such as Mrs. Speen who has a low osmolality and euvolemia, there are several major diagnostic possibilities. Water intoxication, caused by compulsive drinking of water or other hypotonic fluids, can lead to a decrease in the serum sodium concentration. This problem is usually seen in psychiatric patients. It can be exacerbated by some of the medications used to treat psychiatric illness because they can cause a dry mouth. The diagnosis of water intoxication is supported by a very low urine specific gravity (less than 1.003) or a urine osmolality below 100 mosmol/kg. These values are indicative of both a maximally dilute urine and complete suppression of ADH. If the patient is water restricted the urine osmolality will remain low until the serum sodium concentration normalizes. A reset osmostat is a description of a condition where the patient responds normally to changes in osmolality but releases ADH at a lower threshold. It is associated with hypovolemia, quadriplegia, psychosis, malnutrition, and the syndrome of inappropriate ADH secretion (SIADH) (Rose 1994). Patients with a reset osmostat will, with restriction of water intake, stimulate secretion of ADH and have a slight decline in the serum sodium concentration. The urine osmolality will rise as a consequence of the action of ADH.

Adrenal insufficiency can also result in hyponatremia. Hypothyroidism may cause hyponatremia although the mechanism is poorly understood. In younger persons, hyponatremia rarely occurs before there is obvious clinical evidence of hypothyroidism. In older individuals, the symptoms and physical findings of hypothyroidism may be difficult to detect. Therefore, unless an otherwise obvious cause is identified, a thyroid-stimulating hormone (TSH) test should be obtained in screening older persons with hyponatremia.

Various medications may also cause hyponatremia. For example, diuretics are commonly used in the elderly. They increase renal sodium excretion and may cause hyponatremia particularly in individuals on a sodium restricted diet. Non-steroidal antiinflammatory drugs are also commonly used in the elderly. They inhibit the effect of renal prostaglandins. This may limit the ability of the kidney to excrete a dilute urine (Clive and Stoff 1984). As a result, more free water is retained in the intravascular space, and the serum osmolality is decreased. Laxatives, used frequently by the elderly, are also known to cause hyponatremia. They can deplete the extracellular volume. When the ECF volume loss is replaced with drinking water there is a fall in serum sodium concentration.

There are a few other instances when serum sodium concentration may be altered. Occasionally, administration of a hypertonic solution such as mannitol, glycine, or sorbitol may cause hyponatremia. Hypertonic solutions are used in a variety of clinical situations. This includes the use of mannitol for increased intracranial pressure. Hypertonic solutions are also used for bladder flushes after a transurethral resection of the bladder or prostate. These solutions are variably absorbed and may lead to alterations in plasma osmolality. For example, glycine and sorbitol are quickly metabolized, leaving behind a free water load. If an individual's kidneys are diseased and unable to excrete the free water, this is more likely to result in hyponatremia, with a normal total body sodium. This is more likely to occur with the aging kidney. It also occurs in the hospital setting when patients are given free water in intravenous fluids or with intravenous medications, and less commonly with tap water enemas.

SIADH is a common cause of euvolemic hyponatremia. It may be caused by increased production of ADH directly in the hypothalamus. This is seen in some neuropsychiatric disorders, pulmonary disease including pneumonia and tuberculosis, human immunodeficiency virus infection, in post-operative patients, and with many medications (Rose 1994). Some cancers, such as small cell carcinoma of the lung, may secrete ectopic ADH and thus lead to hyponatremia. As mentioned earlier, many medications, including tolbutamide, chlorpropamide, and carbamazepine, may potentiate the effects of ADH. Diagnostic criteria for SIADH includes hyponatremia with hypo-osmolality without evidence of hypervolemia or hypovolemia, and an inappropriately high urine osmolality given the low serum sodium concentration. Normally, with a low serum sodium concentration one would expect the kidney to produce a maximally dilute urine in an effort to excrete excess free water until the serum osmolality normalizes. With SIADH, the high urine osmolality indicates that a concentrated urine is being produced which is inappropriate given the low serum sodium. Diagnostic laboratory findings in SIADH include: (1) a urine osmolality greater than 100 mosmol/kg. In practice, it is usually greater than 250 mosmol/kg, especially in the elderly where the ability to maximally dilute urine maximally is impaired (Crowe et al. 1987); (2) a urine sodium greater than 40 mEq/l; (3) normal adrenal, renal, and thyroid function; and (4) normal acid–base and potassium balance.

Dr. Mobutu reviewed with Ms. Kim the possible causes of normovolemic hyponatremia. They agreed it seemed likely that Mrs. Speen's decreased sodium was caused by SIADH but there was a need to rule out other causes and obtain further laboratory testing. Dr. Mobutu asked Ms. Kim what she thought needed to be done.

'I think that we should check a urine osmolality and urine sodium to make sure they are inappropriately high as we expect. I also think it is worthwhile to obtain a TSH test. There is nothing to suggest that Mrs. Speen is a compulsive water drinker. She is not supposed to be on diuretics but I need to make sure that she does not share medicines with her friends. I will review her use of over-the-counter medications, especially nonsteroidal antiinflammatory agents and laxatives. Her blood urea nitrogen and creatinine certainly do not suggest renal failure. I guess Mrs. Speen could have a reset osmostat but that is a diagnosis of exclusion. I also do not think that Mrs. Speen has glucocorticoid deficiency. We can check a cortisol or do an ACTH [adrenocorticotrophic hormone] stimulation test to be sure. Is that all appropriate, Dr. Mobutu?'

'Yes, I agree with you. In terms of her hyponatremia your plan is excellent. We need to consider what, if any, immediate clinical interventions we need to make for Mrs. Speen.'

The treatment of hyponatremia depends on both the serum sodium concentration and the length of time over which it developed. Clinically, the rate of development of hyponatremia tends to correlate with the severity of symptoms. The more severe symptoms are associated with a rapid decline in serum osmolality. When the serum sodium concentration is between 125 to 130 mEq/l, patients may be asymptomatic although decreased appetite, confusion, muscle cramps, headache, nausea, and vomiting frequently occur. In these patients the first step in treatment is to restrict free water intake to prevent a further decline in serum sodium. This will generally increase the serum sodium concentration no matter what the cause. After the cause of the hyponatremia is determined, treatment should be directed at correcting the underlying etiology. For example, if it is secondary to medications, these should be stopped. If the hyponatremia is secondary to hypothyroidism, the treatment is hormone replacement.

A serum sodium below 120 mEq/l is usually associated with symptoms that begin with confusion and may progress to coma, seizures, and death. The symptoms associated with hyponatremia are related to edema produced in the brain as a result of rapid changes in osmolality. As the osmolality or serum sodium concentration in the extracellular space decreases, fluid moves intracellularly into brain cells as predicted by the osmotic gradient. Patients with chronic severe hyponatremia are over time able to compensate for these changes in osmolality. In general, the elderly are less able to defend themselves against changes in osmolality than are younger individuals. Some studies suggest that when the serum sodium is corrected too rapidly, patients are at risk for central pontine myelinolysis which is a demyelinating disorder characterized by quadriparesis, paraparesis, dysarthria, and coma (Rose 1994).

The treatment of severe hyponatremia is the subject of some controversy. In patients with longstanding hyponatremia, it is generally felt that the brain has partially adapted to the change in osmolality. Depending on the severity of the patient's symptoms, the serum sodium should be increased slowly. A reasonable rate of correction in treating chronic hyponatremia is to increase the serum sodium concentration by 0.5 mEq/l/h. With acute symptomatic hyponatremia, the rate of rise should generally be approximately 1.0 mEq/l/h. However, if the patient is severely symptomatic, many would advocate a more rapid rise until the serum sodium is 120 mEq/l, at which time the rate of rise should be decreased to 0.5 mEq/l/h. The interventions used to correct hyponatremia include water restriction as discussed above and the administration of sodium. Administration of sodium is reserved for those individuals with severe symptomatic hyponatremia. It requires careful monitoring of the patient's volume status to prevent volume overload. Generally, the patient should have vital signs checked as well as a cardiac and lung examination every 2 hours initially, then every 6–8 hours. It is important to check the serum sodium concentration equally as often to assure that the rate of correction is as expected.

In patients with severe symptomatic hyponatremia salt may be given in conjunction with furosemide. The salt may be given either as normal saline, hypertonic saline, or in the form of sodium chloride tablets. The method of administration depends on the severity of the hyponatremia and the patient's symptoms. Again, once the serum sodium is greater than 120 mEq/l, the administration of sodium is generally stopped or slowed and free water restriction continued. The exception to this is in the case of a patient with a true decrease in the amount of total sodium in their body. As stated previously, if the patient has a true decrease in serum sodium content, then there must be evidence that the patient has lost a sodium-containing fluid, for example, through vomiting, diarrhea, or the use of diuretics.

The calculation used to determine the amount of sodium needed to increase the serum sodium to a given level is:

Sodium deficit = 0.6 × weight (kg) × (desired plasma sodium (mEq/l) – measured plasma sodium (mEq/l))

Sodium deficit = 0.6 × weight (kg) × (120 – serum sodium (mEq/l)

Ms. Kim thought for a minute and then said to Dr. Mobutu: 'Given that Mrs. Speen's mild changes in mental status improved with oxygen, I do not believe that she is significantly symptomatic from her hyponatremia. The goal, however, is to keep Mrs. Speen's sodium from dropping further. Mrs. Speen's free water intake should be restricted.

'Good, Ms. Kim. Unfortunately, sometimes restricting free water is easier said than done. You must write the order specifying the fluids that you are calling free water. For example, write that Mrs. Speen is restricted to a specified amount [usually 500–1000 cc/24 h] of free water including coffee, soda, water, and tea. Also specify that, when possible, any medications given intravenously, like her antibiotics, are given in normal saline and not 5% dextrose in water. Here's another challenge for you. Can you summarize the events along with the pertinent diagnostic and treatment plans? It may make sense if you go through Mrs. Speen's different problems one by one.'

'Let's see, Dr. Mobutu. First, with regard to Mrs. Speen's pneumonia, she has not produced a sputum sample but blood cultures have been drawn and she is taking antibiotics for a community-acquired pneumonia. Mrs. Speen is breathing comfortably on two liters of oxygen by nasal cannula. My concern is that her chest X-ray shows a right lower lobe pneumonia in the same spot as that of six months ago. I am worried that she might have a post-obstructive pneumonia and a tumor. Another possibility might be that she has difficulties swallowing and has aspiration pneumonia. I think we should order a chest computerized tomogram scan [CT scan] and if possible obtain sputum samples for cytology. Depending on the results of the CT scan we may need to consider a bronchoscopy. I also need to make sure a PPD is placed. Second, Mrs. Speen's hyponatremia may be due to SIADH caused by the pneumonia. However, I am worried about her ten pound weight loss. She might have a post-obstructive pneumonia which raises the possibility of a lung tumor as the cause of her SIADH and hyponatremia.'

'Ms. Kim, you are a fantastic medical student. Why don't you write the orders in the chart, I will countersign them, and then let's go home. We can meet here at eight in the morning on the patients round.'

Mrs. Speen was comfortable over the next 48 hours.

Her workup proceeded as ordered by Ms. Kim and Dr. Mobutu. Her serum sodium remained at 125 mEq/l despite a moderate fluid restriction. Her urine sodium was greater than 40 mEq/l and her urine osmolality was greater than 100 mosmol/Kg. Blood cultures were negative. A PPD was negative with a positive control. Both hypothyroidism and adrenal deficiency were excluded on the basis of a normal TSH test and ACTH stimulation test. Her serum creatinine and blood urea nitrogen were normal. Unfortunately, Ms. Kim fears about the recurrent pneumonia were all too accurate. A CT scan showed a right lower lobe mass impinging on the bronchus. Bronchoscopy was performed with preliminary cytology results suggestive of small cell carcinoma of the lung.

Dr. Mobutu brought Ms. Kim with her as she told Mrs. Speen the bad news. Mrs. Speen cried a bit when she received the diagnosis. She said that she had been afraid that something serious was wrong. Dr. Mobutu explained that she would have a cancer specialist see Mrs. Speen to consider the best course of action. She also explained that the problem with her serum sodium concentration was likely due to the tumor producing a substance that caused her kidneys to hold on to water. For now, Mrs. Speen should continue to restrict the amount of water and tea that she drank. Most important, Dr. Mobutu told Mrs. Speen that she would continue to care for her, answer her questions, and help her as much as she was able.

Dr. Mobutu finished speaking with Mrs. Speen and went with Ms. Kim to the cafeteria. 'That was hard for me,' she said, 'I have taken care of her since I was in training. I need a few moments to gather my thoughts. Mrs. Speen is a good person.'

Dr. Mobutu's beeper went off, paging her to the emergency room of the hospital. 'Oh great, just what I need,' she commented wryly. 'I guess this is all the time we have for reflection . . . let's go to the emergency room.'

On arrival, the physician in the emergency room, Dr. Luis Gonzalez, told Dr. Mobutu and Ms. Kim that one of Dr. Mobutu's patients, Mr. Ralph Elmore, needed to be admitted. Mr. Elmore, a 90-year-old gentleman who lives alone, had been brought in by ambulance. Normally, he lived independently and was alert, oriented, and able to care for himself except for some shopping. His daughter had not been able to reach him by phone and went over to his apartment as she was concerned. Once there, she had found her father lethargic with vomitus on the floor and immediately called an ambulance. Dr. Gonzalez said that Mr. Elmore appeared very ill and was unable to give a coherent history. He had a systolic blood pressure of 80, a thready pulse of 120, a temperature of 102, and a respiratory rate of 18. Significant physical findings included clear lungs, a normal cardiac exam save for the tachycardia, and tenderness over the bladder with light palpation. Neurologic exam was remarkable for lethargy but was otherwise non-focal. Dr. Gonzalez explained that immediately on the patient's arrival he had begun an intravenous line with normal saline and had placed a catheter through

Mr. Elmore's penis into his bladder. The saline was being infused rapidly until the systolic blood pressure reached above 100 mmHg. The urine catheter drained 150 ml of dark, yellow, cloudy, urine with a foul odor. A sample of urine, examined microscopically, showed numerous white blood cells, too many to count per high power field, and bacteria. On gram stain, the bacteria were gram-negative rods. The urine was sent for culture. Also obtained were blood cultures, a complete blood count, and electrolytes. An electrocardiogram was normal except for a sinus tachycardia. A chest X-ray was clear without infiltrates. Dr. Gonzalez had begun Mr. Elmore on broad-spectrum intravenous antibiotics to cover for sepsis secondary to a urinary tract infection. Dr. Mobutu thanked Dr. Gonzalez for the work he had done.

Dr. Mobutu and Ms. Kim examined Mr. Elmore and confirmed Dr. Gonzalez's findings. His blood pressure had now improved to 120/70 and Dr. Mobutu asked that the intravenous saline be slowed down to a rate of 50 cc/h until the patient was more fully assessed. Arrangements were made for him to be transferred from the emergency room to one of the rooms in the hospital. Dr. Mobutu asked that Mr. Elmore be weighed by the nurses before he was put to bed.

Once in his room, Mr. Elmore remained lethargic. His weight was 66 kg. Dr. Mobutu was paged by the laboratory and told that Mr. Elmore's labs were abnormal. Electrolytes showed a sodium of 160 mEq/l with an elevated chloride, and normal potassium, and bicarbonate. Blood urea nitrogen and creatinine were greatly elevated at 100 and 3.0, respectively. A complete blood count returned with a white blood cell count of 25,000 with 70% neutrophils, 20% bands, and 10% lymphocytes. Dr. Mobutu recalled that on a routine visit last month to the office all of Mr. Elmore's labs had been normal. She turned to Ms. Kim: 'So, Ms. Kim, seeing how well you did with hyponatremia, how do you explain this hypernatremia?'

Ms. Kim answered: 'The greatly elevated serum sodium makes it clear that he has a free-water deficit. Given the history that Mr. Elmore had not answered phone calls for two days, it seems likely that he was ill in his apartment, febrile, vomiting, not eating or drinking. Based on this history together with the physical exam showing significant hypovolemia, he has also become salt-depleted' [i.e., his extracellular volume is depleted].

'I agree with you. There are a few points about his initial management that I would like to emphasize.

'First, when Mr. Elmore showed up in the emergency room, he received normal saline. Dr. Gonzalez responded in this manner because of Mr. Elmore's hypotension. With a blood pressure of 80 systolic, Mr. Elmore was at risk for decreased perfusion of his organs. It would have been inappropriate to sit him up to check for orthostatic changes. Giving saline made sense because it stays in the extracellular space and increases the perfusion of vital organs.'

'Second, saline, although it is very appropriately used in this setting is not without risks. The major potential problem is fluid overload with the risk of congestive heart failure and pulmonary edema. In the elderly, where the chance of underlying cardiac disease is increased, this is a particular risk. Mr. Elmore had been prescribed digoxin four years ago for mild congestive heart failure after a small myocardial interaction. Although there was no recent history of congestive heart failure, fluid needs to be administered with caution. With Mr. Elmore's elevated blood urea nitrogen and creatinine we need to think about how well his kidneys are working. Given the 150 ml of urine found in his bladder it seems likely that his kidneys are still able to make urine. Whenever you give saline, and particularly in an older person who may not have good cardiac function, it is important that you check the blood pressure and listen to the lungs frequently to make sure the patient is not becoming volume-overloaded.'

'Point number three. Sometimes one gives normal saline and the pressure fails to rise. There are three possibilities: (1) the fluid is being lost as you pour it in, which would happen with a hemorrhage or traumatic injury; (2) the blood vessels are dilated and unable to respond to the increased volume, as occurs in sepsis, where patients are hypotensive and can remain that way despite large amounts of intravenous fluids; and (3) there is severe myocardial damage and the heart cannot generate enough force as it pumps.'

'Now, in Mr. Elmore's case, his blood pressure has responded to the saline. I do not expect that he has any bleeding nor is there anything to suggest that he has myocardial injury. I do believe he is infected but he does not seem to be septic, although we need to watch his pressure carefully. Gram-negative rods are bad for blood pressure!'

'A fourth point. The elderly can be especially sensitive to fluid losses compared with younger persons. With aging, there are characteristic changes in body composition. There is a decrease in total body mass, an increase in the percentage of body fat, and a decrease in total body water. For example, in a young man, the total body water is about 60% of the total body mass. In an older man, the total body water is about 55% of the total mass. For women, the figures for total body water are 53% of total body mass in young women versus 45% in an older woman. Therefore, the same amount of fluid loss in an older person compared with a younger person is more serious as there is a greater loss in the percentage of fluid volume.

'A fifth point is that older individuals have more difficulty in maintaining sodium balance than younger individuals. Again, on a day-to-day basis the older kidney is able to meet the needs of salt homeostasis and maintain a normal intravascular volume. However, under physiologic stress their kidneys' response is slower than younger kidneys. As an example, when elderly and young individuals are placed on low sodium diets, the kidneys of the older persons take significantly longer to decrease sodium excretion [Epstein and Hollenberg 1976]. Data from clinical studies have shown that this is secondary to decreased sodium reabsorption at the distal tubule [Nunez et al. 1978]. The pathophysiology behind this observation is not completely understood. When an ill elderly person is ordered to receive nothing by mouth for many hours prior to an invasive procedure, this can lead to a decrease in the intravascular volume due to a failure to conserve sodium by the kidney. This can be prevented by the careful administration of an appropriate intravenous fluid. In patients' with normal electrolytes, renal function, and euvolemia, a 5% solution of dextrose in 1/2 normal saline [$D_5\frac{1}{2}NS$] is used. Just as the aging kidney is slow in decreasing sodium excretion when faced with a need to conserve, so the aging kidney is slower in excreting a sodium load than its younger counterpart. Again, the pathophysiology behind these changes remains the subject of active investigation.'

'A sixth point is that hypernatremia is not uncommon in the elderly. Data collected on hospitalized patients older than 60 showed that 1.1% had an elevated serum sodium concentration [Na greater that 148 mEql] (Synder et al. 1987). Of these patients, 57% developed hypernatremia in the hospital, and the remaining were hypernatremic at the time of admission [Synder et al. 1987]. Of note, the mortality rate among the hypernatremic patients was seven times that of age matched controls and was not related to the severity of the hypernatremia [Synder et al. 1987]. Data from other clinical studies [Prospective Cohort Analytic Study] have shown that the incidence of hypernatremia in febrile nursing home patients is 25% [Weinberg et al. 1994]. Therefore, it is important to learn not only how to treat hypernatremia safely but also how to prevent it from occurring. In the hospital setting, for patients on intravenous therapy and no oral intake, it is important to include free water in the replacement fluids. You can measure whether your replacement is adequate by following the serum sodium concentration. In patients both at home and in the hospital, it is important to make sure they drink enough water or other hypotonic fluid. This is not as easy as it sounds.'

Thirst is an important component in regulating plasma osmolality. It is regulated centrally but sensed peripherally. The patient responds to the sensation of a dry mouth. In the elderly, the sensation of thirst is diminished. Along with decreased thirst, the ability of many elderly to obtain free water may be limited because of decreased mobility, poor visual acuity, cognitive impairment, swallowing disorders, or alterations in thirst from medications such as cardiac glycosides. Older persons with urinary incontinence sometimes limit their water intake to avoid urinary incontinence.

'So, just to summarize Mr. Elmore's presentation, it appears that he is depleted of both salt and free water. He has a sodium of 160, blood urea nitrogen of 120, and creatinine of 3. Last month his blood urea nitrogen and creatinine were normal. I suspect that Mr. Elmore's urine became infected because it stayed in the bladder for too long a time. He became ill and stopped taking fluids. He lost free water from his skin (500 ml/day), respiratory tract (400 ml/day), and stool (200 cc/day). These losses combined with vomiting and poor oral intake resulted in severe salt and water depletion. Not only is his intravascular or plasma volume severely depleted of fluid but the amount of water present relative to sodium is also severely depressed. Therefore, he needs not only normal saline to replete his intravascular space but water to correct his elevated plasma osmolality. In addition, Mr. Elmore's kidneys were not perfused adequately. He did not excrete creatinine or urea and they accumulated in the blood. Now Ms. Kim, to solve this problem we need to work out how much fluid Mr. Elmore has lost and replace it.'

'Dr. Mobutu,' Ms. Kim asked, 'how do you know how much fluid to give and how fast?'

In treating patients with fluid deficits like Mr. Elmore's, the immediate goal is to replace the intravascular or plasma space as this determines the effective circulating volume. As was discussed, this was done appropriately in the emergency room with normal saline until the systolic blood pressure was greater than 100 mmHg. However, the free water deficit is generally replaced more slowly. A reasonable goal is to replace one-half of the free water deficit over the first 24 hours. However, this is a point of some controversy. In cases of severe hypernatremia (greater than 170 mEq/l), some nephrologists would advocate a more rapid repletion initially with slowing later as the serum sodium approached 150 mEq/l. The first step is to calculate the fluid deficit in liters.

Fluid deficit (liters) = desired TBW − current TBW.

The desired total body water is calculated as follows:

Desired TBW = measured serum sodium (mEq/l) ×current TBW / desired serum sodium (mEq/l)

The current total body water is calculated as follows:

Current TBW = 0.6 ×weight in kg [for men].
Current TBW = 0.50 ×weight in kg [for Women].

Once the free water fluid deficit is calculated, it is essential that replacement therapy be carefully monitored. Serum sodium should be checked every hour for two hours then every four to six hours for the next 12 hours. After that time, further measurements are made based on the patient's response. For example, if the serum sodium is falling at a predicted rate, then it is probably not necessary to check it more frequently than every eight hours. The desired rate of decline is 0.5 mEq/l/h. The purpose of such meticulous evaluation is to make certain that the patient is responding to the intervention as predicted and, if not, to re-evaluate.

There are a number of different ways in which the intravascular and free water fluid deficit can be repleted. One option is to use two intravenous lines. In one intravenous line, a solution of 5% dextrose in water is given to replace the free water deficit. In the second, one can give normal saline to replace the depleted intravascular volume. The advantage of two intravenous lines is that over time you can alter either the rate of free water or saline administration independently of each other.

In providing free water and intravascular fluid replacement, careful attention must be paid to urine output. The goal is to have a urine output of at least 50 ml per hour. In Mr. Elmore's case there are two reasons to leave the indwelling foley catheter in place. First, given his altered mental status, he is unable to urinate reliably into a receptacle. Sometimes an external catheter may be used in place of an indwelling catheter. However, until Mr. Elmore demonstrates that he can urinate without a large post-void residual, the indwelling catheter should be left in place. Since there is an increased risk of infection with an indwelling foley catheter, it should be used only when absolutely necessary. It should be taken out as soon as possible, preferably within the first twenty-four hours.

Ms. Kim then calculated the free water deficit and wrote to replete one-half this amount in the first 24 hours. She was aware that she was repleting fluid that had already been lost and not ongoing losses from the skin, lungs, and gastrointestinal tract. If necessary she would increase the free water administration later. There was no evidence for significant ongoing free water losses as can be seen in diabetes insipidus. She ordered the serum sodium to be drawn every hour for the first two hours then every four hours for twelve hours. As Mr. Elmore was clearly volume-depleted on admission, she chose to use normal saline in the second intravenous line for intravascular volume replacement. Mr. Elmore's blood pressure was now back to his baseline. Ms. Kim elected to begin with normal saline at 50 cc/h. She would continue to regularly check Mr. Elmore's blood pressure, pulse, lung exam, and urine output to be sure he was receiving the correct amount of fluid. As she had two intravenous lines it would be easy to increase or decrease the amount of normal saline or D5W over time.

Ms. Kim asked Dr. Mobutu to check her calculations for free water repletion to assure that they were correct. She used the post-hydration weight of 66 kg.

Current total body water = 0.6 ×wt./kg. = 0.6 ×66 = 39.6 liters

Desired TBW = measured serum sodium × TBW / desired serum sodium
= 160 (mEq/l) ×39.6 l/140 (mEq/l) = 45.3 liters

Free water deficit = desired TBW − current TBW = 45.3 l − 39.6 l
= 5.7 liters free water deficit.

One-half of the TBW deficit is 2.85 liters.

This calculates to 2850 cc/24 h = 118 cc/h of D5W for 48 hours.

Ms. Kim was interested in the low dose of digoxin the patient was on for congestive heart failure. Other patients she had seen were always on 0.25 mg/day of digoxin. Dr. Mobutu went on to explain that many medications depend on the kidney for excretion either directly or via tubular secretion. Other drugs have active metabolites that depend on renal excretion. Digoxin is an example of a medication that is excreted by the kidneys and may accumulate to toxic levels if not properly dosed in renal failure. Therefore, it is important to estimate the creatinine clearance in an elderly patient to assist with medication dosing. Almost every medication book will give information on the correct dosing of medication in the presence of altered renal function. The Cock–Gault formula provides an estimate of creatinine clearance (Cockroft et al. 1976):

Creatinine clearance for males = 140 − age × Weight in kg / serum creatinine mg/dl × 72.

[For females, the result is multiplied by 0.85]

It is important to remember that this is just an estimate of creatinine clearance. Recent data suggest that the correlation between the measured and calculated creatinine clearance is only moderate (Malmrose et al. 1993). However, when an adequate urine collection for creatinine clearance is not available, this formula should be used to estimate drug doses. Serum drug concentrations should be monitored whenever possible. It is preferable to collect a 24-hour urine to measure the creatinine clearance. This is often impractical in the acutely ill. However, some physicians advocate a timed urine collection (12 hours) that is standardized for the diurnal variation in plasma creatinine levels (Malmrose et al. 1993).

Mr. Elmore's creatinine clearance is calculated below. Note that this calculation was done after hydration and return of renal function to Mr. Elmore's baseline or steady state. When he was initially admitted with a creatinine of 3.0 mg/dl and a weight of 64 kg, his creatinine clearance was only 14.8 ml/min. Although this may be used as a rough guide to initial medical therapy it must be remembered that the estimated creatinine clearance is not accurate in the setting of a changing serum creatinine.

$$\frac{140 - 90 \text{ years} \times 66 \text{ kg}}{1.3 \text{ mg/dl} \times 72} = 35.2 \text{ ml/min}$$

It is important to remember that with aging, as muscle mass decreases, so does creatinine production. Therefore, the age related fall in renal function is not usually associated with a rise in the serum creatinine. For example, a creatinine of 1.0 mg/dl in a 20-year old man weighing 70 kg corresponds to an estimated creatinine clearance of 116 ml/min. A creatinine of 1.0 mg/dl in a 90-year-old man weighing 70 kg corresponds to an estimated creatinine clearance of 48.6 ml/min.

Data from various studies suggest that the creatinine clearance tends to decline with each decade of life over the age of 40 (Rowe et al. 1976). Of note is that follow-up studies on 254 of these individuals without hypertension, urinary tract disease, or diuretic use demonstrated that approximately 30% had no significant fall in creatinine clearance with aging (Lindeman et al. 1985). The reason for this observed fall in creatinine clearance in approximately 70% of aging individuals is unclear, but is likely to be multi-factorial. Some of the proposed factors contributing to this decline include age associated changes in renal anatomy, vasculature, and tubular function. In addition, age related changes in the hormones that regulate intravascular volume are also likely contributors. The influence of diet and metabolism may also be substantial.

Dr. Mobutu, finishing up after reviewing Ms. Kim's orders on Mr. Elmore, realized it was time to head to the office to see other patients. She made a mental note to make sure that she took more time this evening to speak again with Mrs. Speen and to give Mr. Elmore's daughter a call. He should be better in a couple of days. Dr. Mobutu also realized how much she had enjoyed having this medical student along.

Ms. Kim, I want to thank you for your help. It's been good to have your help and company taking care of some difficult cases. Now tell me, have you ever thought about pursuing a career in geriatrics?'

Questions for further reflection

1. What triggers the secretion of anti-diuretic hormone?
2. What steps should be performed in the evaluation of a patient with hyponatremia? What laboratory tests are useful in addition to the serum sodium?
3. How does one calculate a fluid deficit in a patient with hypernatremia?

References

Clive, D.M. and Stoff, J.S. (1984). Renal syndromes associated with nonsteroidal anti-inflammatory drugs. *New England Journal of Medicine*, **310**, 563–72.

Crockcroft, D.W. and Gault, H.M. (1976). Prediction of creatinine clearance from serum creatinine. *Nephron*, **16**, 31–41.

Crowe, M.J., Forsling, M.L., Rolls, B.J., Philips, P.A.,

Ledingham, J.G.G., and Smith, R.F. (1987). Altered water excretion in healthy elderly men. *Age and Ageing*, **16**, 285–93.

Epstein, M. and Hollenberg, N.K. (1976). Age as a determinant of renal sodium conservation in man. *Journal of Laboratory and Clinical Medicine*, **87**, 411–17.

Helderman, H.J., Vestal, R.E., Rowe, J.W., Tobin, J.D., Andres, R., and Robertson, G.L. (1978). The response of arginine vasopressin to intravenous ethanol and hypertonic saline in man: The impact on aging. *Journal of Gerontology*, **33**, 39–47.

Kleinfeld M., Casimar M., and Borra S. (1979). Hyponatremia as observed in a chronic disease facility. *Journal of the American Geriatrics Society*, **27**, 156–61.

Lindeman, R., Tobin, J., and Shock, N.W. (1985). Longitudinal studies on the rate of decline in renal function with age. *Journal of the American Geriatrics Society*, **33**, 278–85.

Malmrose, L.C., Gray, S.L., Pieper, C.F., Blazer, D.G., Rowe, J.W., Seeman, T.E., *et al.* (1993). Measured versus estimated creatinine clearance in a high functioning elderly sample: MacArthur Foundation Study of Successful Aging. *Journal of the American Geriatrics Society*, **41**, 715–21.

Nunez, J.F.M., Iglesias, C.G., Roman, A.B., Commes, J.L.R., Becerra, L.C., Romo, J.M.T., *et al.* (1978). Renal handling of sodium in old people: A functional study. *Age and Ageing*, **7**, 178–81.

Nunez, J.F.M., Roman A.B., and Commes, J.L.R. (1987). Physiology and disorders of water balance and electrolytes in the elderly. In *Renal function and disease in the elderly*, (ed. J.F.M. Nunez, and J.S. Cameron), pp. 67–93, Butterworth, Boston, MA.

Rose, B.D. (1994). Hypoosmolar states—hyponatremia. In *Clinical physiology of acid-base and electrolyte disorders*, (4th edn), pp 651–694, McGraw-Hill, New York.

Rowe, J.W., Andres, R., Tobin, J.D., Norris, A.H., and Shock, N.W. (1976a). The effect of age on creatinine clearance in men: A cross sectional and longitudinal study. *Journal of Gerontology*, **31**, 155–63.

Rowe J.W., Shock N.W., and DeFronzo, R.A. (1976b). The influence of age on renal response to water deprivation in man. *Nephron*, **17**, 270–8.

Rowe, J.W., Minaker, K.L., Sparrow, D., and Robertson, G.L. (1982). Age-related failure of volume–pressure-mediated vasopressin release. *Journal of Clinical Endocrinology and Metabolism*, **54**, 661–4.

Synder, N.A., Feigal, D.W., and Afieff, A.I. (1987). Hypernatremia in elderly patients. *Annals of Internal Medicine*, **107**, 309–19.

Weinberg, A.D., Pals, J.K., Leveque, P.G., Beal, L.F., Cunningham, T.J., and Minaker, K.L. (1994). Dehydration and death during febrile episodes in the nursing home. *Journal of the American Geriatrics Society*, **42**, 968–71.

15

Problems in voiding and diseases of the prostate

Catherine E. DuBeau

Dr. Irina Anisimov, a medical resident, was sitting in her office in the out-patient clinic waiting for her new patient, when she heard a commotion in the hallway outside her door. She stepped out and saw two ambulance drivers attempting to negotiate a gurney around the corner. Finally successful, they wheeled the stretcher down the hall and stopped outside Dr. Anisimov's office. 'Are you Dr. Anisimov?' one of them asked. 'Here's Mrs. Maida Havel for her three o'clock appointment.'

Peaking out from the blankets on the stretcher was a diminutive elderly woman, her chin tucked against her neck and her hands gnarled from arthritis. 'Oh, Doctor Anisimov, I am so happy to meet you. This problem is very vexing!' She smiled as the ambulance drivers transferred her to the examination table. Dr. Anisimov wondered which problem was so vexing: 'Is she talking about her severe rheumatoid arthritis,' or, she noted how Mrs. Havel's left arm and leg flopped and then drooped as the attendants positioned her, 'is she talking about her stroke?'

Mrs. Havel gave a dry cough and looked up at Dr. Anisimov. 'There it goes again, this is just miserable.'

'What's miserable, Mrs. Havel, are you uncomfortable or in pain? Do you have a cold?' Dr. Anisimov asked as she unwrapped the blankets.

'Oh no, no, no, dear.' Mrs. Havel responded, 'Oh, this makes me want to die.'

At that moment, Dr. Anisimov noted that the sheets surrounding Mrs. Havel were soaked with urine. She felt herself both irritated and overwhelmed. Dr. Anisimov hoped that this was not the problem with which Mrs. Havel wanted help. How come this rarely happened with her younger patients? How was she supposed to deal with this?

Introduction to voiding problems

Urinary incontinence and voiding symptoms are extremely prevalent among older persons, and are the cause of substantial morbidity as well as substantial health care spending. Almost one-third of women over age 65 experience difficulty with bladder control at least monthly, and up to 15% of all community-dwelling elderly are incontinent at least weekly. In long-term care institutions, over 50% of residents are incontinent (UIGP 1992). The medical and psychological morbidity from voiding dysfunction is striking. Cellulitis and pressure ulcers can result from constant skin irritation by urine and wet clothing. Partial or complete urinary retention may necessitate urinary catheters that predispose to urinary tract infections and urosepsis. Falls and fractures can occur from slipping on urine. The associated embarrassment caused by incontinence may lead to social withdrawal, depression, and sexual dysfunction. Caregivers of persons with voiding dysfunction can easily become overwhelmed by the related inconvenience, smell, furniture wetting, and mounds of laundry. Incontinent elders often end up institutionalized. In the United States, incontinence-related health care costs—for evaluation, management, and associated morbidity—currently run up to $10 billion every year (Hu 1990).

Despite the prevalence and importance of the problem, incontinence remains neglected by patients and providers alike (Mitteness 1990). Almost half of incontinent patients never mention their symptoms to their doctors out of embarrassment, a belief that incontinence is a normal part of aging, or the fear that their physician will recommend surgical treatment. At the same time, many health care providers hold the nihilistic view that incontinence is not only inevitable (especially in the frail, institutionalized, and demented), but also difficult to treat in the elderly. As a consequence, many incontinent patients remain undertreated.

This chapter will review the pathophysiology and multiple etiologies of voiding symptoms, including prostate disorders, in the elderly, and provide a framework for evaluation and management. Continued emphasis will be placed on understanding the fact that symptoms of voiding dysfunction are never normal and that relief is almost always possible.

Mechanisms of micturition and continence

To understand how voiding symptoms develop, it is necessary to understand the physiology of normal micturition and continence. (For a review of micturition physiology, readers are referred to overview chapters in current textbooks of urology and neuro-urology.) Micturition requires the storage of urine in the bladder, and the subsequent emptying of bladder contents in a complete, timely, and socially appropriate manner. To achieve this, the bladder must store an adequate urine volume at low pressure and sense when the volume is high. During bladder filling, the urethral sphincter mechanisms must remain closed and then open when voiding occurs. The bladder smooth muscle, called the detrusor, contracts with cholinergic stimulation via parasympathetic nerves that arise from the spinal cord at the S2–S4 level. The urethral sphincter mechanisms include a proximal smooth muscle component and a distal striated muscle component. The smooth muscle portion of the urethral sphincter constricts with alpha-adrenergic stimulation via sympathetic nerves from spinal cord levels T11–L2. The striated muscle component constricts via cholinergic stimulation from somatic nerves arising from the S2–S4 cord levels. The central nervous system plays an important role in modulating and coordinating micturition functions. The parietal lobes and the thalamus are instrumental in sensing the 'urge' to void from detrusor afferent stimuli. The frontal lobes and the basal ganglia inhibit micturition until a person is in the appropriate setting to void (i.e., a toilet is available and the person is prepared). The pontine micturition center integrates inputs from the central nervous system and the periphery, permitting socially appropriate and coordinated voiding.

The sequence of events for voiding thus begins with sensation of bladder fullness by detrusor sensory afferent nerves, which relay this information up the spinal cord to the micturition center in the pons. Availability of toileting facilities is noted and cortical areas relay the release of frontal inhibition to the pontine center. The pons then co-ordinates the shift from sympathetically mediated sphincter constriction and inhibition of the detrusor to parasympathetic sphincter relaxation and detrusor contraction. The detrusor contracts and urine exits through the relaxed urethral sphincter.

The ability to remain continent depends on a number of mechanisms other than bladder and sphincter control modulated by the central nervous system. Adequate cognition is required to comprehend stimuli indicating that voiding is imminent and to respond to such stimuli by locating a toilet or receptacle, or seeking assistance. Mobility is necessary to get to the toilet and adequate manual dexterity is needed to remove clothing and clean after voiding. Motivation also is required for an individual to carry out socially appropriate voiding. Older persons are more likely than younger individuals to have difficulties with cognition and ambulation and may suffer from affective disorders. Dementia, delirium, impaired mobility, and depression all can be causes of voiding symptoms and urinary incontinence regardless of the integrity of lower urinary tract mechanisms.

Changes in lower urinary tract function with age

Several lower urinary tract components and micturition functions undergo changes with aging (Resnick 1992). 'Uninhibited' bladder contractions, which can be difficult or impossible to forestall or interrupt, become more prevalent. The ability to postpone voiding with normal voluntary bladder contractions diminishes slightly. The total capacity of the bladder also decreases somewhat, which may make the need to void more frequent. Although older persons may need to urinate more frequently, the stream tends to be less strong and it takes longer to empty the bladder. The urinary flow rate decreases, possibly due to an age-related decrease in detrusor contractility. The bladder is also less likely to empty completely. The amount of urine remaining in the bladder after voiding, the post-void residual, tends to increase with age, although residuals larger than 50 ml should not be ascribed to age alone. Most elderly men develop prostatic hyperplasia, and in a subset the prostate enlargement can cause partial or complete obstruction of voiding. In women, low estrogen levels after the menopause may cause thinned and inflamed urethral and vaginal mucosa, a condition called atrophic urethritis and vaginitis. Atrophic and inflamed urethral tissues lack the mucosal seal necessary to keep the urethra closed. Not only may urine seep out but bacteria can enter and cause urinary tract infections. The associated irritation may trigger uninhibited contractions. In addition, both the bladder and the urethra become less compliant in older women.

Changes in the diurnal pattern of fluid excretion may also affect voiding and continence in elderly persons. Younger persons excrete the bulk of their daily ingested fluid during the hours they are awake. Many older persons, however, experience a shift in this pattern, with an increasing proportion—and sometimes the majority—of ingested fluid excreted later in the day and during the night. A number of factors create this change in fluid excretion. Age-related changes in renal filtration rate and/or circadian changes in atrial natriuretic peptide levels occur in healthy older adults. In many older persons, a variety of conditions and medications that predispose to pedal edema (e.g., congestive heart failure, venous insufficiency, ibuprofen) contribute to increased nocturnal urination: when edematous legs are elevated during sleep, the retained fluid re-enters the circulation and is subsequently excreted.

All of the above age-related changes may predispose an older person to develop voiding symptoms and incontinence. Many older persons who undergo these changes, however, remain asymptomatic and continent. Age-related factors do not necessarily cause voiding problems. Voiding symptoms and incontinence are not 'normal aging'.

Evaluation: the history and hypothesis formation

Once Mrs. Havel was safely settled on the exam table with the nursing assistant taking her vital signs, Dr. Anisimov went to find her preceptor, Dr. Sherman. He reminded her that problems with voiding dysfunction should be assessed in the same way as any other symptom, like chest pain: determine the onset, quality, severity, associated symptoms, and the exacerbating and ameliorating factors.

Reassured, Dr. Anisimov returned to the exam room and began: 'Mrs. Havel, is the problem you wanted to talk with me about bladder control?'

'Oh yes, Doctor. I am so embarrassed! I hate it! Can anything be done to help me?'

'Well, yes, Mrs. Havel, I do think we can help you. Can you tell me when you first noticed a problem controlling your urine?'

'It was over a year ago . . . maybe even two years.'

'How long ago did you have the stroke?'

'That was a long time ago. Maybe fifteen years.'

'What happens when you lose control of your urine . . . do you have a sudden urge to go that you can't control?'

'No . . . not at all. I leak when I cough.'

'How about when you laugh or bend over or lift things?'

'I don't do too much bending over or lifting, and I haven't laughed much since I got this problem.'

'Well, when you cough, does the urine squirt out the exact moment you cough or do you cough and then the urine leaks out several seconds later?'

'Oh, well, it comes out right when I cough.'

For all her nervousness at dealing with incontinence, Dr. Anisimov made a good start with her questions of Mrs. Havel. Dr. Anisimov brought up the topic of incontinence and offered reassurance so Mrs. Havel could relax, concentrate, and discuss the important details of her voiding symptoms. Patients may find voiding symptoms difficult to discuss. Older men may not feel comfortable talking about urination difficulties with younger women physicians, and older women may feel inhibited bringing up and discussing incontinence with younger male physicians. Patients may have had previous negative interactions with health care providers who were uncomfortable discussing voiding problems or who were nihilistic about their management. Expressing interest and a desire to help a person with a voiding problem is important when caring for older persons.

Exploring the onset of incontinence, its relationship to other illnesses, and the use of medications is crucial. As noted above, many older persons will have comorbid conditions outside of the lower urinary tract that may predispose to the development of voiding symptoms. As was the case with Mrs. Havel, however, such conditions may be comorbid but not causal; a stroke that occurred in the distant past is unlikely to be directly related to incontinence of more recent onset.

One of the main goals in taking the history from a patient with urinary incontinence is to tease out the details that will allow the determination of the clinical type of lower urinary tract dysfunction. There are four major clinical types of voiding dysfunction: urge incontinence (due to an 'overactive' bladder with uninhibited contractions), stress incontinence (due to sphincter insufficiency), overflow incontinence or incomplete bladder emptying (due to either outlet obstruction or detrusor weakness), and functional or transient incontinence (largely due to factors outside of the lower urinary tract). This classification is simple yet pathophysiologically descriptive. A common previous classification, 'neurogenic bladder', should be avoided; it is imprecise and can be used to describe situations in which the bladder may be either overactive or underactive. The causes of transient incontinence are summarized in Table 15.1; note that the causes can be recalled using the mnemonic DIAPPERS.

The type and quality of voiding symptoms provide important clues to the underlying cause of voiding dysfunction. The sudden onset of an overwhelming urge to void suggests uninhibited contractions. Patients may tell you that the urge comes on 'like a bolt out of the blue'. The key factor is the abruptness of the onset of the urge. Some individuals may be able to forestall a sudden urge. 'Holding' or forestalling an urge

Table 15.1 Causes of transient incontinence (mnemonic DIAPPERS)

D	Delirium/confusional state
I	Infection, urinary (symptomatic)
A	Atrophic urethritis/vaginitis
P	Pharmaceuticals
P	Psychological, especially depression
E	Excessive urine output (e.g., heart failure, hyperglycemia)
R	Restricted moblility
S	Stool impaction

Adapted from Resnick (1984).

depends on functions other than the detrusor, such as mentation, sphincter control, or the speed with which one can find and run to a toilet. Frequent precipitants of sudden urges to void include dishwashing or washing one's hands, stepping out into the cold, and even the sight of the garage or front door after returning from shopping.

Urine leakage with maneuvers that increase intraabdominal pressure—coughing, laughing, bending over, running, changing position—suggests stress incontinence. Typically, leakage will occur instantaneously with the increase in abdominal pressure. A delay between the provocative maneuver and the leakage, or the occurrence of a sudden urge before or coincident with the leakage, suggests that the maneuver provoked an uninhibited contraction, and that the patient has urge and not stress incontinence. That is why it is so important to ask if a patient notes leakage immediately with a sneeze or a cough or if the leakage is delayed a few seconds. This differentiation is especially important in older women in whom both types of incontinence are common.

Overflow incontinence symptoms may include continual or frequent leakage of small amounts of urine associated with a weak, hesitant, and/or interrupted urine stream. Occasionally, patients with overflow incontinence will have a sense of incomplete emptying. Such symptoms, however, are not as specific as urge and stress symptoms and they cannot differentiate between the two causes of overflow incontinence, outlet obstruction, and detrusor weakness.

Dr. Anisimov was feeling better about the case. Given the symptoms she had elicited, she was fairly sure that Mrs. Havel had stress incontinence. A quick look at Mrs. Havel's chart, however, brought up new dilemmas. Mrs. Havel had a long history of severe, deforming rheumatoid arthritis, punctuated with multiple joint replacements; a right middle cerebral artery stroke with a persistent dense left hemiparesis; hypertension; anemia; and insomnia. The medication list included enalapril, naproxen, hydrochlorothiazide, atenolol, diphenhydramine, trazodone, propoxyphene, iron, enemas, and estrogen.

How could she manage stress incontinence in such a complicated patient, especially since she thought she remembered that stress incontinence required surgical correction?

Dr. Anisimov paused for a moment. She recalled that functional and other factors could cause or exacerbate incontinence in the elderly. She mentally went down the checklist. Mrs. Havel did not seem delirious, she was attentive and quite oriented. Infection? Dr. Anisimov made a note to order a urinalysis, and asked Mrs. Havel about frequency, dysuria, and suprapubic discomfort, all of which she denied. Atrophic vaginitis was a possibility, despite the estrogen prescription. Medications could be part of the problem, although Dr. Anisimov was not sure how to begin to sort out which one might be the culprit. Motivation to stay dry was not a problem here. Excessive urine output? Dr. Anisimov could check for pedal edema, but beyond that she was uncertain as to what else to do. Restricted mobility was also an issue: a review of ADLs and IADLs revealed that Mrs. Havel could do little except feed herself and use the telephone. The rest of Mrs. Havel's care was provided by her daughter with assistance from the Visiting Nurse Association and a home health aide. Stool impaction? Dr. Anisimov asked Mrs. Havel about her bowel function and frequency of enema use. In addition, she asked about perineal sensation, previous urinary tract infections, and prolonged labor or difficult deliveries.

Transient incontinence often coexists with the other main types of dysfunction, and is especially common in the elderly, who are more likely to have comorbid conditions and take medications. Even in patients with definite lower urinary tract abnormalities, treatment of 'transient' incontinence factors alone can be sufficient to improve or cure voiding symptoms. A thorough investigation of all possible factors causing transient incontinence should be done in every older person with voiding problems.

The evaluation and treatment of the many causes of transient incontinence are relatively straightforward, although not always easy. A person who is incontinent and delirious needs attention to the cause of the delirium and appropriate steps taken to return the person to baseline cognition. Likewise, an individual who is incontinent in the setting of a urinary tract infection should have the infection treated. Some persons are incontinent because of impaired mobility. A careful functional assessment and proper institution of assistance and adaptive aids (walker, commode, raised toilet seats, etc.) may greatly lessen or eliminate episodes of urinary incontinence. Medications are among the most common causes of transient incontinence. Any medication that impairs cognition, mobility, fluid balance, bladder contraction, or sphincter function can affect continence. The potential effects of medication

Table 15.2 Effect of drugs on functions necessary for continence

Function necessary for continence	How drugs impair function	Examples
Cognition	Cause confusion, sedation, decrease motivation	Antipsychotics Benzodiazepines Antidepressants alcohol
Mobility	Induce rigidity Induce orthostatic hypotension Cause sedation	Antipsychotics Antihypertensives, nitrates, tricyclic antidepressants (See above for drugs that impair cognition)
Fluid balance	Cause fluid retention/pedal edema with increased nocturnal diuresis Induce excessive diuresis	Non-steroidal antiinflammatory agents Calcium channel blockers Diuretics, alcohol, caffeine
Bladder contraction	Impair contractility Cause fecal impaction	Anticholinergics, calcium channel blockers, narcotics Anticholinergics, calcium channel blockers, narcotics
Sphincter function	Prevent adequate relaxation Prevent adequate closure Increase stress maneuvers (cough)	Alpha-adrenergic agonists Alpha-adrenergic antagonists Antiotensin-converting enzyme inhibitors, nicotine

on these functions and examples of drugs from each category are given in Table 15.2. Many medications may impair several of these functions. As an example, haloperidol can produce confusion, cause muscle stiffness and make ambulation difficult, and, through its anticholinergic effects, impair detrusor contraction and cause stool impaction. It is easy to see why stopping one such medication, when possible, can profoundly improve voiding function!

Another common cause of transient incontinence, and one that can exacerbate any form of established incontinence, is excessive fluid output. Occasionally, the etiology of the increased output may be as simple as increased fluid input! The 'eight to ten glasses of water every day' recommended in some articles and diet plans can flood a small-sized elderly person. Some individuals may be drinking more in response to a dry mouth induced by anticholinergic medications such as tricyclic antidepressants. Excessive fluid output also can be a consequence of diuretic medications. There also may be a direct diuretic effect from common beverages such as coffee, tea, and alcohol. In other instances, excess fluid output is caused by edema fluid returning to the circulation at night when the person assumes a recumbent position. Examples of conditions where edema may accumulate include congestive heart failure, hypothyroidism, renal failure, severe protein malnutrition, and venous insufficiency. Some medications, such as non-steroidal antiinflammatory drugs and some calcium channel blockers, also can lead to the accumulation of edema fluid.

A voiding record or diary can demonstrate the existence of excessive output and provide important clues to its etiology. Ask the patient to record the time and volume of every episode of voiding over a 48-hour period, including the hours of sleep. The record should include the estimated volume of incontinent losses as well as all continent voids. If the patient is in a hospital or nursing home and unable to create a voiding diary, the staff can chart a patient's continence status (e.g., dry, damp, soaked) every two hours. The modal voided volume (i.e., the most frequently recorded volume of urine voided over the 48 hours of recording) gives an estimate of the functional bladder capacity; add this to the post-void residual to obtain an estimated total bladder capacity. Calculate the nocturnal output by adding the first morning void to the total volume voided (continent and incontinent) during the hours of sleep. This simple information can be very useful. For example, if the nocturnal output is 800 ml and the functional bladder capacity is 200 ml, then the patient has to void about four times each night (800/200 = 4), and the nocturia is driven by the high urine output. In such cases, search for causes of the diuresis such as pedal edema, congestive heart failure, or an alcoholic 'nightcap'. If, in reviewing the voiding record, it appears that incontinence tends to occur at the same time each day, check to see if it is related to a medication dosage or fluid intake.

Examination: physical examination and ancillary testing

The nursing assistant came in to assist Mrs. Havel in undressing while Dr. Anisimov quickly went to check again with her preceptor, Dr. Sherman.

'Dr. Sherman, on the exam I'll need some help with the pelvic.'

'No problem, Dr. Anisimov. But before you get to that part of the exam . . .'

'I know. I will do a good general exam and check for all the factors outside the lower urinary tract that can affect continence.'

Returning to the exam room Dr. Anisimov considered how working up incontinence was well within her primary care skills. She did not need to do extra training in gynecology or urology to get through this evaluation.

While the physical examination of older persons with voiding symptoms must concentrate on the genitourinary system, a comprehensive general exam is needed to evaluate the possibility of important contributing factors or serious comorbid conditions (e.g., previously undetected Parkinson's disease, a rectal mass). Table 15.3 provides a systematic review of important points on the physical exam in evaluating a person with voiding symptoms.

Dr. Anisimov completed the general physical exam. She found that Mrs. Havel had mild atrophic vaginitis in addition to the findings associated with her rheumatoid arthritis and previous stroke. She then asked Dr. Sherman's assistance in performing a urinary stress test. With Mrs. Havel in the lithotomy position, Dr. Sherman checked to see that her perineum was relaxed, and then instructed her to give a single, vigorous cough. A spurt of urine instantaneously flew out of the urethra. Dr. Anisimov was glad she was standing off to one side! Dr. Sherman pointed out that with a more mobile patient, he would have asked the patient to stand and cough. The examiner can then stand behind the patient, offering a modicum of privacy, and also make sure that the gluteal folds are not tightened, which would indicate inadequate perineal relaxation. If the patient is not adequately relaxed, the stress test can be falsely negative. Another reason for a false negative urinary stress test is an inadequate bladder volume (e.g., less than 100 ml).

Dr. Sherman then asked Mrs. Havel to void. With help from the nursing assistant Dr. Sherman and Dr. Anisimov moved Mrs. Havel into the bathroom, where a collecting 'hat' was placed in the toilet to catch and measure the voided urine. Mrs. Havel voided 230 ml. Next, they took Mrs. Havel back to the examining room and catheterized her bladder, finding a post-void residual (PVR) volume of 100 ml.

A witnessed void can be an extremely helpful part of the incontinence assessment. It provides information on the amount voided, offers an opportunity to check the PVR, and—from the ratio of the voided volume to the total bladder volume (voided volume plus any residual)—gives an estimate of the voiding efficiency. In addition, peak urine flow rate can be measured using special urine flow meters. The average flow rate can be calculated by dividing the voided volume by the voiding time in seconds. Measuring the PVR is relatively easy and only requires a simple sterile catheterization. An elevated PVR (greater than 50 ml in the elderly) can contribute to urinary frequency and urgency (an inadequately emptied bladder more quickly refills to the volume at which it tends to empty) as well as worsen stress incontinence (by increasing the usual bladder volume). A very large amount of residual urine in the bladder may lead to hydronephrosis and potential renal damage. While such extra testing may be time consuming, especially when one is first learning the techniques, they are well worth the effort.

Laboratory tests that should be checked in every patient include blood urea nitrogen; creatinine; and urinalysis (followed by culture if indicated). Additional tests to consider are glucose and calcium levels (when elevated, both of these cause a diuresis and may be associated with confusion) and vitamin B_{12} level. Urine cytology should be checked in any patient with hematuria, pelvic pain, or pain on urination.

Treatment

Dr. Anisimov sat down with Dr. Sherman as Mrs. Havel was getting dressed. 'I know that Mrs. Havel definitely has stress incontinence, atrophic vaginitis, and a relatively high PVR which can make the stress incontinence worse.'

'Good,' said Dr. Sherman, 'now could she also have excessive urine output?'

'I don't think that's the issue here, Dr. Sherman, but I will give her a voiding diary to check.'

'That's fine. Let's try to figure out why the PVR might be elevated.'

'Medications can do it . . .'

'Any suspects on her medication list?'

'She is taking diphenhydramine and propoxyphene.'

'Does she need them?'

'Well, let's see. The diphenhydramine is not a great anti-insomnia agent and she also has trazodone to take. I would like to see if she could use acetaminophen rather than propoxyphene for her pain.'

'What about the rectal exam? Impaction can impair bladder emptying, was she impacted?'

'No, there was a lot of stool in her vault and it was pretty black, but tested negative for occult blood.'

'Well, the dark stool is probably from the iron that she takes. Iron is very constipating, too. I wonder if she really needs iron. Iron deficiency anemia can be a difficult diagnosis to make without a bone marrow biopsy in someone with active rheumatoid arthritis. Has she had any iron studies?'

'Let me see . . . yes, six months ago the iron and the TIBC [total iron binding capacity] were both very low.'

'She could just have anemia of chronic disease related

Table 15.3 Key points in the physical examination of patients with voiding symptoms

General	What is the level of alertness?
	Obvious cognitive deficits?
	Is there an odor of urine?
Vital signs	Check for orthostatic hypotension that could impair mobility.
HEENT (head, eyes, ears, nose, and throat)	Limited cervical range of motion suggests cervical spondylosis—with secondary impingement of detrusor efferents in the spinal cord that could cause detrusor instability.
	To check for cervical spondylosis: Ask the patient to turn head to the left and then the right (lateral rotation) and then touch each ear to the respective shoulder (lateral flexion). Note the degree of turning with each maneuver. Significant cervical spondylosis is often accompanied by atrophy of the interossei muscles of the hands due to compression of roots C7–T1.
Back	Are there signs of occult dysraphism (incomplete spinal bifida) such as dimpling or a hair tuft at the base of the spinal cord? Is there evidence of previous laminectomy that may have involved spinal roots with detrusor efferents?
Chest	Auscultate lungs for evidence of congestive failure.
Heart	Check jugular venous distention and heart sounds for evidence of congestive failure.
Abdomen	Is the bladder distended? A distended bladder can be detected by palpation or 'scratching out' as is done to evaluate liver size.
Extremities	Is there pedal edema and, if present, how severe? Evidence of arthritis that could impair mobility?
Genital	*In women:* The vaginal mucosa should be checked for atrophy (thinning, paleness, loss of rugae, narrowing of the introitus by posterior synechia, vault stenosis) and any associated inflammation (redness, erosions, petechiae);
	Check for masses on the bimanual.
	To assess pelvic support, remove the top blade of the speculum. Hold the bottom blade firmly against the posterior vaginal wall. Ask the patient to cough (and be ready to avoid a spurt of urine!). A quick anterior forward swinging of the urethra indicates *urethral hypermobility*. Bulging of the anterior vaginal wall suggests a *cystocele*. Next, turn the speculum to support the anterior vaginal wall and have the patient cough. Posterior vaginal wall bulging with coughing is evidence of a *rectocele*.
	In uncircumcised men:
	Make sure that the foreskin can be retracted and that there is no phimosis or paraphimosis.
Rectal	Check for rectal masses and fecal impaction.
	Assess resting and volitional tone of anal sphincter: Ask the patient to squeeze the anus around examining finger, making sure that patient is not performing a Valsalva maneuver.
	Check integrity of sacral roots S2–S4 by testing anal 'wink' reflex and *bulbocavernosus* reflex. To test these reflexes: Anal wink: lightly scratch the perineal skin on each side of the anus, which should contract ('wink').
	Bulbocavernosus reflex: discuss the procedure with patient first. Lightly squeeze the tip of a man's glans penis or a woman's clitoris and look for the same anal contraction seen with the 'wink.'
	In men: Check prostate consistency (boggy? asymmetric firmness?) and whether there are any nodules present. Do not worry about estimating prostate size as the rectal exam is very inexact in this regard.
Neurological	Assess cognitive status and affect.
	Check motor strength and tone, especially with regard to mobility. Note any evidence of previous stroke.
	Check vibratory sense and peripheral sensation for possiblity of peripheral neuropathies (e.g., from diabetes or vitamin B_{12} deficiency—detrusor nerves may be affected by same process.

Table 15.4 Treatments for established incontinence

	Urge	Stress	Overflow
Behavioral	Bladder training Prompted voiding Biofeedback	Pelvic muscle exercises Biofeedback Electrical stimulation Weighted vaginal cones	For underactive detrusor: Crede maneuver Valsalva voiding
Pharmaceutical	Bladder relaxants: oxybutynin propantheline tricyclic anti-depressants calcium channel blockers	Alpha agonists: Phenylpropanolamine Estrogen	For obstruction: Alpha blockers Finasteride For underactive detrusor: Bechanecol (not reliable)
Surgical	Augmentation cystoplasty (rare)	Bladder neck suspension 'Sling' procedures Periurethral bulking injections Artificial sphincter	For obstruction: Prostatectomy Prostatic incision Urethral stents

to her rheumatoid arthritis. We can't be entirely sure without a bone marrow biopsy. However, given that her hematocrit is stable and not very low we can just stop the iron, follow her hematocrit, and see what happens. What other reasons could Mrs. Havel have for an elevated PVR? What about bladder outlet obstruction? I doubt it in a woman. But, in a man we would have to worry about benign prostatic hyperplasia or prostate cancer. Any reason why her detrusor shouldn't work? Did you find any neurological disease other than the previous stroke?'

'No, Dr. Sherman,' said Dr. Anisimov, 'but let me mention something else from the medication list. I was just thinking ... some people get a cough on enalapril, right?'

'Yes. When did Mrs. Havel start on the enalapril?'

'About two years ago, according to her chart.'

'And when did her incontinence start?'

'Hmmm ... about two years ago!'

'Dr. Anisimov, I think you are going to make this woman feel a whole lot better!'

While understanding and making a diagnosis of the lower urinary tract pathology causing voiding symptoms and incontinence is important, determination of any exacerbating factors or precipitating factors is often the most critical component in developing a treatment plan. Because age-related changes in lower urinary tract function—and other physiological systems as well—render the elderly more vulnerable to developing symptoms from additional insults such as medications, correction of these insults alone often can relieve symptoms. In Mrs. Havel's case, age-related atrophic urethritis and decreased urethral length impair urethral closure. When the additional insults of an increased bladder volume (from the elevated PVR) and a medication-induced cough, stress incontinence resulted. Surgical correction of urethral function is not needed, however, since the actual precipitants of incontinence were the high

PVR and the cough. As is often the case with problems in older persons, Mrs. Havel requires a multi-factorial approach to correct the precipitating factors: replacement of enalapril with another antihypertensive medication; an increased estrogen dosage to relieve the atrophic urethritis; and decreasing the PVR by stopping the diphenhydramine, propoxyphene, and iron and improving constipation by adding fiber to the diet. While Mrs. Havel has other conditions that may affect voiding function—notably impaired mobility and manual dexterity—it is clear that these deficits do not play a major role in her voiding dysfunction. In the actual case on which Mrs. Havel's story is based, continence was restored by the medication changes and the improvement in bowel function; surgery and improvement in mobility never had to be addressed. Her story underscores that the primary care provider has a pivotal role in a multi-factorial approach to voiding symptoms in the elderly that focuses on precipitating factors.

In some cases, correction of precipitating factors may not relieve symptoms sufficiently and additional treatment will be necessary. Table 15.4 outlines management strategies for specific types of urinary tract dysfunction. (For a more complete discussion of treatment, see UIGP 1992.) In all cases, treatment should be tailored to the individual. This will usually involve a stepped approach moving from least to more invasive types of therapy: for example, behavioral treatment should be tried before pharmacological, and both should be tried before surgery (UIGP 1992). Before considering surgical correction of incontinence in the elderly, both primary care providers and specialists should appraise the accuracy of the diagnosis, and weigh the potential risks of surgery against the desired gain. Specialized testing may be necessary to confirm diagnoses.

Catherine E. DuBeau

Prostate diseases in elderly men

Dr. Anisimov was now on her in-patient medicine rotation, and her second admission Tuesday evening seemed straightforward. Mr. Kelso, age 72, had come to the emergency room for evaluation of a hot, painful, and swollen right knee. The emergency room doctor had tapped the knee joint, examined the cloudy fluid obtained, and found birefringent urate crystals indicating acute gout. Since Mr. Kelso lived alone in a third floor walk-up apartment, and could not walk with his painful knee, he was admitted for treatment and pain control. He had few other medical problems; his only medications were occasional ibuprofen for osteoarthritis in his hips and knees, and furosemide for pedal edema. When asked if anything else was bothering him, he had bellowed 'No sirree fellas!' around the cigar clamped in his teeth. A retired barber, Mr. Kelso was a master of opinionated small talk, and regaled Dr. Anisimov and her team with stories that evening and again on ward rounds the next day.

Thursday morning, however, was an entirely different story. On prerounds, Dr. Anisimov found Mr. Kelso bleary eyed and cranky, with Dr. Smith, the cross-covering intern from the previous evening, at the bedside splashed with iodine disinfectant solution and surrounded by opened catheter trays. From the bits of story the two related as they continually interrupted each other, Dr. Anisimov pieced together the following sequence of events:

Around midnight, Mr. Kelso had called his nurse complaining that he could not pass his urine. The nurse had paged Dr. Smith. Dr. Smith felt that perhaps the problem was that Mr. Kelso had taken in too little fluid that day, and set him up with intravenous fluid. Despite the additional fluid, several hours later Mr. Kelso still could not pass any urine and now was writhing in pain. Dr. Smith then decided to insert a urethral catheter, but could not get it to pass into the bladder. Dr. Smith called the urology resident for assistance but found he was unavailable because of an emergency case in the operating room. Dr. Smith tried to catheterize Mr. Kelso with a smaller catheter, but to no avail. Now there was blood oozing from the tip of Mr. Kelso's penis, and he was refusing to let anyone near him again with 'one of those damn rubber hoses'. Dr. Anisimov was taken aback. What had happened to Mr. Kelso? Why did a complication occur in a man who otherwise had seemed so healthy?

The evaluation of voiding dysfunction in elderly men in general is approached as discussed in the case of Mrs. Havel, based on an understanding of age-related lower urinary tract changes and with a focus on possible precipitating factors. With men, however, the provider must add to the differential diagnosis the possibility of prostatic diseases. Benign prostatic hyperplasia (BPH) and prostate cancer both increase in prevalence with age. Nearly 80% of men by age 80 will have BPH. Approximately half of these men will have

secondary prostate enlargement, and approximately one-third will develop voiding symptoms (Berry et al. 1984). Other than skin cancers, adenocarcinoma of the prostate is the most common form of cancer in men, and the second leading cause of cancer death (Boring et al. 1994).

Although commonly assumed to cause symptoms by urethral blockage and bladder outlet obstruction, prostate diseases also become symptomatic through non-mechanical mechanisms. Contraction of the smooth muscle in the prostate and prostatic urethra is mediated by alpha-adrenergic receptors. These receptors are increased in BPH, and can be stimulated by exogenous sympathomimetics such as the decongestants found in many over-the-counter cold medications. Thus, an elderly man with BPH taking nose drops for allergies or a cold can find himself unable to void. BPH also is associated with changes in bladder function. Approximately two-thirds of men with BPH develop uninhibited bladder contractions; although the exact mechanism of this phenomenon is not clear, the causal association is suggested by the disappearance, in most cases, of uninhibited contractions following prostatectomy (Abrams 1985). The most feared effect of prolonged prostate enlargement is detrusor decompensation with subsequent urinary retention and possible renal insufficiency. Longitudinal natural history studies have shown, however, that the actual risk of urinary retention is quite small, less than 10% over five years. In addition, such studies—along with many placebo arms from BPH treatment trials—have also shown that prostatism voiding symptoms wax and wane, and are not inevitably progressive (DuBeau and Resnick 1992).

The most important fact about 'prostatism' symptoms, however, is that they are non-specific and can occur in the absence of a prostate! Hesitancy, nocturia, and slow stream are present in many elderly women because of age-related lower urinary tract changes, comorbid conditions, medications, or urinary tract pathophysiology. Despite the increased prevalence of BPH and prostate cancer with age, one should not assume that voiding symptoms in elderly men are necessarily due to prostate diseases. Age-related lower urinary tract changes (including BPH) render elderly men more vulnerable to precipitating factors outside of the lower urinary tract. Symptoms in an older men are more likely to be due to factors outside of the lower urinary tract.

How can we use this information to sort out what has happened to Mr. Kelso? Three points are helpful. First, one must recognize that Mr. Kelso has developed acute urinary retention, the new onset of an inability to pass urine. The fact that the retention is acute should not lead one to presume that Mr. Kelso previously was healthy and free of voiding symptoms. Like many older persons, he may have accepted a gradual onset of symptoms as a normal part of aging, never mentioned the symptoms to his physician, or was never asked by his doctor about voiding problems. Second, recall that incomplete bladder emptying has two possible causes, outlet obstruction and detrusor weakness. The fact that Mr. Kelso is male and that the covering intern had difficulty passing a catheter through the urethra does not imply that Mr. Kelso has outlet obstruction. Inability to pass a catheter often can be due to striated sphincter spasm from the discomfort of catheterization. Impaired detrusor contractility, however, may be a consequence of exogenous factors like bed rest and, especially, narcotic analgesics. Third, Mr. Kelso's plight finally was provoked by the rapid diuresis caused by the intravenous fluid bolus which overwhelmed the weakened detrusor and led to a quick and painful distention of the bladder.

At that point, the urology resident entered the room. 'Sorry, I was stuck in the operating room. I got the message that you were having a tough time passing a catheter in this gentleman. Mr. Kelso,' he said, quickly palpating the patient's very distended bladder, 'give me a minute and I think I can make you more comfortable.'

The urology resident asked for a new catheter tray with a larger catheter. He instructed Mr. Kelso to roll on his side and, after explaining what he was going to do, performed a digital rectal exam. 'Four plus enlarged prostate, no nodules, probably BPH.' Mr. Kelso was asked to return to lying on his back. With the liberal use of anesthetic jelly, a catheter was finally inserted, and 800 ml of urine immediately flowed out. The team decided to leave the catheter in for several days. Later that morning, Dr. Anisimov asked Mr. Kelso about his bowel function and ordered a bulk laxative. He was able to tolerate a decrease in his pain medications and began limited ambulation with the help of physical therapy.

The next day brought new developments. Dr. Anisimov found the urology resident back in Mr. Kelso's room. 'He needs a prostate biopsy to rule out cancer, his PSA [prostate specific antigen] is 63.'

As noted above, prostate cancer is a potential concern in elderly men with voiding symptoms. For further evaluation of these men—as is the case with any type of early cancer detection or screening program in the elderly—the first step is to determine whether the patient is a candidate for cancer treatment. If the elder is very frail or has extensive comorbid diseases that would preclude treatment, one can question the utility of early detection or screening. Age alone is a poor proxy for deciding whether or not to screen: functional status, frailty, and life expectancy are more dependent on the presence or absence of comorbid conditions than on chronological age. For example,

Mr. Kelso, although he is 72 years old, is in excellent general health, and, based on 1974 data, has an average life expectancy of 11 years (Catalona 1993). If Mr. Kelso were agreeable, an important caveat, he would be a candidate for either radical prostatectomy or radiation therapy, the treatments for potentially curable organ-confined prostate cancer.

In the past, the only method for prostate cancer detection was the digital rectal examination (DRE). DRE testing is subjective. Prostate size estimates have marked intra-observer variability, and correlate poorly with whether the prostate is obstructing the urethra (DuBeau and Resnick 1992). DRE is only fairly sensitive for cancer detection; nearly 70% of prostate cancers detected by DRE have already spread beyond the prostate (Catalona 1993). The advent of PSA testing, however, has radically changed prostate cancer screening. PSA is more sensitive than DRE, and widespread PSA testing has increased the numbers of prostate cancers detected nationwide. In two-thirds to three-quarters of asymptomatic men with cancer detected by PSA, the cancer will be organ confined (Catalona 1993).

The most commonly used PSA assays have a normal range of 0–4 ng/ml. While abnormal levels are associated with prostate cancer, other conditions can also elevate PSA, and result in a false positive test for cancer: BPH, prostatitis, prostate infection, and urinary retention. Because these conditions, especially BPH, are common in older men, 55%–80% of men with an elevated PSA will not have prostate cancer (Partin and Oesterling 1994). PSA levels also increase with age, largely as a function of increased prostate size caused by BPH. Indeed, about 40% of men over age 70 will have either an abnormal DRE or an elevated PSA (Partin and Oesterling 1994). The change in PSA over time, the PSA velocity, greater than 0.75 ng/ml/yr may offer greater specificity (Partin and Oesterling 1994). PSA levels greater than 20 ng/ml most likely indicate cancer, but such levels have also been found with benign conditions (Partin and Oesterling 1994). Mr. Kelso's PSA value of 63 in the setting of urinary retention should not lead to the assumption that he has cancer. In his case, postponing PSA testing until the retention had resolved and the effects of related prostate edema or infarction were allowed to subside may have increased the specificity of the test. A normal PSA does not mean that a man does not have prostate cancer. PSA fails to detect cancer in an appreciable number of cases. Among men with cancer confined to the prostate, approximately 35% will have a normal PSA level (Partin and Oesterling 1994).

PSA testing and prostate cancer screening in general is controversial because of the lack of definite data showing that detection and treatment of prostate cancer improves survival. Prostate cancer screening is best done when there has been sufficient education of the provider and the patient regarding the reliability of testing methods, the potential impact of cancer detection on morbidity and mortality, and the appropriateness of anticipated treatment for the individual patient.

The prostate biopsy was performed and Mr. Kelso given a follow-up appointment with the urologist. Mr. Kelso was discharged home with a visiting nurse to assist him with his bladder catheter. After five days at home, the catheter was removed and Mr. Kelso had no difficulty voiding on his own. After spontaneously voiding, a post-void residual of 50 ml was obtained.

Dr. Anisimov ran into Mr. Kelso in the hospital later the next week. He explained to her that he had gone to see the urologist for his follow-up appointment. He had been very upset about the PSA result and could not get it out of his mind that he might have cancer. The urologist had told Mr. Kelso that because of the episode of urinary retention, he probably could benefit from a transurethral prostatectomy, which also would yield more tissue that could be examined for cancer. Mr. Kelso was very reassured, however, by the urologist's explanation that the prostate biopsy done in the hospital showed only benign tissue and some areas of tissue infarction, with no cancer. Following the transurethral prostatectomy, the formal pathology report again described BPH with some areas of infarction and no evidence of malignancy.

Questions for further reflection

1. What are the causes of transient incontinence?
2. In persons with established incontinence, what factors are amenable to intervention to alleviate or end the amount of incontinence?

3. Discuss the potential benefits and pitfalls of screening for prostate cancer.

References

Abrams, P. (1985). Detrusor instability and outlet obstruction. *Neurourology and Urodynamics*, 4, 317–28.

Berry, S.J., Coffey, D.S., Walsh, P.C., and Ewing, L.L. (1984). The development of human benign prostatic hyperplasia with age. *Journal of Urology*, 132, 474–9.

Boring, C.C., Squires, T.S., Tong, T., and Montgomery, S. (1994). Cancer statistics, 1994. *CA: Cancer Journal for Clinicians*, 44, 7–26.

Catalona, W.J., Smith, D.S., Ratliff, T.L., and Basler, J.W. (1993). Detection of organ-confined prostate cancer is increased through prostate-specific antigen-based screening. *Journal of the American Medical Association*, 270, 948–54.

DuBeau, C.E. and Resnick, N.M. (1992). Diagnosis and management of benign prostatic hypertrophy. *Advances in Internal Medicine*, 37, 55–83.

Hu, T. (1990). Impact of urinary incontinence on health care costs. *Journal of the American Geriatrics Society*, 38, 292–5.

Mitteness, L.S. (1990). Knowledge and beliefs about urinary incontinence in adulthood and old age. *Journal of the American Geriatrics Society*, 38, 374–8.

Partin, A.W. and Oesterling, J.E. (1994). The clinical usefulness of prostate specific antigen: update 1994. *Journal of Urology*, 152, 1358–68.

Resnick, N.M. (1984). Urinary incontinence in the elderly. *Medical Grand Rounds*, 3, 281–90.

Resnick, N.M. (1992). Urinary incontinence in older adults. *Hospital Practice*, 27, 139–42; 147; 150.

UIGP (Urinary Incontinence Guideline Panel) (1992). *Urinary incontinence in adults: Clinical practice guideline.* AHCPR Pub. No. 92–0038. Agency for Health Care Policy and Research, Public Health Service, U.S. Department of Health and Human Services, Rockville, MD.

16

Pneumonia in the elderly

David F. Polakoff

It was a few minutes past 2 a.m. on a Tuesday morning in late January, and Dr. Robin Tomazewski was getting very tired. But she was quite gratified by what she had learned this night on-call. Dr. Tomazewski was a senior resident on an emergency department rotation, and had been assigned early in the evening to see Mr. Archie Santinga, an 83-year-old nursing home resident who had been transferred for evaluation and treatment of a high fever and lethargy. When she entered the treatment room, she found an elderly man, lying on a stretcher apparently oblivious to his surroundings. His breathing was rapid and somewhat labored, and his color was a bit dusky. He was able to answer most questions only with single word responses that were not very informative. Dr. Tomazewski had had to rely on the transfer forms from the nursing home for most of his history. They had been carefully completed, and revealed that Mr. Santinga had resided at the Atwood Manor Nursing Home for the past two years, having been admitted there five months after the death of his wife of 51 years. His problem list was long, and encompassed a total of 14 items. Among them were coronary artery disease, peripheral vascular disease, venous stasis, a number of surgical procedures in the remote past, gastroesophageal reflux with chronic heart burn, and modest impairment of memory. Despite this apparently long list, his primary nurse reported that on most days he was quite functional and independent. With only occasional reminders, he was capable of attending independently to his own hygiene and personal care. He also walked without assistance, was continent, and took his meals without any assistance. But on Saturday afternoon when his daughter and grandchildren had visited him, they had first noted that he wasn't 'quite himself'. He had been less attentive than usual to the grandchildrens' stories, and had shown little interest in the cookies that his daughter had brought from his favorite bakery. On Sunday, he had eaten very little. The nurse's notes also indicated that, quite uncharacteristically, he had napped on and off all day long. Finally, on Monday, he had been so tired, that he had been allowed to stay in bed. He slept most of the day, and the nursing staff found him somewhat difficult

to arouse. When he did wake, he did not speak with his usual coherence. His physician, Dr. Carter, had been called, and by the time he made rounds that afternoon, Mr. Santinga was breathing rapidly and had a fever of 103°F. The doctor had ordered that Mr. Santinga be transferred to the emergency department of the local hospital for evaluation of what he felt was probably pneumonia, and possibly sepsis.

Dr. Tomazewski had immediately recognized that Mr. Santinga was very ill. She conducted a quick physical examination, noting that despite his warmth, her patient had very dry skin that was pale but greyish in color. His cheeks were a bit sunken and his mouth was dry. The pulse was rapid and diminished in volume, but regular, and his breathing rapid (she counted 28 breaths per minute) and labored. When she placed her stethoscope on his chest, Dr. Tomazewski heard noisy rattling breath sounds (coarse râles) which she judged to be the result of secretions in the patient's pharynx. But as she continued to listen, she also thought she heard finer rales over the right lower lobe. She requested that a pulse oximeter be placed on Mr. Santinga's index finger, and that blood be sent off immediately to the laboratory for a complete blood count with differential, electrolytes, blood, urea, nitrogen (BUN), and creatinine. She also asked for a urine specimen for urinalysis with microscopic examination and culture, and that an attempt be made to obtain a suctioned sputum specimen for gram stain and culture. A chest X-ray was also to be performed in the emergency department. The first available result was the pulse oximetry, which showed mildly diminished oxygen saturation at 88%. On this basis, Dr. Tomazewski performed a radial artery puncture to obtain a specimen for arterial blood gas analysis. These results returned quickly, and showed that Mr. Santinga was indeed hypoxemic. The po_2 was 58 mmHg, with a normal pH and pco_2. The chest X-ray was performed next, and showed an ill-defined infiltrate involving the right lower lobe. A few minutes later, some of the other laboratory test results returned, and showed that Mr. Santinga had only a mild leukocytosis (10 800 white blood cells), but a marked leftward shift of the differential count, with 12% bands and 70% neutrophils. There was also confirmation of the volume depletion that Dr. Tomazewski had suspected: the serum sodium, BUN, and creatinine were all significantly elevated.

While there was still much to be done, Dr. Tomazewski was now confident that Mr. Santinga was indeed suffering from a right lower lobe pneumonia. She completed the necessary paperwork to admit him to the hospital, and commenced treatment with intravenous normal saline, supplemental oxygen delivered by facemask, and two antibiotics. All of this was accomplished in a manner that had become almost automatic over the past two and a half years of her residency. In that time she had seen countless patients very much like Mr. Santinga. Elderly people, some residing in nursing facilities, others in their own homes, often arrived in the emergency department in significant distress as a result of the rapid onset of symptoms associated with pneumonia. After finishing her work that evening, Dr. Tomazewski took the opportunity to think a bit more about this case. She realized that her treatment of Mr. Santinga and other patients like him was based mostly on what she had been taught about pneumonia as they occur in younger adults. But she also realized that there were some very notable differences between the younger and older adults whom she had treated for pneumonia. Since it was still early in the evening, and she felt driven by her curiosity, she decided (time and workload permitting) to spend the rest of the evening in the hospital's medical library reading about pneumonia in the elderly. The available sources included a computerized search of the literature, several relevant textbooks, and the article files maintained by the chief resident. She was fascinated by what she found, and decided to prepare a conference presentation for her colleagues on the topic.

While the advent of the antibiotic era of modern medicine has brought about vast changes in the management of pneumonia, this disease remains the fifth leading cause of death among Americans over the age of 75 (Fein and Nederman 1994). It is the most common fatal infection in the elderly, representing about half of all deaths caused by infectious diseases (Fein and Nederman 1994). The elderly are hospitalized for pneumonia at ten times the rates of young adults (Fein and Nederman 1994; Granton and Grossman 1993). The incidence of pneumonia is two to four times higher for elderly residing in nursing facilities than for elderly individuals residing in the community (Garibaldi et al. 1981; Granton and Grossman 1993). Viewed cross-sectionally, up to 2% of all nursing home residents may be suffering from pneumonia at any given time (Garibaldi et al. 1981). Mortality rates associated with pneumonia also rise dramatically with age, doubling from the fifth to the sixth decade of life, and reaching a level over 80% among those above 90 years of age (Fein and Nederman 1994; Garibaldi et al. 1981).

The ominous sound of this basic epidemiology alarmed Dr. Tomazewski, and spurred her to read on. She began to wonder what predisposed the elderly to such high attack rates and mortality rates from this disease. She decided to begin by reading about the changes that occur in the lungs as a part of usual, healthy aging to see how these might alter susceptibility to pneumonia.

A variety of structural and mechanical changes in the chest are seen with aging, even among the healthiest of older individuals. The ribs are subject to decalcification, and at the same time the costal cartilages calcify; there are frequently mild arthritic changes in the costovertebral joints. Some degree of dorsal kyphosis is also seen in about two-thirds of those over the age of 75. In sum, these changes result in a chest that is increasingly barrel-shaped and has an increased anteroposterior diameter. At one time this radiographic finding was commonly referred to as 'senile emphysema' because it is also seen in patients with emphysematous disease. This term has fallen into disfavor, however, because elderly individuals with these changes do not necessarily have any of the clinical features of emphysema.

The basic architecture of the lungs is preserved with normal aging. The dry weight changes little. The only alteration seen in the large conducting airways is calcification of the cartilaginous rings resulting in a slight increase in the anatomic dead space. The small bronchioles slightly change in the opposite direction, diminishing in diameter as they progressively lose the connective tissue tethering that helps to maintain their shape.

The tendency of the lungs to recoil inward is balanced by the opposing tendency of the chest wall to recoil outward. The system balances at its resting volume, called the functional residual capacity (FRC), a value that increases slightly with advancing age. There is a progressive age-related loss of lung elastic recoil after the second decade of life, which results in decreased lung compliance. These changes seem to be attributable to alterations in the location and orientation of elastin fibers in the lung parenchyma. While these changes are relatively small in magnitude, they are sufficient to result in moderate alterations in pulmonary function. They are most prominent at lower lung volumes, where the elderly become increasingly dependent on muscle effort (in the absence of elastic recoil) to achieve flow. With decreased compliance, the older person performs 20% more work to achieve a given level of ventilation than the does the younger adult. Fully 70% of this work goes toward moving the chest wall, as compared to 40% in the younger person. The impact of these changes in workload is mitigated slightly by the relatively greater abdominal contribution to breathing in the elderly; more than half of the work is done by the abdominal muscles, the remainder by those in the chest.

The structural and mechanical changes in the chest produce modest changes in some of the standard measurements of lung volume. The residual volume (that which remains in the lungs at the end of a maximal voluntary expiration) increases slightly with healthy aging, as does the vital capacity (the amount of air one can expire following a maximal inspiration). These two changes balance one another to result in a total lung capacity that is essentially unchanged with aging. Forced expiratory volume in one second (FEV1, the most commonly used clinical measurement of airflow) diminishes only slightly with healthy aging.

Dr. Tomazewski also began to wonder about the arterial blood gases she had obtained on Mr. Santinga. Almost all of the elderly patients she treated with pneumonia were hypoxic, as he was. She also knew from prior reading that the oxygen content of the blood diminishes with normal aging. But she knew little of the basis for this change.

Gas exchange in the lungs is dependent on close regional matching of ventilation and perfusion. In younger adults, the lung as a whole achieves such matching, although it is based on relative perfusion deficits in the upper lung zones and relative perfusion excess at the bases. This perfusion differential persists with aging. But ventilation, which in youth is relatively constant throughout the lungs, undergoes some regional changes with aging. The loss of connective tissue tethering that affects the smaller airways results in occasional closure of some airways during normal tidal breathing. The lung volume at which this occurs is referred to as the 'closing volume' or 'closing capacity' of the lungs. Closing volume rises with age after the fourth decade and by the sixth decade the closing volume exceeds the functional residual capacity. Some lung units may be closed during normal, tidal breathing. The effects of gravity dictate that these units are predominantly located in the lung bases, the areas which receive preferential perfusion. An exaggeration of the physiologic ventilation/perfusion mismatch and a decline in the normal partial pressure of oxygen (pO_2) are the end results of these changes. Such changes have been demonstrated in several large population studies of screened, healthy adults. It is consistently found that pO_2 falls, by about 4.5 mm Hg each decade, from a high of about 110 mmHg at age 20 (Sorbini *et al.* 1968).

Dr. Tomazewski thought for a minute about this. Based on this formula, Mr. Santinga should have a pO_2 of about 80 mmHg. While this is significantly lower than the levels seen in healthy younger adults, it still corresponded to an arterial oxygen saturation of about 95% (well within the normal range). One should not become hypoxemic on the basis of age alone.

Although normal aging changes do not cause hypoxemia, they leave the older individual with far less reserve capacity to buffer the impact of insults such as pneumonia.

Increases in closing volume are important for other reasons. The predominant impact of changes in this parameter is on the lower lung zones. When a person assumes a supine posture, the abdominal organs shift upwards, and exert pressure on the diaphragm. Total lung capacity (TLC) and functional residual capacity (FRC) are both decreased, and the gap between FRC and closing volume (CV) widens even further, resulting in another decrease in arterial oxygen concentration. A sick, older person whose oxygenation is in the lower range of normal (but quite adequate) when upright, may find himself or herself precariously close to desaturation when lying down.

As Dr. Tomazewski considered this information, many of her clinical experiences came into sharper focus. She thought of the many older patients she had treated during the years of her residency experience. Patients who had complained of symptoms clearly representing orthopnea, but who despite careful examination showed no evidence of congestive heart failure. Perhaps at night, when sleep led to decreases in ventilation, lying flat was the last straw that tipped them over into desaturation. No wonder they felt more comfortable sleeping on several pillows, or in a reclining chair. She also remembered the many older pneumonia patients she had cared for whose hypoxemia and requirements for supplemental oxygen improved when they were mobilized, even if that just meant spending a substantial portion of the day sitting up in a chair, or in bed.

In contrast, the arterial concentration of carbon dioxide does not change significantly with age. This concentration is determined by the balance between metabolic production of carbon dioxide and and the minute ventilation. The latter remains unchanged with advancing age, but alveolar ventilation decreases a bit because an increasing proportion of the constant minute ventilation goes to physiologic dead space. Elderly people typically have diminished muscle mass, producing less carbon dioxide. These two opposing effects balance each other, and the net result is no change in the pO_2.

She thought again about Mr. Santinga. The information she had uncovered so far gave her a much clearer understanding of why he had been hypoxic, despite a relatively modest area of infiltration that would have left a younger adult with a much higher oxygen saturation. But none of this explained why Mr. Santinga had developed pneumonia, nor why so many older individuals are afflicted with and die of such infections. What factors, she wondered, predispose the elderly to an increased incidence of pneumonia?

The lower respiratory system is protected by four levels of defenses. The epiglottis and upper airway serve as a mechanical barrier to prevent aspiration. While studies have shown diminished levels of ciliary function and mucociliary clearance, no direct connection has been made between these observations and the occurrence of disease.

The second level of defense consists of the gag, cough, and laryngeal reflexes. The intensity and responsiveness of these reflexes become attenuated with age, and decline even further with disease. For instance, the response to inhaled cough stimulants such as ammonia have been shown to be significantly lower in older individuals than in younger adults. These factors represent an increased risk for aspiration pneumonia, and perhaps even for other types of pneumonia. Such risk is further exaggerated by the presence of feeding tubes. The relative contribution of this factor has not been elucidated.

Humoral and cellular immunity represent the third and fourth levels of defense. Again, there are demonstrable declines in immune function with advancing age. Secretory immunoglobulin (IgA) on the nasal and respiratory mucosal surfaces functions as a major deterrant to viruses. The concentration of IgA in these locations has been shown to be lower in older than in younger persons. Similarly, cellular immune function (as measured, for example, by skin tests) declines somewhat with advancing age. But there is little direct evidence that such immune factors play a significant role in predisposing the elderly to pneumonia. Despite these measurable declines, there is no increase in the incidence of infections due to pathogens usually associated with compromised cellular immunity. Thus, while the absolute levels of function are diminished in the elderly compared to younger adults, they appear to be quite adequate, in the absence of other clinical factors such as therapy with steroid hormones or antineoplastic agents.

Dr. Tomazewski began to wonder about pathogens. She knew that the epidemiology of pneumonia varied considerably with host factors. Which pathogens were most common in older adults like Mr. Santinga? Her reading revealed that, for this purpose, elderly patients are classified according to the place where the pneumonia was acquired. The categories she found were community-acquired, nursing home-acquired, and hospital-acquired pneumonia.

The most common pathogen in community-acquired pneumonia remains *Streptococcus pneumoniae*, accounting for about half of such cases (Granton and Grossman 1993). In the pre-antibiotic era, streptococcal pneumonia was considered highly virulent and often took the life of debilitated and very ill older people. It still remains quite virulent, with a 20% mortality from bacteremic pneumonia, a figure which has not changed with the advent of antibiotics. In the elderly, mortality rates are much higher (Fein and Niederman 1994).

Haemophilus influenzae and *Moraxella catarrhalis* represent the two next most common pathogens in a community-dwelling elderly population. Both organisms commonly colonize the upper respiratory tracts of patients with chronic obstructive lung disease, and those of smokers. They are also frequent colonizers in the period following a bout with influenza.

Legionella pneumophila is a surprisingly common cause of pneumonia in community-dwelling elderly. Because it occurs both sporadically and in epidemic fashion, various surveys have found it accounting for as little as 1% and as much as 16% of all cases of pneumonia in elderly individuals. Both the attack rate and the mortality rate increase with age. Earlier studies had emphasized a specific, but somewhat atypical presentation of *Legionella*-caused pneumonia. This presentation included a non-productive cough, and a variety of extrapulmonary manifestations such as neurologic and gastrointestinal symptoms and renal insufficiency. More recent studies have shown that these presentations are no more common in *Legionella* pneumonia than in that caused by other pathogens.

Among various adult populations, the incidence of pneumonia caused by *Chlamydia pneumoniae* may be highest in the elderly. This was previously known as the TWAR agent, and is also responsible for a great deal of upper respiratory illness in the elderly. Although the disease is usually mild, recovery is frequently very slow. The diagnosis is usually made on clinical grounds, as effective diagnostic methods are not yet available.

Despite twenty years of public health programs aiming to vaccinate the high risk population, large portions of the community-dwelling elderly population remain susceptible to influenza pneumonia. Attack rates as well as morbidity and mortality rates remain very high. Influenza incidence is seasonal, with the annual peak occurring in late autumn and winter, and lasting into the spring. Primary influenza pneumonia is most often characterized by a relatively sudden onset of fever, cough, and dyspnea. The chest X-ray usually shows a diffuse interstitial infiltrate. Support is the primary aim of treatment, but other more specific methods are available. Amantadine or rimantadine may be helpful, and in severe cases, ribavirin may be tried. Bacterial superinfection following influenza pneumonia is quite common, and vigilance is necessary to determine whether it is necessary to add antibacterial therapy (Gross *et al.* 1988).

The spectrum of typical pathogens for nursing home-acquired pneumonia includes the list of those seen most commonly in the community, and a few others as well. Gram-negative bacilli are found more frequently colonizing the oropharynx in nursing home residents, and consequently account for a higher proportion of the pneumonia in this group. A variety of factors predict higher rates of colonization with this group of organisms, including chronic illness, inability to walk or perform activities of daily living, recent use of antibiotics, recent stay in a hospital, and urinary incontinence (Valenti *et al.* 1978). Coverage for gram-negative pathogens mandates a different choice of antibiotics for this population.

Staphylococcus aureus is a relatively uncommon cause of community-acquired pneumonia. But among nursing home-acquired and hospital-acquired cases, it rises in importance, causing 10–30% of all cases among the elderly in these settings (Fein and Niederman 1994; Gross *et al.* 1988). It is also the second most common cause of post-influenza pneumonia. The appearance on chest X-ray is more that of a bronchopneumonia, without findings of segmental or lobar consolidation.

Respiratory syncytial virus (RSV) is frequently thought of as an infection affecting school-age children. But it also causes outbreaks of rapidly spreading disease in nursing homes. A variety of clinical syndromes are seen, including a flu-like syndrome, bronchitis, and pneumonia characterized by high fever and dyspnea. Epidemics in the nursing home setting show a peak incidence in the early spring, a few months behind the midwinter peak seen in school-age children. Recent exposure to young children (such as visiting grandchildren) represents a risk factor for the occurrence of this disease. Nursing home outbreaks should be treated with population control measures, including cohorting of patients, contact isolation, and strict handwashing procedures. Therapy is again primarily supportive, but aerosolized ribavirin is used in patients with severe disease.

Hospital-acquired or nosocomial pneumonia is caused by the same list of pathogens as nursing home-acquired pneumonia, with the addition of anaerobic organisms. The most important additional consideration in this group of patients is antibiotic resistance. Isolation of specific pathogens, and sensitivity testing assume added significance.

Dr. Tomazewski found herself feeling reassured. In choosing the initial antibiotic regimen for Mr. Santinga, she had given consideration to the correct list of potential pathogens. She had chosen ceftriaxone and erythromycin as initial antibiotics, providing good coverage for the list of organisms which she now knew would likely include the one responsible for her patient's pneumonia.

In her reading she had run across a quote from Dr. William Osler, attributed to *Principles and practices of medicine*, his classic 1892 textbook:

'In old age, pneumonia may be latent, coming on without chill. The cough and expectoration is slight, physical signs ill defined and changeable and the constitutional symptoms out of all proportion ... Of fever, there may be none. Fever is higher in healthy adults than in old persons ... [In] senile pneumonia the temperature may be low and yet brain symptoms are very pronounced.'

Dr. Tomazewski's thoughts shifted to the description of Mr. Santinga in the notes from his nursing home, in the days preceding the transfer to the hospital. His presentation, until the time of transfer, had been much more subtle than that which she was accustomed to seeing in younger patients. There had been very few symptoms directing attention to the respiratory tract as the source of his trouble. She decided to continue her reading with attention to the presentation of disease in older adults.

More recent observations confirm those made by Dr. William Osler a century ago. The elderly are more likely to present atypically with acute infections, including pneumonia. While many will indeed display the classic signs of fever, productive cough and leukocytosis, a higher proportion (than in younger adults) will lack one or even all of these signs. Among elderly patients, those most likely to present in such an atypical fashion are those with cognitive impairments (who are often unable to communicate accurately any symptoms that they may have experienced), those with functional impairments in their ability to perform basic activities of daily living, and those of very advanced age. As many as one in three older adults with pneumonia will fail to demonstrate either fever or leukocytosis. Even larger numbers will not complain of cough, sputum production or chest discomfort. Tachypnea may be the only sign associated with the respiratory system. The initial chest X-ray may not show any evidence of infiltrate, or if one is present, it may be described as having an 'atypical' appearance. More than half of elderly patients with pneumococcal pneumonia demonstrate a radiographic pattern of bronchopneumonia, rather than the classically described lobar infiltration. The diagnosis of pneumonia may also be obscured by the presence of other comorbid conditions, such as congestive heart failure and chronic obstructive pulmonary disease, producing respiratory symptoms, an abnormal chest X-ray and hypoxemia.

Unfortunately, the absence of classical features is not helpful in arriving at a diagnosis of pneumonia in an elderly person. The presentation most often includes non-specific features, as seen in Mr. Santinga. Patients will often lose their appetite, even refusing food and liquids entirely for several days. As a result of this diminished oral intake, significant dehydration often occurs before the pneumonia is recognized. For this reason close attention should be paid to the oral intake and urine output of elderly patients, particularly those residing in long-term care facilities. By the time the underlying infection and resultant dehydration are recognized, they are often at an advanced stage where this secondary condition poses a hazard as great as the primary pneumonia. There may also be an abrupt decline in the individual's cognition, ranging from mild irritability and forgetfulness, to frank delirium with alterations in level of consciousness and hallucinations. Abrupt changes in cognition always deserve close investigation, as they often serve as the harbinger of incipient acute illness.

Therapy

Optimal therapy is always guided by culture and sensitivity results, and accordingly an effort should always be made to obtain a sputum specimen for gram stain, culture, and sensitivity. Oropharyngeal contamination, along with the fastidious constitution of some of the organisms, results in a relatively low accuracy for such cultures; but an effort should be made to obtain sputum, nonetheless. If the gram stain reveals a good specimen (more than 25 polymorphonuclear leukocytes and less than 10 epithelial cells per high power microscopic field), and one predominant organism, then the accuracy is likely to be quite high and should be used to guide treatment.

Several factors must be considered in choosing an empiric therapeutic regimen where gram stain results cannot be relied upon. Gram-negative flora must be well covered if the infection is nosocomially acquired, if the patient resides in a long-term care facility, or if he or she suffers from an impaired functional status or advanced frailty (Mylotte *et al.* 1994). Staphylococcal coverage is also essential in residents of long-term care facilities, as well as in patients whose pneumonia follows an episode of influenza (Gross *et al.* 1988; Mylotte *et al.* 1994). Patients who have poor dentition require antibiotics providing adequate coverage for anaerobic organisms, and those with specific immunocompromise (treatment with corticosteroids or other immunosuppressive agents) should be given consideration of coverage for opportunistic infections (such as *Nocardia* and *Pneumocystis*). Smokers and those suffering from chronic obstructive lung disease have higher rates of infection with *Streptococcus pneumoniae* and *Hemophilus influenzae*. Consideration of *Legionella pneumophilia* depends on the degree to which it has been found to be endemic in the geographic region where the patient resides. Finally, patient-specific issues such as renal and hepatic function, other medications and drug allergies must be considered in choosing an antibiotic regimen.

With these considerations in mind, the initial regimen for a patient such as Mr. Santinga, who lives in a nursing home but who was quite functional and ambulatory prior to his acute illness, should include coverage for *Streptococcus pneumoniae, Hemophilus influenzae, Neisseria catarrhalis,* and *Chlamydia pneumoniae*. Coverage for *L. pneumophila* must be considered a variable factor, depending on local endemic levels. For these organisms, a simple regimen would include either cefuroxime or ceftriaxone, with the addition of erythromycin (ATSCC 1993). This initial choice would be adjusted if culture results demonstrated a singular pathogen that was either highly sensitive to a single agent, or not adequately sensitive to the initial choice, or if the patient failed to respond as expected to initial therapy. Had Mr. Santinga been more debilitated, additional coverage would have been required for *Staphylococcus aureus*. In this instance, the empiric choice of drugs would likely have included nafcillin or oxacillin, with the addition of a second agent such as either ceftazidime or aztreonam to provide coverage for gram negative organisms. A growing number of institutions around the world are suffering from widespread problems with methicillin resistant *Staph. aureus*. Local experts must be consulted in this instance to discuss precautions and containment policies, as well as to choose an antibiotic regimen appropriate to the circumstances.

Pneumonia in the older individual cannot, however, be treated with antibiotics alone. Since such patients are almost invariably dehydrated at the time of presentation, vigorous but careful rehydration is essential, along with close monitoring of renal function and electrolytes, and frequent reassessment of volume status. The associated delirium and general level of acute illness frequently preclude adequate oral rehydration, and another method must usually be chosen. The choices, depending on location and available resources, include administration of fluids by intravenous infusion, nasogastric tube, or hypodermoclysis.

Supplemental oxygen is usually required for reasons previously outlined, and initially it is wise to monitor oxygen saturation to assure adequate supplementation. It is also prudent to check arterial blood gases at least once after beginning supplemental oxygen to ensure that the patient has not begun to retain carbon dioxide. The effect of posture on oxygenation in older adults suggests that elderly patients should be mobilized as soon as they are capable; this will also have salutory effects on their general sense of well-being, and possibly on any delirium that may accompany the pneumonia.

Other elements of treatment will depend on the patient's symptoms and comorbid conditions. Those with chronic obstructive lung disease or asthma may require the continuation or institution of bronchodilator therapy, and may also require inhaled or systemic corticosteroids if bronchospasm becomes prominent. Sputum production may require management with respiratory therapy techniques such as chest physiotherapy and postural drainage.

Prevention of pneumonia

The most important preventive measure, on both an individual and a population basis, is the annual administration of influenza vaccine in the autumn. This vaccine has a moderately high efficacy in preventing infection with influenza virus (Centers for Disease Control 1991). But even in those who do become infected, it appears to attenuate the severity of the ensuing illness, and to lower the incidence of complicating illnesses such as primary influenza pneumonia and secondary viral and bacterial pneumonia. The well-described phenomena of antigenic shift and drift mandate that vaccination be repeated each year, prior to the onset of the winter influenza (flu) season (CDC 1991).

The other vaccine to be considered in the elderly is the polyvalent vaccine against *Strep. pneumoniae*. The efficacy of this vaccine has been the subject of some debate, but even the lowest estimates suggest a rate of effectiveness greater than one-half, and approaching two-thirds of all immunocompetent elderly individuals vaccinated (CDC 1991; Shapiro *et al.* 1991). Because pneumococcal pneumonia has a very high mortality rate in the first two to three days (even when treated with appropriate antibiotics), it is wise to provide this vaccination to all elderly individuals. The accepted indications for vaccination include age 65 and over, asplenia, immunocompromise, renal, hepatic, lung or cardiac disease, and diabetes mellitus. Some have argued that because the response rate to the vaccine is higher, vaccination should begin at a younger age, such as 55.

Dr. Tomazewski had now been reading for several hours, and the sun was beginning to rise. Although she found her reading fascinating, she was being summoned back to the emergency department to see a patient who had just arrived with chest pain. She knew that she would be busy until her night shift was complete, but promised herself she would return to this topic at the next available opportunity.

Fourteen hours had passed. Dr. Tomazewski had completed her night's work, had been home to sleep, and now had a few hours remaining before the next night's work would begin. She thought again about Mr. Santinga, and

wondered how he was doing. She also contemplated his presentation again, and recalled a lecture on tuberculosis that she had attended earlier in her training. She recalled that nursing home patients were considered to be at high risk for infection with tuberculosis, and that these institutions were also considered to represent reservoirs for this disease. This looked like a good avenue along which to continue her case-based learning related to Mr. Santinga.

Up to a quarter of all reported cases of tuberculosis in the United States occur in elderly individuals, and elderly nursing home residents suffer tuberculosis incidence rates that are at least twice as high as their community-dwelling counterparts (Stead *et al.* 1985; Stead 1981). There are several reasons for these extraordinarily high incidence rates. The elderly have higher rates of both age and disease-related immune dysfunction than younger adults. Epidemiologic studies have revealed that, when the current cohort of elderly were young adults (in the 1930s and 1940s), the majority of the American adult population was exposed to tuberculosis. Since tuberculosis was at that time under-recognized, and treatment with modern effective therapies was not available, many of these individuals were either never treated or incompletely treated. Consequently, this generation represents an enormous reservoir of dormant tuberculosis, awaiting the opportune circumstances for reactivation. Finally, nursing homes and other long-term care facilities represent precisely the sort of closed environment in which tuberculosis has been demonstrated to spread rapidly and in epidemic fashion (Stead *et al.* 1985; Stead 1981).

For these reasons, active skin-testing programs are considered essential for all chronic care facilities (Stead and To 1987). Prevalence surveys have revealed that 10–15% of all patients are PPD (purified protein derivative) skin test-positive at the time of admission to nursing homes (Stead *et al.* 1985). Even more alarming is the fact that cross-sectional surveys of nursing home populations have shown that 30–40% of all residents exhibit positive PPD skin tests (Stead *et al.* 1985; Stead 1981). This suggests very high nosocomial transmission rates. Conversion rates in nursing homes have been shown to be as high as 3.5–5% of the previously PPD negative population (Stead *et al.* 1985).

Skin testing should be performed for several groups of elderly persons, in addition to those in whom active tuberculous disease is suspected. These groups include individuals who have had close contact with patients diagnosed with active disease, those with comorbid conditions that elevate the risk of developing active disease (poor nutrition, immunocompromise as a result of disease or concomitant therapy, hematologic and lymphoreticular diseases, diabetes mellitus, silicosis, end-stage renal disease, and gastrectomy, among others), and those at increased risk for exposure to tuberculosis. The latter is the category most applicable to residents of nursing homes. It also should not be overlooked that while HIV (human immunodeficiency virus) disease is commonly thought of as a condition of young people, there are a significant and growing number of afflicted elderly individuals who are at high risk for tuberculosis and should be screened with skin testing.

The most recent guidelines from the Centers for Disease Control (CDC) suggest the use of three different cut-off values for interpretation of the standard 5-tuberculin unit purified protein derivative (5–TU PPD) (CDC 1990). The choice of cut-off value depends on the individual patient's risk of developing active disease should they become infected, and the probability of exposure to tuberculosis. The most sensitive cut-off, 5 mm of induration at the site of intradermal testing, applies to those at the highest risk of exposure and development of disease. This includes recent contacts of patients with documented cases of pulmonary tuberculosis, HIV-infected individuals, and those whose chest X-ray shows evidence of old untreated infection with tuberculosis (parenchymal scarring). The 10 mm cut-off, representing intermediate sensitivity, applies to persons whose risk of exposure is elevated. This includes residents of nursing homes and other long-term care facilities, as well as health care workers and those originating in areas of world where tuberculosis is endemic. Those with comorbid conditions exposing them to increased risk of development of active disease, if exposed, are also subject to this intermediate cut-off. All other persons are subject to the higher cut-off of 15 mm of induration to be considered 'PPD positive'.

Patients with unknown or negative PPD status at the time of admission to a nursing home, or at the time that another risk factor for the development of tuberculosis is identified, should undergo PPD testing with an initial 5–TU test, using the two-step procedure to define a baseline status (CDC 1990). If the initial test is negative, it should be repeated, in identical fashion, after a one to three week delay to determine whether a positive reaction can be elicited by recall of a T-cell response (or boosting reaction). Such a reaction indicates tuberculosis infection in the remote past, and should be considered full evidence of PPD positivity. No subsequent PPD testing should be performed in individuals found to be positive after the first or second step of this procedure. Nursing home residents with negative initial tests should be retested annually to detect those with recent infection (CDC 1990).

Tuberculosis well-deserved reputation for complexity arises in part from its ability to present in several different ways. All of these presentations may be seen in the elderly. The classic features include a persistent or chronic cough, fever, night sweats, and weight loss. *Reactivation* of an old and previously dormant infection with tuberculosis is more common in the elderly than in any other group, and usually arises from healed lesions in the apex of the lung. Cavitation of these lesions is a common, but not an essential feature. *Primary tuberculosis* often goes unrecognized in nursing home patients because the illness may be mild and may subside without specific treatment. However, when it is progressive it often results in infiltrates in the middle and lower lobes that may be mistaken for common bacterial pneumonia. *Tuberculous bronchitis* or *endobronchial tuberculosis* manifests as segmental atelectasis and hemoptysis, and may be mistakenly identified as malignant disease. *Miliary tuberculosis* is also more common in the elderly than in any other age group, may occur with either recent infection or reactivation, and is caused by the rupture of a caseating lesion into the bloodstream.

Diagnosis of tuberculosis requires the demonstration of acid-fast bacilli in a stained sputum smear. In a patient who is not coughing, and who cannot be stimulated to produce a sputum specimen, bronchoscopy with bronchoalveolar lavage may be required to achieve a diagnosis.

Treatment of tuberculosis has become considerably more complex in recent years with the widespread emergence of drug resistance. Fortunately, most isolates from elderly patients are still drug-sensitive. The major exception is seen in older individuals who were previously given antituberculous therapy. In locales where the prevalence of isoniazid resistance is thought to be less than 4%, a three-drug initial treatment regimen is recommended. If the prevalence of resistance is higher than 4%, a four-drug initial regimen should be chosen (ATS 1994; CDC 1993). In a nursing home with an epidemic of tuberculosis and the isolation of resistant strains, it is necessary to provide initial treatment with five to six drugs. This almost invariably mandates expert consultation. The addition of the third drug in the treatment of sensitive isolates has allowed for a shortened course of therapy from the previously standard duration of nine months to six months, without loss of effectiveness or significant increase in toxicity (ATS 1994; CDC 1993). The predominant form of toxicity seen with all regimens is drug-induced hepatitis, and all patients should be monitored monthly during the course of treatment for loss of appetite, nausea and vomiting, and abdominal pain, as

Table 16.1 Antituberculosis drugs: dosage and frequency

Drug	Daily dosage	Twice-weekly dosage
Isoniazid (INH)	5 mg/kg (usually 300 mg)	15 mg/kg (usually 900 mg)
Rifampin (RIF)	10 mg/kg (usually 450–600 mg)	10 mg/kg (usually 450–600 mg)
Pyrazinamide (PZA)	15–30 mg/kg (usually 2 g)	45–70 mg/kg (usually 3–4 g)
Ethambutol (EMB)	15–25 mg/kg (up to 2.5 g)	50 mg/kg (up to 2.5 g)
Streptomycin (SM)	10–15 mg/kg (usually 0.5–1 g)	20–30 mg/kg (usually 1–1.5 g)

Adapted from CDC (1993).

well as abnormalities in liver function tests (Stead *et al.* 1987).

It is thought that the recent increase in drug-resistance among tuberculosis isolates is related in large measure to noncompliance, or incomplete compliance with prescribed treatment regimens. For this reason, the CDC began in 1993 to recommend direct observation of treatment for all patients (CDC 1993). All patients should be carefully questioned regarding prior drug therapy for tuberculosis, and those previously treated should be started on a regimen of at least four agents untill susceptibility test results are available. Drug susceptibility testing should now be performed on all patients with documented tuberculosis, and should be repeated in those patients whose sputum fails to convert to negative after three months of treatment.

Initial treatment refers to that given until susceptibility testing is complete on the specific isolate from the patient in question. Once this is accomplished, treatment may be tailored to these susceptibilities. Accepted regimens and dosing are summarized in the Tables 16.1 and 16.2.

The following modes of administration are considered equally efficacious options:

1. All drugs given daily for the entire six months.

2. All drugs given daily for the first two months, followed

Table 16.2 Initial antituberculosis drug regimens

Two-drug regimen:
INH for 9 months
RIF for 9 months

Three-drug regimen
INH for 6 months
RIF for 6 months
PZA for first 2 months

Four-drug Regimen:
INH for 6 months
RIF for 6 months
PZA for 2 months
EMB *or* SM for 2 months

Adapted from CDC (1993).

by twice or three times weekly for the remaining four months, with direct observation of therapy.

3. (a) four drugs given daily for two weeks, then
   (b) four drugs given three times per week with direct observation for six weeks,
   (c) four drugs given twice per week with direct observation for four months.

4. Four drugs given three times per week with direct observation for six months.

Two days later Dr. Tomazewski arrived half an hour early for her night on duty in the emergency department, and decided to visit Mr. Santinga on the medical ward where he had now been for four days. She found him sitting up in a chair in his room, looking alert, and just finishing his dinner. His daughter was sitting across from him. Dr. Tomazewski introduced herself, and Mr. Santinga indicated that he did not recall meeting her before. In striking contrast to their first encounter, Dr. Tomazewski found him to be quite lucid about all matters except the days of his acute illness. He recalled having felt poorly for several days at Atwood Manor, but did not recall being transported or admitted to the hospital. He was now feeling reasonably well, he said. But on more specific questioning, he admitted to being very tired, and his usual appetite had still not returned. He was also still coughing frequently, and was producing purulent-appearing sputum. When he coughed, the right side of his chest ached. When not coughing, however, he was able to speak easily and fluently, without any apparent dyspnea. He still wore a nasal oxygen cannula.

Dr. Tomazewski expressed her pleasure at seeing Mr. Santinga so much improved since their first meeting, and proceeded to the nurses' station where she found the patient's hospital record and read about the course of his illness and treatment over the first four days in the hospital. His rehydration had been accomplished without complication, and had resulted in normalization of his electrolytes within 36 hours of admission. However, over the same interval he had become more alert, but very confused and agitated. This had been recognized as a delirium associated with his acute illness, and for this he had been treated only in supportive fashion. No sedatives were employed, and he had not been physically restrained. By the end of the third day his agitation had resolved, but he had remained extremely fatigued and somewhat lethargic. His fever had continued for two days following the initiation of antibiotic therapy, and had been suppressed with antipyretic medication. But after 48 hours he had become afebrile. His white blood cell count had also returned to the normal range after four days. A follow-up chest X-ray was taken on the fourth day, and confirmed the same infiltrate seen on admission, which had become more clearly defined , but had not yet begun to resolve, and had also not progressed to involve other lung zones. He remained free of complicating congestive heart failure. His oxygenation had also improved somewhat, allowing a change from a facemask to the nasal cannula for his oxygen supplementation. His sputum culture had grown only mixed oral flora, and the blood cultures drawn at the time of admission had not shown any growth. So the choice of initial empiric antibiotic therapy had been appropriate, and he had indeed not been septic.

The team of physicians and nurses caring for Mr Santinga had consulted extensively with both his family, and his caregivers in the nursing home. They had asked questions about his usual daily habits, abilities, and preferences. They had also discovered that Atwood Manor did indeed screen all residents for tuberculosis on admission and annually thereafter. Mr. Santinga had always tested negative, with positive controls indicating adequate immune function to mount a positive skin test in the presence of exposure. His last test had been just six weeks earlier. On this basis it was concluded that tuberculosis was an unlikely cause of the present acute illness.

With improvement in Mr. Santinga oral intake on the fourth hospital day, plans were in place to switch him to oral antibiotic therapy on the fifth day. If a day of observation demonstrated that this was well tolerated, he would be returned to Atwood Manor Nursing Home after five full days in the hospital. It was felt that he would return to his baseline level of function more rapidly in the familiar environment that had become his home.

Dr. Tomazewski felt gratified that despite her perceived knowledge deficit in this area she had been able to establish an appropriate and effective course of treatment for Mr. Santinga. She was also pleased with his prompt return to good health, and was confident that the remaining difficulties would also resolve in the weeks to come.

Questions for further reflection

1. Why does arterial oxygenation decline with increasing age?
2. What are the most common pathogens that cause pneumonia in an older population?
3. Discuss methods of screening for tuberculosis in elderly persons. Does this differ depending on the setting (e.g., nursing home vs living at home)?

References

ATSCC (American Thoracic Society Consensus Committee) (1993). Guidelines for the initial management of adults with community-acquired pneumonia: Diagnosis, assessment of severity and initial antimicrobial therapy. *American Review of Respiratory Disease*, 148, 1418–26.

ATSCC (American Thoracic Society) (1994). Treatment of tuberculosis and tuberculsosis infection in adults and children. *American Journal Respiratory Critical Care Medicine*, 149, 1359–74.

CDC (Centers for Disease Control) (1990). Prevention and control of tuberculosis in facilities providing long-term care to the elderly. *Morbidity and Mortality Weekly Report*, 39 (RR-10), 7–20.

CDC (Centers for Disease Control) (1991). Update on adult immunization. *Morbidity and Mortality Weekly Report*, 40, (RR-12), 1.

CDC (Centers for Disease Control) (1993). Initial therapy for tuberculosis in the era of multidrug resistance. Recommendations of the advisory council for the elimination of tuberculosis. *Morbidity and Mortality Weekly Report*, 42 (RR-7), 1–9.

Fein, A.M. and Niederman, M.S. (1994). Severe pneumonia in the elderly. *Clinical Geriatric Medicine*, 10, 121–43.

Garibaldi, R.A., Brodine, S., and Matsumiya, S. (1981). Infections among patients in nursing homes: polices, prevalence and problems. *New England Journal of Medicine*, 305, 731–5.

Granton J.T. and Grossman, R.F. (1993). Community-acquired pneumonia in the elderly patient. *Clinics in Chest Medicine*, 14, 537–53.

Gross, P.A., Rodstein, M., LaMontagne, J.R., Kaslow, R.A., Saah, A.J., Wallenstein, S., et al. (1988). Epidemiology of acute respiratory illness during an influenza outbreak in a nursing home. *Archives of Internal Medicine*, 148, 559–61.fi(#fi

Mylotte, J.M., Ksinzed, S., and Bentley, D.W. (1994). Rational approach to the antibiotic treatment of pneumonia in the elderly. *Drugs and Aging*, 4, 21–33.

Shapiro, E.D., Berg, A.T., Austrian, R., Schroeder, D., Parcells, V., Margolis, A., et al. (1991). The protective efficacy of polyvalent pneumococcal polysaccharide vaccine. *New England Journal of Medicine*, 325, 1453–60.

Sorbini, C.A., Grassi, V., Solinas, E., and Muiesan, G. (1968). Arterial oxygen tension in relation to age in healthy subjects. *Respiration*, 25, 3–13.

Stead, W.W. (1981). Tuberculosis among elderly persons: An outbreak in a nursing home. *Annals of Internal Medicine*, 94, 606–10.

Stead, W.W. and To, T. (1987). The significance of the tuberculin skin test in elderly persons. *Annals of Internal Medicine*, 107, 837–42.

Stead, W.W., Lofgren, J.P., Warren, E., and Thomas, C. (1985). Tuberculosis as an endemic and nosocomial infection among the elderly in nursing homes. *New England Journal of Medicine*, 312, 1483–7.

Stead, W.W., To, T., Harrison, R.W., and Abraham, J.H. III. (1987). Benefit–risk consideration in preventive treatment for tuberculosis in elderly persons. *Annals of Internal Medicine*, 107, 843–5.

Valenti, W.M., Trudell, R.G., and Bentley, D.W. (1978). Factors predisposing to oropharyngeal colonization with gram-negative bacilli in the aged. *New England Journal of Medicine*, 298, 1108–11.

17

Ethical concerns in the care of the older person

Myles N. Sheehan

Sorting through her mother's things in her old flat was proving to be less of an ordeal than Rita had imagined. She had inherited her mother's practical sense and the emptying of drawers and closets proceeded rapidly. As she sat down in an armchair in the small study, she looked out with satisfaction at the boxes and bags of belongings she would give to the homeless shelter. Now there was a bit of time to have a cup of tea and rest. Rita reached over to the top of the desk in her mother's study. She picked up the picture of her mother, herself, her father, and her half brother taken during a visit to her brother while he was attending college in the United States. The picture was faded and the styles of 1955 looked peculiar forty years later. Rita looked at the picture of herself at age 10 and recalled how she was nervous with this mysterious older brother of hers. James, her brother, looked strained next to his stepfather. The whole visit had been a bit awkward, but given the possibility for disaster, it had gone well. Quite unlike this past visit to the United States when her mother died. Rita felt her eyes fill with tears. She certainly was still grieving her mother's loss and her ability to be businesslike when it came to emptying the apartment did not keep her from feeling the deep pain caused by her mother's death.

Rita looked again at the old picture and saw her mother and father. At 40, her mother looked much like Rita had at that age. Her father seemed content and happy in the picture. He was 65 at the time of the picture. Rita's mother and he had married in the autumn of 1944. Both had lost their first spouse. Her mother's first husband, James' father, had been killed in North Africa. Rita's father had lost his first wife in 1943 when their London flat was hit by a bomb. He had been working at the War Office late that evening. His skills as a translator kept him busy and there had been some important documents that had to be translated that night. Rita's mother and father had been introduced to each other by a mutual friend who knew their loneliness and sadness. Rita looked again at the picture. James had been, in some ways, another casualty of the war. Her mother had sent him to the United States in 1940. Many English children had been sent abroad or to the countryside to avoid the dangers of the Blitz. James had lived with a family in New Hampshire. The couple were childless and

fell in love with the little boy. James had thrived in the countryside of New Hampshire. At the end of the war, when he returned to his mother, he moped about. At 10 years old, he missed his American foster parents more than he remembered his English mother. The death of his father and the presence of a stepfather were confusing to him. After a miserable year, and multiple letters across the Atlantic, James returned to New Hampshire to live with his foster parents. He was raised as their son. He developed, over the years, a cordial but distant relationship with his mother in England. As he became an older man, and grew in understanding of the multiple strains of the war and the difficulties his mother had faced, a new warmth had grown between James and his mother. Rita and he began to correspond. She visited him in Boston and came to know his family. The fated trip for her mother was an eightieth birthday present from Rita and James to her.

Rita's mother, Anne Winters, had recovered well from a heart attack two years before this trip. She had occasional chest pain but did well. At the time of her myocardial infarction, the doctors at the hospital in London had been quite gloomy and her course had been stormy. Over time, however, Anne had rallied and the period of hospitalization was a distant memory. She was excited by the prospect of the trip. They flew to Boston and arrived on a hot and humid summer night early in July. James had met Rita and her mother when they cleared customs. Anne had been excited and happy to see her son.

As they stepped out of the terminal, James had turned to his mother and half-sister: 'Why don't you wait here, I'll get the car from the parking garage, and pick you up at the curbside.' Rita noticed how his face froze. She turned to see what was wrong and saw her mother begin to collapse. Both she and James were too startled to do more than exclaim as Mrs. Winters fell backward, hitting her head heavily on the sidewalk. Rita began to scream and yell for help. Several people rushed to their aid. A small crowd gathered about Mrs. Winters. A security guard came to help. He placed his hand on Mrs. Winters' neck, did not feel a pulse nor see respirations. He began to perform cardiopulmonary resuscitation, after ordering a bystander to go to a nearby ticket counter and ask for an ambulance for a woman in cardiac arrest.

Rita could barely remember what happened over the

next few hours. An ambulance came and whisked her mother off, with emergency medical technicians pumping on her chest. Somehow, she and James had stumbled to his car and he had driven to the hospital where the drivers said they would take Mrs. Winters. They were directed to the waiting room of the emergency department and told that a doctor or nurse would speak to them as soon as possible.

After a long wait, a young but tired looking physician came out and asked for the family of Mrs. Winters. Rita and James jumped up and introduced themselves. The doctor introduced herself and began: 'Your mother seems to have had a large heart attack. When she was brought to the hospital she was in full cardiac arrest. We have been able to stabilize her heart rhythm for the time being but her blood pressure is erratic and she is still very unstable. Your mother remains unconscious. We will be taking her up to the intensive care unit on the fourth floor and you will be able to see her there.'

Mrs. Winters' disastrous trip to Boston may not seem to be filled with great ethical conflict. The details, however, set the scene for common problems in caring for older patients: a severe illness suddenly causes a life-threatening crisis. Mrs. Winters' clinical condition makes it unlikely that she will be able to discuss with her physicians how she wants to be cared for. In this setting, family members often find themselves trying to decide how to respond to the recommendations of physicians. An important caveat is that physicians should not turn to family members when an older person is capable of participating in decision making.

The lengthy introduction to this case may seem somewhat peculiar. After all, who cares about Mrs. Winters' first marriage, the events of World War II, or the unsettled relationship between the various family members? Details like these, however, are often at the heart of ethical conflicts when a family member becomes ill. Tensions and misunderstandings are part of every family. They surface again in times of crisis. When a mother is critically ill, the fabric of the family is torn and decision making between siblings can become very strained.

The immediate response to Mrs. Winters' cardiac arrest reveals a key ethical principle in medical care: one acts to preserve life in an emergency unless there is a clear reason to withhold life-preserving therapy. An example of such a clear reason would be a lengthy illness in which the emergency is simply the terminal event. Even in these situations, it is best that discussion occur prior to the emergency so that the person with the illness can help decide what sort of treatments are consistent with his or her values.

There are a number of different ways to approach ethical analysis and problem solving in medical practice. Much of this discussion assumes a standard in which the desires of the person who is being treated are at the center of therapeutic decision making. What are the values that underlie ethical decisions? How does one become aware of and learn how to manage the ethical considerations? How are conflicts dealt with in the clinical setting? What is the role of the physician? Is there something different about decision making with regard to older patients?

Many authorities favor a principle based ethics where certain principles serve as rules to follow in clinical decision making. Five principles are considered essential: autonomy, beneficence, non-maleficence, non-paternalism and justice (Beauchamp and Childress 1989). Patient autonomy may well be the central value. Autonomy, as the word itself suggests from the Greek *auto* (self) and *nomos* (law), refers to the ability of an individual to be a law unto himself or herself, to choose and decide what is best for the person without undue interference from others. Physician respect for patient autonomy is considered a crucial part of ethical behavior, especially in the United States. Concern for autonomous decision making by the patient lies behind the concepts of informed consent and competence. Informed consent is the process by which a physician assists a patient with a decision regarding proposed treatments or procedures. Informed consent requires that the physician and other caregivers explain the treatment, describe the risks and benefits associated with the treatment, discuss alternative courses of action, and detail possible outcomes. An informed consent, or refusal, is obtained when the patient demonstrates understanding of the treatment, risks, benefits, outcomes, and alternatives and then chooses (Applebaum and Grisso 1988). Patients are allowed to refuse what the doctor recommends! What can be very difficult is when it is unclear if the patient understands what the physician has explained. Many conflicts about informed consent are concerned with determinations if the patient is competent to decide. Competence is a legal concept. In the United States, an individual is considered competent until a court has ruled that the person is not competent. The role of the physician is to assess the patient's capability to handle information. This assessment is frequently done by psychiatrists and involves formal assessment of an individual's mental status and the person's ability to make judgements.

Particularly in the United States, concerns about patient autonomy, informed consent, and competence become entangled about fears of legal action and claims of malpractice. This is unfortunate as it obscures the role of the physician in trying to protect a person's ability to decide and in facilitating methods to reach agreement and a reasonable care plan when there is a disagreement. Three points may be helpful. First, respecting an individual's autonomy means, for the physician, a conscientious effort to inform the patient about his condition, make a recommendation about treatment based on the physician's expertise and the patient's informed wishes, and be willing to respect the patient's wishes if he disagrees. Second, disagreements between patients and physicians need not require legal intervention. Often, further conversation and explanation can clarify misunderstandings and lead to a mutually agreed upon course of action. Occasionally, when the physician has genuine doubts that the patient understands the information, then it is reasonable to obtain consultation to help assess the patient's decision making ability. Rarely, in life-threatening cases where there is evidence that the person is incapable of decision making, legal consultation may need to be obtained. Third, most of the time discussions about patient wishes for treatment and efforts to obtain informed consent are straightforward. Conflicts represent an exception.

Respecting patient autonomy is often contrasted with paternalistic behavior on the part of physicians. Paternalism describes the situation when the physician assumes a parental role and decides what is best for the patient without fully informing the person about diagnosis and treatment. The previous practice of not revealing to a patient a diagnosis of cancer is an example of paternalism. Most ethical authorities consider paternalism inappropriate behavior as it robs the patient of the ability to decide for him/herself. It sets the physician in an omnipotent role, deciding about the lives of patients without allowing the patient to choose.

Beneficence is a third key principle in medical ethics. Physician behavior should flow from a genuine desire to do what is right for the person who has come seeking medical assistance. There are three elements to beneficent behavior for a physician: technical, human, and ambiguous. The technical element to beneficence requires that the physician possess the appropriate skills and knowledge to practice in a competent manner. The human element of beneficent behavior includes the willingness and ability of the physician to behave in a compassionate manner to the sick. Finally, beneficence can be ambiguous! Although paternalism and respecting autonomy may appear as sharp contrasts, there are situations when it is very unclear how to inform a patient, what to make of an unclear decision making capability, and how to proceed in a manner that is compassionate. Occasionally, patients make decisions that are obviously poor choices and it can be extremely difficult for a physician to assist them. Perhaps the ambiguous nature of beneficence can be synthesized as the honest attempt of a physician to choose and recommend therapeutic options in a way that respects patient autonomy, show concern for the patient as a person, but recognize that the expertise of a physician does not mean that the doctor can independently decide what is best for those in his or her care.

Non-maleficence is, superficially, the easiest ethical principle to explain. Doctors do not do evil things to their patients. Specifically, doctors do not lie, cheat, abuse, or deliberately hurt those in their care. Doctors do no commit fraud and bill for services they have not performed nor do they charge exorbitant rates. Obviously, the ease of definition depends on the grossness of what is considered evil. Traditionally, the direct taking of life has been considered a clear example of evil. Physicians have been considered pariahs who participate in euthanasia or assist the suicides of patients. This understanding, however, is currently being questioned.

The principle of justice mandates that physicians treat patients in the same manner, despite differences in race, creed, national origin, or, in an emergency, the ability to pay. Just behavior for the individual physician is interwoven with what a society decides is just in health care. In countries like the United States, where there is no program of universal health coverage, physicians are forced to struggle with decisions about how to care for patients who lack the resources to pay. Practically speaking, for people who are critically ill, decisions about treatment are to be made without reference to ability to pay. What is more problematic are non-emergency and preventive services. It can be very difficult for a physician to earn a livelihood caring for people who cannot pay. In countries where there exists a program of national health coverage, physicians may be confronted with bureaucracies and rationing decisions that are unreasonable for the patient. Justice may require the physician to advocate for the patient so that he or she is treated fairly. As with most ethical principles, justice provides gross guidance for behavior but there remains the potential for multiple conflicts and much uncertainty.

In the care of older persons, ethical decision making may be especially problematic. There are two major reasons for this. First, in an ethical system that places great emphasis on autonomy, decision making can be very difficult when individuals are unable to express their own wishes because of illness, delirium, or dementia. Unfortunately, alterations in cognitive ability are a frequent occurrence with sick older persons. Second, ethical dilemmas are often associated with death and dying. No matter how much geriatricians talk about aging successfully, death comes for everyone. Care of older persons involves dealing with the inevitable death of patients. Part of this care is consideration of the patient's wishes, values, and the medical options. Ethical decision making thus is a common and necessary part of caring for the elderly. Because the death of a person is a time of great stress and tension for the person, family members, and caregivers it is not a surprise that conflicts can develop.

A principle-based ethics may not always be of much guidance in conflictual situations. Principles serve as external guides, rules that suggest what to do but reveal little about how to behave when there is little clarity on what is the right course. What is often frustrating for physicians is that dealing with ethical dilemmas can take up a lot of time, discussing issues with a patient, family, and other caregivers. It is important to recognize that difficult and abusive patients usually remain that way no matter how hard the physician tries. Families that had deep-seated conflicts prior to a health care crisis will likely display those conflicts during the illness of a family member. Physicians and other caregivers can often be caught up in long-standing dysfunctional relationships. Recognizing what is happening will not necessarily yield a solution, but it might help the physician to maintain equanimity and the ability to provide consistent care, rather than be driven to distraction by a family in disarray. Some rules of thumb for clinical practice may be of help.

First, make the effort to prevent crises. In discussions with primary care patients, physicians should speak with them about their wishes for care regarding both life-sustaining treatment as well as who the patient would wish to speak for them in the event of an illness that rendered the patient incapable of participating in decision making.

Second, recognize that decision making capability is a continuum and not 'all or nothing.' An individual may be confused about the date or other details, but quite clear on his or her wishes not to be intubated and placed on a ventilator. With patients who are confused more important than absolute clarity in every cognitive domain is the recognition of clear, consistent, and repeated preferences for care.

Third, conversations regarding patients' choices should be documented in the medical record. This provides a reference during a crisis, gives guidance to any covering physicians, and can help settle disputes among family members.

Fourth, recognize the need to be patient. The most difficult decision-making processes involve sudden illnesses with an unclear clinical prognosis and a family that is suddenly plunged into a nightmare. If the older person who made a clear determination of

his or her wishes prior to the illness, then these wishes should be respected or the person named as proxy should speak for the patient. If there is no clear sense of the patient's wishes nor a proxy decision maker, then the family may need time to adjust to the sudden illness. The time frame for decision making may be different for health care providers than for patients and their families. Although this can be awkward, a recognition that the family may need a day or two to come to a decision can lower tensions and permit better decision making.

Perhaps the most important part of medical ethics, and one that is rarely commented on, is the need for reflection by the physician. As mentioned, principles serve as external guidelines but the interpretation of principles in conflictual situations can be far from straightforward. Physicians are forced to come to decisions that are difficult and may leave them feeling uncomfortable. In dealing with the care of patients, considering three questions can give insight as to how to proceed: 'Who am I? What am I doing? Who am I becoming by these actions?' Rather than seeking external validation from principles, this process of questioning considers medicine as a profoundly moral activity, one where practitioners are changed by the way they behave. Ethics is more than rules, it is systematic reflection on the meaning of behavior. A physician concerned about the meaning of his or her decisions and the ways he or she interacts with patients will strive to grow in excellence, both in clinical skills and in behaving correctly. Principles set the framework for understanding the boundaries of moral decision making. But without individual reflection on the meaning of one's practice, a physician can not grow as a moral person (Sheehan 1994a).

After several hours of waiting, James and Rita were brought into their mother's room in the intensive care unit. Mrs. Winters was intubated and mechanically ventilated. A Swan–Ganz catheter had been introduced via her right internal jugular vein. An arterial catheter was present in her right radial artery. Urine was draining from her bladder by a Foley catheter. Mrs. Winters remained unconscious. Rita began to cry as she saw her mother. The combination of the shock of her sudden illness and the sight of the variety of tubes and monitoring devices left her feeling helpless and overwhelmed. James awkwardly attempted to comfort his half-sister, but he too felt overwhelmed and horrified.

Over the next two days, Mrs. Winters' condition showed improvement in some areas but worsening in others. Neurologically, she began to have periods of wakefulness and purposeful movements. Her respiratory status improved to the point that she was extubated. No meaningful conversation was possible, however, with Mrs.

Winters either sleeping, moaning, or answering questions with an unintelligible mumble. Unfortunately, her cardiac condition remained precarious. She had had a large anterior wall myocardial infarction and remained relatively hypotensive and dependent on intravenous medications for blood pressure support. There were multiple episodes of ventricular arrhythmias that were only partially controlled with intravenous lidocaine. Her renal function was not good with a poor urine output. There were several episodes when Mrs. Winters' temperature would spike to 101°F without a clear source.

Rita and James began to discuss what would happen to their mother. James seemed reluctant to consider any possibility but a full recovery. Rita feared that her mother would linger and not return to an active lifestyle. She felt James was angry with her when she would express her fears or her doubts as to recovery. They asked the attending physician in the intensive care unit if it would be possible to speak with him about their mother.

Dr. Coughlin met with Rita and James the next morning. Mrs. Winters had had a rough night with an episode of hypotension and pulmonary edema that required her reintubation. Her urine output was also quite low. She was, once again, unconscious and had a continuous fever.

'Mr. Williams, Ms. Winters, we need to make some decisions about your mother's care. She is not doing well and I need to hear from you what you feel is appropriate. First, she has had a large myocardial infarction complicated by arrhythmias and congestive heart failure. As a consequence of her episodes of low blood pressure, her kidneys are not working well and we will need to consider dialysis, at least on a temporary basis. Her chest X-ray from this morning shows a pneumonia developing in her right lower lobe. We have begun antibiotics to treat this. Her prognosis is unclear. I have a couple of questions for you? Do you want us to continue to do everything or do you want us to pull back? We could make her a DNR and let her be comfortable.'

Rita asked what was a DNR.

'It means "Do Not Resuscitate", Dr. Coughlin answered. 'If your mother were to have another cardiac arrest then we would not attempt to restart her heart and she would pass away.'

James looked pained and began to speak: 'Look, Doctor, I want you to do everything for my mother.'

Dr. Coughlin seemed relieved at getting an answer: 'Well, we will try. Your mother is very sick but I appreciate your wishes. I will keep you posted as to what happens.'

As Dr. Coughlin left the room, Rita felt a combination of confusion and anger. 'What had he been talking about? What was all that information? I still don't have a sense of whether my mother will get better. And why did he take James' word as the final word?'

Later that evening, as she sat alone in the hospital cafeteria, Rita thought about the conversation with Dr. Coughlin. She remembered when her mother had been sick in England that the doctors in the hospital had

been quite different than Dr. Coughlin. There had not been the avalanche of information that Dr. Coughlin had conveyed. Neither had the physicians in London ever asked Rita or her mother what type of treatments they desired. Mrs. Winters had told Rita that tests and procedures were simply done and the doctors were not very communicative. Although Rita appreciated being told what was happening, she thought that the information could have been conveyed better than the way Dr. Coughlin spoke with her and her brother. Rita felt herself annoyed that Dr. Coughlin had looked to her and her brother to decide what treatments to use in the care of their mother. 'Isn't that the doctor's job to decide?' she thought.

Dr. Coughlin did not do a very good job in dealing with Rita and James. Discussions about the provision, withholding, and withdrawal of life-sustaining therapy are not easy. Physicians, like any human being, are often uncomfortable at discussing the possible death of a patient. Doctors are good at prolonging life and trying to beat death, even if only for a short time. Admitting that a patient may die can make a physician feel, in some ways, like a failure and incompetent. It may be easier to proceed with vigorous treatment rather than face the reality of the clinical situation and the need to discuss with the patient and family the possible courses of action.

Rita was surprised at the amount of information communicated by Dr. Coughlin. Although she appreciated being informed about her mother's condition, she was overwhelmed by the sheer detail and unable to assimilate the meaning of the medical facts. Her recollection of the differing level of information communicated by her mother's physicians during the earlier hospitalization in London and the American practice of seemingly telling every detail may overstate the differences between English and American habits in informing patients and families. Clearly, however, the emphasis on autonomy in the United States has lead practitioners to provide patients and family members with large amounts of information. Keeping patients informed about their condition and prognosis is an important part of medical care. It is a concrete way for the physician to show respect for the patient as a person rather than an object to be treated. It also allows the physician the opportunity to understand the goals and values of the patient so that the doctor can make responsible recommendations to the patient about various care options. Unfortunately, not all American physicians recognize that simply telling patients and their families a wealth of information and then expecting a decision is irresponsible. It may be that other physicians need to recognize that critical decisions in caring for patients involve more than medical expertise and it is essential to humane care to consult with the patient about his or her values and desires (Sheehan 1994*b*).

Rita's annoyance at Dr. Coughlin is well founded. It is not the job of family members to decide how a relative is to be cared for. The appropriate role of family members is to provide the physician with insights into how that family member would want to be treated. A son or daughter should not be asked: 'Do you want us to do everything for your mother or do you want us to pull back?' A more appropriate question would be: 'Did your mother ever discuss with you how she would have wanted to be treated if she was seriously ill?'

Particularly in decisions regarding life-sustaining therapy, the discussion about treatment must be contextualized. The obligation to preserve life is part of the role of the physician. Life is generally considered a great good. It has been the tradition that doctors normally sustain life and never directly take life. This tradition has always been tempered by a recognition of mortality. Everyone dies and a death need not represent a failure of the medical profession. The therapies available to maintain life are medical treatments. They have indications and contraindications, like any other types of treatments. A rule of thumb for discussions about life-sustaining treatment is to avoid focusing the conversation on a specific treatment. Dr. Coughlin made this error in concentrating on the issue of resuscitation. Especially in a complicated case like Mrs. Winters', where there is multi-organ failure and a very unclear prognosis, the physician should present a care plan that attempts to meet patient goals, limit suffering, and ensure attentive care that does not abandon the patient. Patients and their families do not have the expertise to make medical judgements. It is the role of the physician to discuss the medical condition of the person, inquire about what the patient expects and hopes for, as well as what the patient is afraid of, and then attempt to provide a care plan that respects the patient's values.

After a week of hospitalization, Mrs. Winters continued to do poorly. Although she did not require intravenous medications to maintain her blood pressure, she still had poor urine output, was in congestive heart failure, had a right lower lobe pneumonia, and remained intubated and mechanically ventilated. Dr. Coughlin asked to meet with Rita and James in the conference room next to the coronary care unit. He announced that he was leaving on vacation and introduced them to Dr. Hunt, the physician who would be caring for their mother while he was away. Dr. Coughlin excused himself while Dr. Hunt asked Rita and James to remain as he wanted to take advantage of the opportunity to speak with them.

'Ms. Winters, Mr. Williams, as you are likely aware, your mother's case is quite complicated. I would find it very useful if we could talk together about how she is

doing and then discuss some plans for her care. First, I find it very difficult to know how to proceed in a case like your mother's. She is clearly critically ill. There are a variety of decisions that have to be made. She cannot participate in those decisions because she is so sick and unable to speak with us. It also is difficult because I have no idea what she would want. What I need to know from you is a sense of what your mother would want. Did she ever talk to you about how she would want to be treated if she became seriously ill?'

Rita and James looked at each other. Rita began: 'Well, Doctor. Particularly after she had the heart attack in England a few years ago, she began to mention that she did not want to be on tubes and in bed being kept alive. She made it clear she had had a good life and that when her time came she was ready to go.'

Dr. Hunt continued: 'That is very helpful. But, you know, very few people would want to be intubated. If she could be treated aggressively for a brief period of time and then recover, how do you think she would feel?'

'Oh, my mother was always very practical,' Rita answered, 'I am sure she would be willing to have a go at anything for a while if it might work, but I am quite certain she would not want to be kept going just to become an invalid.'

'I have reviewed your mother's case very carefully with Dr. Coughlin and I have just finished examining her. Although there are a number of problems, the critical issue currently is her kidney function. For a variety of reasons, your mother's kidneys are not functioning properly. Although there is the possibility of performing temporary dialysis, I do not recommend such an effort. First, given your mother's marginal blood pressure it is not uncommon for people to have a very low blood pressure during some of the dialysis period and that could lead to other problems. Second, I am not certain it would really change much. I am afraid your mother's prognosis is quite poor. I am sorry to be the bearer of such news. But what you have told me about her wishes is very helpful. What I would suggest is that we keep up our current level of care over the next few days and see what happens. It is possible that kidney function will return spontaneously over the next few days. If it does that is fine. Should your mother survive, she will have a long road ahead of her as she will be quite debilitated. I also will write in the chart an order that your mother not be resuscitated in the event she suffers another cardiac arrest. In the current situation, it is unlikely that our efforts to resuscitate her would be successful. At best, they would only prolong her life for a few days. My plan is to keep your mother as comfortable as possible. As I mentioned, we will continue to support her breathing and treat her pneumonia with antibiotics. I also will meet with you to let you know about any changes and answer your questions. Do you have any questions now?'

'Yes, Dr. Hunt, I want to make sure I understand. The plan is to continue her treatment, not begin dialysis, and not resuscitate if her heart stops?

'Yes, Ms. Winters, that is the plan.'

James had appeared anxious while Dr. Hunt had been speaking. Now he could barely control himself: 'Well I don't like that plan and you are going to dialyze my mother and resuscitate her. She is going to get better and you are not going to let her die! Do you understand, Doctor?!'

'Well, Mr. Williams, I understand that you are very upset about your mother's illness and I also understand that it is not in my power to keep her alive. She has had a massive heart attack and several complications. Your mother made it clear that she did not want to be kept alive in a situation where it was unlikely she would recover. My recommendation is that we maintain support for the time being. Mr. Williams, I am sorry that your mother is so sick. But I must also recognize that she is my patient and I need to do what she would have wanted. Do you disagree with your sister's recollections of your mother's wishes about what to do if she became seriously ill?

'No, Dr. Hunt, I never spoke with her about it. But I want you to do everything for her.'

'Mr. Williams, I am doing everything I believe your mother would want.'

'Then why would you let her die if her heart stops?

'Mr. Williams, if your mother has another cardiac arrest, there is some chance that I may be able to restore a heart beat, but given that she is in renal failure, has a serious pneumonia, and is also in congestive heart failure there is almost no chance she will recover. Do you understand? There are lots of things that I can do. But they would not do much for your mother except increase the potential for suffering. I know that you do not want that. I want to preserve your mother's comfort and dignity, while allowing the possibility that she may recover.'

'James, stop being so impossible,' Rita shouted at her half-brother, 'Mom does not want to be kept going so we can feel content that we insisted on every damn torture this hospital has to offer. Leave her alone and let the doctor try his best to take good care of her!'

Dr. Hunt's method of discussing Mrs. Winters' case with her children emphasized Mrs. Winters' wishes, her medical condition, and his recommendations about the course to be followed. He attempted to present a care plan that described what treatments would be continued and those that he recommended against. Ascertaining Mrs. Winters' wishes, through the recollections of her daughter, provided the rationale for the clinical plan. It is important to realize that if Mrs. Winters was able to communicate herself, then the decisions would be between her and her doctor. Too often, especially with older patients, physicians will avoid conversations with the patient and turn to family members. *Although the family may be a tremendous support, if a patient is capable of decision making, then he or she is the one with whom decisions about care should be made.* In the United States, many patients have 'advance directives' for their health care. An advance directive can take a number of forms, but it provides a way for an individual to communicate his or her wishes regarding care in the event that he or she is too ill to participate in decision making (Orentlicher 1990). One form of advance directive is called a 'living will' and it specifies the treatments that an individual would or would not want to receive. Other advance directives will specify an individual who can speak for the patient if the patient is not capable. The appointment of a health care proxy can be of great value in guiding physicians. The role of the proxy is to communicate the wishes of the patient to the doctor and to act for the patient with regard to medical decision making. Mrs. Winters did not have a living will or a health care proxy. In such a situation, the reasonable course is to attempt to ascertain what the individual would have wanted. Unfortunately, there is evidence that the preferences of patients regarding their care are often poorly understood by family members and treatments are chosen that the person would not have desired (Seckler *et al.* 1991; Uhlmann *et al.* 1988). There is also the risk of conflict among family members and physicians may be concerned about legal action if they act against the wishes of some of the family.

Dr. Hunt found James to be upset about the clinical plan and wanting more aggressive medical therapy for his mother than seemed clinically reasonable or in keeping with his mother's previously expressed wishes. Disagreements about care can have a number of different reasons (Wolf 1988). First is the possibility of a genuine disagreement about values. James may feel that every moment his mother is alive is worthwhile, regardless of the condition in which she is existing. It would not be reasonable, however, for a physician to keep Mrs. Winters alive against her previously expressed wishes simply to gratify her son's feelings. Second, the emotional tension surrounding the illness of a close relative can bring up a number of strong feelings. James' desire that everything be done may be as much an expression of his unresolved feelings about his mother, given their years of separation. Third, family members and patients may have unrealistic expectations about the ability of physicians and technology to prolong life. Many people do not comprehend that resuscitation (outside of the setting of an acute myocardial infarction) does not automatically result in restoring function, or that many individuals who survive the initial resuscitation die a short time thereafter (Murphy 1988; Schiedermayer 1987; Blackhall 1987). Likewise, other items of medical technology may simply postpone the dying process, and bring with them the potential for great suffering. Gently attempting to educate patients and families can result in improved understanding. Fourth, patients and families may be concerned that setting limits to life-sustaining therapy or agreeing to 'do not resuscitate'

(DNR) orders may mean the attention and care that the patient receives will be compromised. Physicians and other care providers must make certain that these fears are not true and that a decision to treat in a less aggressive manner does not imply decreased attention to the comfort and dignity of the person.

In the case of Mrs. Winters, where there is argument between the children and a need for medical decision making, the physician faces a quandary. Should he forego writing a 'DNR' order, concerned about the absence of a health care proxy, relying on the recollections of Mrs. Winters' daughter, and nervous about the reaction of Mrs. Winters' son? Or, given the evidence that Mrs. Winters is doing poorly, simply continue with the course that was outlined in the discussion with the family? There is no clear answer and the solution depends on the style of the physician, the policies of the hospital, and the particular realities of each case (Tomlinson and Brody 1988). Given Mr. Williams' anguish over his mother's condition, one could respect a decision to forego writing a DNR order for the time being, meanwhile keeping in close contact with him and his sister, with frequent discussions. That course of action would allow for the most support for the family and provide the physician the opportunity to come to a consensus decision that combines the medical aspects of the case with the wishes of Mrs. Winters, as far as they can be ascertained, and the desires of her children. Such a course, however, is very time consuming and it runs the risk of placating Mr. Williams at the expense of Mrs. Winters' desires. This is why ethical decision making is hard in the practice of medicine, and why it requires reflection on the part of the physician as to the meaning of his or her actions and the need to consider what is prudent in each situation.

Over the next 24 hours, Mrs. Winters continued to do poorly. Her urine output fell to only a few milliliters an hour, her blood pressure become unstable with episodes of hypotension, and her oxygenation worsened with increasing heart failure and no clearing of the pneumonia. She began to have bouts of arrhythmias.

Dr. Hunt met Mr. Williams at his mother's bedside. Rita Winters was taking a nap in the waiting room.

'How is my mother doing, Doctor?'

'Not very well, I'm afraid. How are you doing, Mr. Williams?'

'Oh, this is very hard. I just want her to get better.'

'I know you do, but I do not think that is very realistic.'

'Yes, I am beginning to get that impression ... but I still want you to do everything ... You did not write that "do not resuscitate" order, did you?'

'No, I haven't yet, but I do not think it will matter much pretty soon, Mr. Williams. Can I ask you a question?'

'Sure, Doctor, what is it?'

'What do you think I can do by "doing everything"? What will it accomplish?'

Mr. Williams gripped the handrails of his mother's bed tightly. His eyes began to fill with tears. He reached into his pocket and pulled out a handkerchief, trying to stop the tears. 'Oh, Doctor Hunt, I am not sure. I wanted to have as much time with my mother as I can. I still feel cheated out of the time we lost because of the war ... it's a long story, but I was sent over here when I was very young, during World War II, to avoid the Blitz. I was very lucky. My American foster parents were wonderful parents to me, so wonderful I ended up staying with them after the war. But now that I have had a chance to know my own mother, I regret the time we did not have. I am jealous of my sister and the childhood she had with our mother. It makes me angry when I realize how much better Rita knows my mother. Frankly, I do believe my mother would not want to be kept alive. She lived a full life. But I had such hopes. I was looking forward to this visit of my mother's. I wanted to spend time with her and just enjoy her company. I wanted to tell her so many things ...'

Dr. Hunt stood quietly as Mr. Williams fought to maintain his composure. Mr. Williams stepped away from the bedside and sat down in the chair in the corner of the room. He put his face in his hands for a few moments, and then looked up.

'Doctor Hunt, what is going to happen to my mother?'

'I think that she is dying. I will make sure that she is well treated and kept comfortable. I do think that you have some time to be with her now. What you have told me has been very helpful. As a doctor, I cannot give you your mother back in the way you wish. But, as her doctor, I can make sure that she is treated in the way it seems likely she would want to be treated. Resuscitation and dialysis are not the point. What is the point is your strong feelings about your mother ... You should spend as much time with her as you feel comfortable. That might be the best thing for both of you, right now. I am going to write the DNR order now, because we are all clear on what we believe your mother would want. I promise you I will continue to pay very close attention to your mother. I am only a doctor, I am not God. She could get better on her own. Stranger things have happened, but there is really little more that medicine has to offer than just continue to support her the way she is currently. I also know, Mr. Williams, that this is a very hard time for you. If you want to talk some more with me let me know, and I will be available or help you find some other assistance as you need it. Do you have any questions?'

'No, Doc, thanks for putting up with me. I appreciate the time you have taken. I will be all right, it just is hard.'

Often taking the time to speak again with family members or a patient can bring resolution when the initial discussion seemed laden with conflict. As Dr. Hunt discovered, Mr. Williams had put a meaning on resuscitation and aggressive treatment that had little to do with the medical realities. He wanted to make up for the time he had lost with his mother as a youngster and have a chance to spend more time with her, speaking with her and getting to know her better. Mr. Williams had very human reasons for not being able to agree with the initial discussions to limit treatment: his mother's death seemed doubly painful as he felt he was cheated out of their relationship when he was a child. Dr. Hunt could be criticized for not writing the DNR order once he felt reasonably certain that Rita Winters was conveying her mother's wishes. At the same time, he realized that a bit of time might allow another opportunity for discussion. The new discussion allowed Dr. Hunt the opportunity to reaffirm his commitment to not abandon Mrs. Winters, to do everything to limit her suffering, and to provide some relief and comfort for Mr. Williams. Mr. Williams's desire for a new close relationship with his mother was not in the reach of medicine. Once he came to grips with his feelings, Mr. Williams was able to confront the realities of his mother's condition. Should he and Mr. Williams have continued to disagree and Mrs. Winters' condition continued to worsen, then the ethical course would have been to follow Mrs. Winters' wishes despite her son's demands. Simply looking to rules and principles, however, without considering the context would be a mistake. Dr. Hunt, by his firm attention to the medical realities of the case, his desire to respect Mrs. Winters' wishes, and his concern for the anguish of the family, demonstrated why medicine is filled with moral challenges. He considered a number of options, kept in mind ethical principles as a guide to handling the situation, but also showed personal strength and courage in working with the family and caring enough to listen to Mr. Williams' pain.

Mrs. Winters' arrhythmias began to increase in frequency as her heart failure worsened. The arrhythmias were accompanied by more hemodynamic instability and worsening oxygenation.

After another few days, Rita Winters raised the issue of withdrawing therapy. She discussed with her brother and with Dr. Hunt the possibility of removing the ventilator and endotracheal tube. Mr. Williams spoke frankly with his sister that he did not want this to happen, not because he felt that their mother was going to get better, but because he would find it too painful to remove treatment. It had been hard enough to reconcile himself to the decision to limit treatment.

Dr. Hunt once again raised the question of what Mrs. Winters' children thought their mother would want. Neither was entirely sure. Rita felt certain her mother would not want to be supported indefinitely but she admitted she was a bit uneasy about withdrawing therapy at this point.

Dr. Hunt explained that withdrawing the ventilator would clearly result in Mrs. Winters' death due to her heart failure and pneumonia. If there were a more clear indication of Mrs. Winters' wishes, then he would respect those wishes. Given, however, the ambiguity of the situation, he was more comfortable continuing the present course. He reiterated his commitment to limit Mrs. Winters' suffering and make sure she was well cared for. He also told Rita and James that he felt that it was likely their mother would not survive much longer.

Decisions about withholding and withdrawing life sustaining therapy remain painful for patients and caregivers. The end of life is a time that frightens individuals in two ways. First is the fear of death and the sadness that a life is over. Second, conversely, is the fear of a life unnaturally prolonged, maintained by machines in a manner that most would find undignified. Wrapped up with the difficulty in deciding about life-sustaining therapy are fears of causing a death and killing another person. Although there is growing acceptance by some individuals of assisted suicide and voluntary euthanasia, many disagree with the direct taking of life and will not participate in such decisions.

Recognizing the possibilities for confusion and ambiguity, four points can provide some assistance in considering withholding or withdrawing life-sustaining treatments.

First, patients have the right to refuse treatment, even if it will result in their death. It is important for a physician to ascertain that such refusals are informed refusals, and made out of a genuine understanding of what is at stake and the consequences. In cases where there is doubt about a patient's decision because of depression, decision making capability, or fears of coercion, then consultation should be obtained. The refusal of a patient to agree to what seems like ordinary therapy to the physician may be morally repugnant to some caregivers. In such a situation, physicians and others may find it difficult to continue to care for the patient and will need to find others to take over that person's care. Prior to such a drastic step, however, caregivers need to make the effort to understand the patient's understanding of the procedure and the reasons for the refusal. Repeated efforts at explanation and communication may lead to truly informed consent or refusal, rather than misunderstandings.

Second, there is a difference between killing and letting die, even if the boundaries may seem blurred. Medical treatments may be withheld or withdrawn if the patient finds the treatment burdensome and the treatment is unlikely to restore a person to health. This is not the same as intending death by withholding or withdrawing treatment. Although death is the outcome whether one intends it or not, many believe there is a morally relevant difference between deliberately intending the death of a patient, even at the patient's request, by withholding or withdrawing therapy versus recognizing that a treatment has the potential to maintain life but is too painful or otherwise unacceptable to be tolerated, and thus allowing death to occur (Callahan 1990).

Third, ethicists argue that withholding and withdrawing therapy are morally equivalent. In the situation where a critically ill person receives aggressive and intensive therapy but it becomes clear the patient is not responding, treatment can be withdrawn after discussion with the patient or proxy, if the patient is incapable of participating in the discussion. The traditional caveat would be that treatment is withdrawn not to bring about the person's death, which would be morally wrong, but because of a recognition that the treatment is burdensome for the patient and unlikely to result in restoration of the person to his or her previous level of function. It can be, however, quite difficult for caregivers and family members to cease life-sustaining treatments (e.g., removing a patient from a ventilator), as it may feel like actively bringing about the patient's death. It may seem less difficult to withhold aggressive treatment (e.g., by not providing intubation and mechanical ventilation), and thus allowing death to occur. In both the case of withdrawing and withholding, however, the person is allowed to die from the disease process. Death is not occurring from the action of the physician (Callahan 1990). Although this makes rational sense, there remains the emotional pain of decisions to withdraw therapy. The need for careful reflection, concern about the value of human life, diligent effort to understand as much as possible the wishes of the patient, and painstaking review of the medical realities of the situation are required in all decisions regarding life-sustaining therapy.

Fourth, cost is not to be used by the physician as a criterion for bedside ad hoc *rationing.* Health care costs are valid concerns. Most industrialized countries have made decisions about the distribution of health care resources as part of a commitment to health coverage for their citizens. Recognizing that health care costs can overwhelm the resources of a society, there is a need for decisions to be made about the allocation of resources. Physicians who practice in countries where decisions have been made on a national level about the allocation of health care resources can ethically operate within that system, recognizing the need, on occasion, to advocate for the needs of an individual patient. In a nation like the United States, however, where there is no agreement on what is the appropriate allocation of health care resources, it is unjust for a physician to

decide based on personal arbitrary criteria about how resources are to be rationed. An example would be a physician who decides all people over the age of 70 should be denied access to the intensive care unit because the care of these individuals is too expensive. In the absence of a national consensus as to how societal resources are to be allocated, the physician's job is to care for the patient with the resources available. The physician has a responsibility to steward resources based on what is medically reasonable and likely to produce a benefit. This responsibility should not be conflated with bedside rationing decisions (Churchill 1988; ACPEC 1989; AGSPPC 1989).

Rita packed away the last few things into boxes in her mother's apartment. She remembered the final day in the hospital. James had stayed with her at their mother's side. Her blood pressure had steadily decreased. Dr. Hunt said he thought it likely she had had another small heart attack. All in all, their mother had seemed comfortable and Dr. Hunt was as good as his word in keeping an eye on her. The end had come peacefully and she and James were glad for each other's company. During the final day, James had the chance to tell Rita how hard he found their mother's dying and how much he had longed to make up for the lost time of their relationship.

Rita smiled for a moment as she thought how perfectly horrid she found Dr. Coughlin with all his questions and his nervous habit of inundating her and James with unintelligible information. Her opinion of American medicine was redeemed by their encounter with Dr. Hunt. He seemed to share the American obsession for telling everything, but that was actually helpful in small doses, and she was grateful for his attention and concern, as well as the resolute way he worked with James.

Rita prepared to leave her mother's apartment for the last time. She looked around the room and wiped a tear from her eye. 'Oh, time to get out of here,' she thought to herself, 'I will go home and call James, and see how he is doing.'

Questions for further reflection

1. Discuss the elements of a principle based approach to ethical decision making and some of the difficulties.
2. How would you approach a healthy older person regarding the development of advance directives for health care.
3. What factors must be considered in decisions to withhold or withdraw life-sustaining therapy?

References

ACPEC (American College of Physicians Ethics Committee) (1989). The American College of Physicians ethics manual. Part 2: The physician and society; research; life-sustaining treatment; other issues. *Annals of Internal Medicine*, **111**, 327–35.

AGSPPC (American Geriatrics Society Public Policy Committee) (1989). Equitable distribution of limited medical resources. *Journal of the American Geriatrics Society*, **37**, 1063–4.

Applebaum, P.S. and Grisso, T. (1988). Assessing patient's capacities to consent to treatment. *New England Journal of Medicine*, **319**, 1635–8.

Beauchamp, T. and Childress, J.F. (1989). *Principles of biomedical ethics*, (3rd edn). Oxford University Press.

Blackhall, L.J. (1987). Must we always use CPR? *New England Journal of Medicine*, **317**, 1281–5.

Callahan, D. (1990). *What kind of life: The limits of medical progress*, pp. 231–7. Simon & Schuster, New York.

Churchill, L.R. (1988). Should we ration health care by age? *Journal of the American Geriatrics Society*, **36**, 644–7.

Murphy, D.J. (1987). Do-not-resuscitate orders: Time for reappraisal in long-term care institutions. *Journal of the American Medical Association*, **260**, 2098–101.

Orentlicher, D. (1990) Advance medical directives. *Journal of the American Medical Association*, **263**, 2365–7.

Schniedermayer, D.L. (1987). The decision to forgo CPR in the elderly patient. *Journal of the American Medical Association*, **260**, 2096–7.

Seckler, A.B., Meier, D.E., Mulvihill, M., and Cammer Paris, B.E. (1991). Substituted judgment: How accurate are proxy predictions? *Annals of Internal Medicine*, **115**, 92–8.

Sheehan, M.N. (1994a). Why doctors hate medical ethics. *Cambridge Quarterly of Healthcare Ethics*, **3**, 289–95.

Sheehan, M.N. (1994b). Technology, older persons, and the doctor-patient relationship. In *Health care ethics: Critical issues*, (ed. J.F. Monagle and D.C. Thomasma), pp. 374–83. Aspen, Gaithersburg, MD.

Tomlinson, T. and Brody, H. (1988). Ethics and communication in do-not-resuscitate orders. *New England Journal of Medicine*, **318**, 43–6.

Uhlmann, R.F., Pearlman, R.A., and Cain, K.C. (1988). Physicians' and spouses' predictions of elderly patients' resuscitation preferences. *Journal of Gerontology (Medical Science)*, **43**, M115–21.

Wolf, S.M. (1988). Conflict between doctor and patient. *Law, Medicine, and Health Care*, **16**, 197–203.

18

Disease prevention, health promotion, and health protection

Claus Hamann and Jeanne Y. Wei

'Well, Doctor Poulos, I feel just fine. I am eighty years old and I came to see you for a check-up. I have high blood pressure. Over the last year, my blood pressure readings have been good but I am concerned about my pressure and I would like to go over my medication. I was also hoping that you would have the time to talk to me a bit about things that I could do to stay healthy.'

Lillian Shipper, a healthy older woman, comes to your office for the first time. Her desires for a thorough check-up, evaluation of her blood pressure medications, and advice on how to stay healthy are appropriate concerns. Like many of her peers, she recognizes that health is a dynamic process requiring active effort by her and her physician to prevent disease and to promote her health. Also, like many of her peers, she relies mostly upon her physician for ongoing education in the important areas of disease prevention, health promotion, health protection, and preferences for future medical care. This patient–doctor collaboration is a key feature of maintaining and enhancing Mrs. Shipper's ability to function in her daily activities and her continued enjoyment of the highest possible quality of life.

The initial office visit with a new patient challenges the primary care physician to establish rapport and understanding, examine the stability of existing conditions, ascertain the risk of future conditions, and discuss a plan of possible interventions. In an older person with multiple medical conditions and impairments, it would be extremely difficult if not impossible to accomplish all of these goals during a single visit. They should therefore be prioritized and addressed over several visits during the early phase of the new patient–doctor relationship. Since Mrs. Shipper appears healthy and her only medical condition (hypertension) is stable, the issues of disease prevention, health promotion, health protection, and preferences for future medical care can be initiated on this first visit (Herman and Robertson 1993).

Disease prevention

Heart disease and stroke

The annual economic burden of cardiac and cerebrovascular disease in the United States is in excess of $135 billion (HCFAOA 1988), greater than any other major disease. As a preventive maneuver, antihypertensive treatment reduces the major risk factor for heart disease and stroke in the elderly, two of the three most frequent causes of death and major contributors to morbidity in older adults. Aside from hypertension, other preventive measures are directed towards the other major alterable risk factors for cardiovascular and cerebrovascular disease, when present: smoking and high blood cholesterol (see Tables 18.1 and 18.2, respectively). In addition, increased physical activity, reduction of weight and of sodium, fat, and alcohol intake, as well as stress management are essential.

Mrs. Shipper brought with her a list of blood pressure readings that she had taken over the last few months on a home sphygmomanometer. The pressures were stable, being in the range of 150/90 mmHg. Mrs. Shipper said that when she began her blood pressure medication, her blood pressure had been around 180/100 mmHg. Mrs. Shipper said that she took her enalapril, 5 mg, each morning. She mentioned that she puts a check mark on the calendar after she takes the medication each day.

Recent clinical trials of antihypertensive treatment in older adults suggest that appropriate treatment of hypertension likely lowers the risk of developing left ventricular hypertrophy, congestive heart failure, acute myocardial infarction, and stroke (Kupersmith *et al.* 1995; Massei 1994). Medications that have been used in these studies include a diuretic (chlorthalidone or other), a central adrenergic inhibitor (alpha-methyldopa), beta blockers (propranolol, metroprolol, or

Table 18.1 Annual preventive evaluation for older adults (above 65 years)

Service/Maneuver	Comment
All older adults	
Weight	>15% above or below the desired range may confer health risk
Blood pressure	More frequently than yearly if elevated (systolic and/or diastolic)
Visual acuity	Use newsprint text or Rosenbaum card
Intraocular pressure	By eye specialist, to prevent glaucoma
Hearing acuity	Especially if prior exposure to loud noise
Oral examination	Especially with tobacco or alcohol use
Mental status	Screen for cognitive and affective changes
Clinical breast examination	Good evidence for effectiveness, together with mammography
Mammography	Reimbursed by Medicare every 2 years
Total cholesterol	*In men*, non-fasting; every 5 years, more often if coronary risk factors are present. *In women*, if coronary artery disease is present; otherwise controversial
Urine analysis	For bacteriuria (especially in diabetics) and bladder cancer (especially in smokers), by dipstick
Stool for occult blood	Especially if family history of colon cancer or history of inflammatory bowel disease; otherwise somewhat controversial, due to low sensitivity and specificity
Sigmoidoscopy	Controversial: every 3–5 years may be of benefit, or air-contrast barium enema every 5 years
High-risk older adults	
Depression	For any one of the following: evidence of psychiatric disorder, personal or family history of depression/suicide, substance abuse, chronic illness, living alone, recent loss (bereavement, separation, unemployment), sleep disturbance, multiple somatic complaints
Carotid auscultation	History of transient ischemic attack, or cerebrovascular disease risk factors
Skin examination	History of substantial sun exposure, or family history of dysplastic nevi
Papanicolaou smear	Every 3 years if not screened during 10 years prior to age 65, and total number of negative smears is less than three (reimbursed by Medicare)
Hematocrit	In institutionalized older adults, and possibly those in the community who need assistance
Fasting plasma glucose	In obese (if motivated to lose weight), diabetics, or those with family history of diabetes
Thyroid hormone levels	In women, for hypothyroidism
HIV serology	High-risk behavior, history of transfusion 1978–85
Syphilis serology	High-risk behavior
Tuberculin skin test	History of exposure (e.g., nursing homes, shelters, dialysis units), immigration from high-risk areas, immune deficiency (e.g., high dose-steroids)
Resting, stress ECG	If coronary artery disease and/or substantial risk factors and starting vigorous exercise program
Bone mineral analysis	In frail, slender women (especially history of smoking or hyperthyroidism) or those considering estrogen for osteoporosis only
Colonoscopy	If two or more first-degree relatives with colon cancer, history of ulcerative colitis for 10 or more years, history of familial polyposis; otherwise controversial

Adapted from Sox (1994) and McCormick and Inui (1992)

atenolol), calcium channel blockers (diltiazem, vera-pamil, nifedipine or telodipine), and angiotensin con-verting enzyme (ACE) inhibitors (captopril, enalapril, or lisinopril).

At present, many antihypertensive agents are effi-cacious in reducing blood pressure. The choice of medication often involves selecting the agent with the least risk of side-effects for a particular patient. For example, in a diabetic patient, one may wish to avoid masking the adrenergic signs and symptoms of hypoglycemia, making a beta blocker, such as propranolol, a second choice in that case. In choosing to use a calcium channel blocker, care should be exer-cised to avoid orthostatic hypertension, a side-effect to which age-related changes in baroreceptor sensitivity

may predispose older adults (Wei 1989). In an older hypertensive patient whose electrocardiogram shows no significant conduction defects and who is not in severe congestive heart failure, diltiazem would be a reasonable choice among the calcium channel-blocking agents.

Mrs. Shipper takes enalapril, an ACE inhibitor and a reasonable choice in the absence of pre-existing renal insufficiency, when ACE inhibitors would generally be avoided or used only with very careful monitoring of renal function, because of the potential for worsened renal function due to the associated efferent glomerular arteriolar vasoconstriction.

Two additional aspects of monitoring Mrs. Shipper's antihypertensive treatment are important to consider

Table 18.2 Health promotion and health protection counseling for all older adults

Service/Maneuver	Comment
Preferences for future medical care	Identify surrogate decision makers for health care and life-prolonging interventions, discuss values underlying preferences for interventions, document and update decisions
Oral health	Daily oral hygiene
Nutrition	Daily intake of total fat <30% of calories, saturated fat <10%, cholesterol <300 mg, high fiber (5–40 g); low sodium (<6 g); calcium (1 g for men, 1.5 g for women); vitamin D 600–800 IU (twice the recommended daily allowance)
Estrogen	For women at increased risk for osteoporosis, such as those with low body weight for height, smoking, alcohol, inadequate dietary calcium, medications such as corticosteroids and thyroid replacement—and no contraindications such as abnormal vaginal bleeding, active liver disease, thromboembolic disorders, or estrogen-dependent cancer.
Physical activity	Maintain desirable weight, cardiovascular and musculoskeletal fitness, including osteoporosis prevention, through endurance and strength training (e.g., sustained walking for 30 min/day)
Functional assessment	Assess cognitive, emotional, and physical function, including driving; elicit marital/sexual problems
Injury prevention	Avoid falls through exercise, adequate nutrition, improved vision/lighting, and hazard reduction in living environment; safety belts, helmets; other home safety, including smoke detectors, unloaded firearms stored in locked containers
Environmental safety	Past and present exposures, toxins
Food and drug hazards	Dietary and food preparation history; inquiry about non-prescription medications and home remedies
Sexual practices	Provide information on safe sex for all sexually active patients
Substance use	Quit tobacco; limit alcohol intake; stop problem alcohol drinking; if IV drug use encourage to quit, at least stop needle sharing
High-risk older adults	
Conditions to monitor	Abnormal bereavement, symptoms of depression, risk factors for suicide, changes in cognition, multiple medications, signs of abuse or neglect, malignant skin lesions, peripheral arterial disease, loose/decayed teeth or gingivitis

Adapted from Sox (1994) and McCormick and Inui (1992).

at this point: adherence to her medical regimen, also known as patient compliance; and possible adverse effects of her medication, including potential interactions with other medications.

Mrs. Shipper volunteers that she keeps track of her medication by making a note on her calendar. Asking about a person's reminder system can give the physician an idea of how compliant a patient is with the prescribed regimen.

A good open-ended question to begin a discussion about possible drug side effects would be 'How do you feel?' This question could be followed with questions about light-headedness, especially on standing up (checking for orthostatic hypotension), gastrointestinal distress, and mood changes. Symptoms potentially due to enalapril include cough and edema, which may prove refractory to dose changes and necessitate a change to another antihypertensive medication. Serum urea nitrogen and potassium need to be monitored at intervals; elevated levels may signify worsening renal function. In considering the responses to these questions, one needs to keep in mind that the symptoms may not be due to the medication under consideration but could be due to an as yet undiagnosed condition or to other medications, prescribed or non-prescribed.

'Mrs. Shipper, I am glad that you are doing well with control of your blood pressure. You mentioned that you are interested in considering ways in which you can stay healthy and active. There are a number of possible ways to assist you in keeping good health and limiting disability. Let me ask you a few questions.'

'Good, Doctor, that's why I am here for this appointment!'

At Mrs. Shipper's initial visit, you explain to her that the menu of preventive activities is extensive. These activities include history, physical examination, and laboratory tests (Table 18.1); counseling for healthy lifestyles (Table 18.2); and immunizations (Table 18.3). Some tests and activities, such as immunizations, apply to all adults 65 years of age and older. Others apply selectively, depending on the results of previous screening, established diseases and functional status, present risks for future diseases, as well as patient beliefs and preferences. The evidence supporting screening and prevention in the older-old, such as Mrs. Shipper, is less robust than that in the younger-old. The rationale for their performance in the older-old is based on prudent extrapolation of the evidence from the younger-old group.

Table 18.3 Immunizations in older adults (above 65 years)

Service/Maneuver	Comment
Influenza	Annually, ideally in October
Pneumococcal pneumonia 23-valentre	Once, consider vaccination every 6 years
Tetanus-diphtheria toxoid	Every 10 years (e.g., on mid-decade birthdays)
Hepatitis B	For hemodialysis patients, those exposed to blood and/or secretions, and those traveling in endemic areas

In reviewing Mrs. Shipper's antihypertensive regimen, you also ask her about any over-the-counter medications, specifically non-steroidal antiinflammatory drugs (NSAIDs), whose concurrent administration could lead to increased blood pressure or renal insufficiency via inhibition of prostaglandin synthesis and subsequent renal vasoconstriction.

While the focus here is on secondary prevention (i.e., preventing the sequelae of established hypertension), you would also verify that she is following the primary prevention strategies: maintaining a low-sodium diet, reducing stress if present, and engaging in regular physical exercise.

Chemoprophylaxis to prevent ischemic heart disease and stroke is another consideration in Mrs. Shipper's disease prevention plan. Given her advanced age (over 75 years), gender (female), and hypertension, she is a good candidate for primary prevention with aspirin or other antiplatelet medication such as ticlopidine (Albers 1995; ATC 1994). She would also be a candidate for anticoagulation with warfarin if she had left ventricular dysfunction, history of a prior stroke or atrial fibrillation (Albers 1995; ATC 1994).

If on examination the patient had a carotid bruit, even if there were no symptoms, she would likely benefit from further evaluation of the carotid lesion; surgical therapy would be indicated if the carotid stenosis was found to be 60% or greater (NINDS 1994).

Diet and exercise are also very important for cardiovascular disease and osteoporosis prevention in Mrs. Shipper. For her, administration of exogenous estrogen may be important for preventing coronary artery disease as well as osteoporosis and hip fracture (Manson 1994; Geelhoed et al. 1994). Before deciding about estrogen therapy, you carefully inquire about potential contraindications to estrogen replacement such as breast cancer and/or a family history of breast cancer or endometrial cancer. Likewise, Mrs. Shipper should be informed about the possible risks of endometrial cancer (low absolute risk), breast cancer (potential risk), and vaginal bleeding (often absent at low doses).

Cancer

Does Mrs. Shipper regularly perform breast self-examination? Has she had a mammogram recently? The clinical breast examination is most effectively performed in tandem with a mammogram by a radiologist. The incidence of breast cancer rises steadily throughout old age, and the prognosis is better when lesions are detected early (Henderson 1995; see also Chapter 13 on breast cancer). There appears to be no age that is 'too old' for this essential preventive service. Has she noticed any vaginal bleeding? When did Mrs. Shipper last have a pelvic examination with a pap test? Obtaining cervical cytology through pelvic examination would be necessary if Mrs. Shipper had had abnormal or insufficient results, or if no testing was done in the years prior to this office visit. Has there been any change in appetite or bowel habits? Has she lost any weight? There is currently some controversy concerning efforts directed toward the prevention of colon cancer. Recent research has shown that the presence of occult blood in the stool may be insufficiently sensitive or specific in the detection of colon cancer. Expert panels recommend sigmoidoscopy at three to five year intervals, with colonoscopy for those individuals at greater risk.

Throughout the examination you are attentive to any skin changes and potentially malignant skin lesions, from the common basal cell carcinoma to the less common but faster-growing melanoma. Many older persons have sustained considerable sun exposure, and ultraviolet radiation-induced skin damage along with it (Austoker 1994).

Reduction of dietary fat intake, together with increased consumption of fruit, vegetables, and grain products would be vitally important to protecting the elderly person against cancer (Block 1992). Similarly, reducing tobacco use, decreasing sun exposure, and undergoing regular pap tests and mammograms as well as other health screening evaluation and follow-up would be important (Taplin et al. 1994, Sutton et al. 1994).

Diabetes

Does Mrs. Shipper have any symptoms of polyuria, lethargy, weight loss, or visual disturbances? Is she experiencing increased thirst? Has there been a decrease in energy level?

Diet and weight control, physical exercise and careful control of blood glucose levels are important to the prevention of diabetic complications. Routine foot care is also important in the elderly diabetic, in order to avoid possible trauma and infection with ensuing complications. Careful attention to diabetes-related eye changes is essential to maintaining functional independence (Singh *et al.* 1994).

Chronically disabling conditions

These include arthritis, deformities or orthopedic impairments, visual impairments, hearing and speech impairments. Does Mrs. Shipper have any joint pains, limitations in joint motion, or trouble with her feet? When did Mrs. Shipper last have her vision checked and her intraocular pressures measured? On examination of Mrs. Shipper, you ask her to read newsprint using her corrective lenses. (Alternatively, you could have her read the 20/70 line on the Rosenbaum pocket-size vision chart.) Primary care physicians can screen for visual acuity and perhaps, intraocular pressures, while a more thorough exam may be provided by an ophthalmologist or optometrist. How is Mrs. Shipper's hearing? Does she wear hearing aids? Two are usually better than one, but more expensive! The presence of hearing aids is usually a good sign, but it does not necessarily mean that they work. If Mrs. Shipper is wearing hearing aids, you would ask her to demonstrate their function to you. If she does not wear a hearing aid, you would ask if she has difficulty with conversations on the telephone or with hearing conversations when there are other voices in the background. You check her hearing using the finger-rub test which has been shown to correlate well with audiometry, and examine the ear canals for cerumen.

The prevention of chronically disabling conditions includes reducing the impairment where possible through appropriate therapy and assistive devices, increasing physical activity, reducing excess body weight, and accessing self-help resources in the community.

Infection

Because older persons may be more susceptible to infectious agents and to the complications of infections, a major effort should be devoted to prevention of infection through public health, education, and immunization (Fabacher *et al.* 1994; Sox *et al.* 1994). Vaccination of asymptomatic individuals is a cornerstone of primary preventive care in older adults (recommendations for immunizations are given in Table 18.3). Prompt treatment of those afflicted is also vital to preventing serious complications as well as further spread of the diseases. You are not surprised that Mrs. Shipper has never had an influenza shot: only a minority of people at risk for influenza has received the vaccine. It is underutilized at least in part because the patient's physician did not recommend it.

Unless Mrs. Shipper has a rare allergy to egg protein, you strongly recommend that she make an appointment with your office in the fall to receive the flu vaccine. You also plan to send her a reminder. For your patients who are less mobile than Mrs. Shipper, you arrange for in-home vaccination by the visiting nurse. You also discuss with Mrs. Shipper your recommendation that she receive pneumococcal vaccine, which provides protection against up to 85% of the sources of pneumococcal bacteremia (Butler *et al.* 1993). This vaccination is usually given only once in a lifetime, although recent research suggests that revaccination every six years might be beneficial due to the attenuation over time of the immune response (Mufson *et al.* 1991). Pneumococcal vaccine can be given a minimum of one week after the influenza vaccine, so as to avoid the possibility of simultaneous local reactions.

You are even less surprised that Mrs. Shipper cannot recall receiving a tetanus shot for many years. While rare, tetanus is a completely preventable near-fatal disease. Many people receive the vaccine when they are treated for a wound and their vaccination status is unknown. Once inoculated, everyone should have the tetanus-toxoid booster every 10 years; the mid-decade birthday provides a convenient reminder.

Dementia

In the absence of a reliable informant, such as a spouse, close friend, or family member, a person's progressive cognitive changes may be initially difficult to detect. It may also be a challenge to detect cognitive changes in the presence of visual or hearing loss. Without the adequate reception of sensory stimuli and information, subsequent encoding and retrieval are impaired, but may not necessarily be due to impaired integrative or cognitive capacity. Examination of Mrs. Shipper's vision, hearing, and recall ability will help to resolve this potential issue.

Efforts devoted to the maintenance of cognitive function in the presence of stroke and/or other neuro-degenerative diseases are most worthwhile. Recent and ongoing research studies to help elucidate the causes of Alzheimer's disease and of other dementias appear promising and will likely be invaluable to future preventive and therapeutic efforts (Fitten 1994). Educational programs and social support systems designed to maximize the level of function of cognitively impaired persons are also essential (Eisdorfer 1994).

Screening for changes in cognition is potentially delicate, but essential. Mrs. Shipper appears to be cognitively intact and functionally independent. In addition to performing a Mini-mental State Examination (MMSE), you discuss with her several recent national and international events, in order to see that she is fully oriented, that her attention, memory and language skills are good, and that her judgement and ability to synthesize new information are intact. Depending on the results of this initial screen, you may schedule in-depth cognitive testing during a future visit.

Functional status

Mrs. Shipper came to your office visit unaccompanied. How did she get there? Did she drive or take public transportation? Was she dropped off by someone? Mrs. Shipper's answers to these questions give you information about her baseline abilities to carry out activities of daily living. These abilities are essential benchmarks of health in all older adults. Although Mrs. Shipper appears healthy, there may be a professional or recreational activity that she may no longer be able to perform. This change in function could herald undetected illness.

Promoting the use of preventive services is key to the prevention of disease in older persons. The elimination of barriers to clinical preventive services is vitally important. Increasing the delivery of preventive services would also be helpful (Lemley et al. 1994).

At several junctures during the history, examination, tests, and counseling, you have gauged Mrs. Shipper's ability to perform professional, recreational, and daily living activities in order to determine any change in her status caused by potentially remediable illnesses or diseases. This effort is termed 'functional assessment', and embraces the emotional, cognitive, physical, social, and environmental domains. You ask Mrs. Shipper specifically about one basic activity of daily living or self-care which is often overlooked by patient or doctor alike: urinary incontinence. Can she make it to the bathroom on time? Does she ever leak on the way to the bathroom, or when she laughs or coughs? If so, she will benefit from further questioning to ascertain the exact cause of the incontinence and consideration of what can be done to eliminate or decrease the problem.

As part of your review of systems, you ask her about her sexual activity. You inquire more specifically about her partner or partners and any potential risky sexual behaviors. Human immunodeficiency virus (HIV) infection is on the rise in the older population, mainly as a consequence of transfusions prior to 1985. Safe sexual practices are always indicated.

Other preventive services

Several laboratory or special diagnostic tests would complement the menu of preventive services relevant to Mrs. Shipper. These tests include a resting electrocardiogram and a dipstick urinalysis for proteinuria, bacteriuria, or hematuria (which may presage renal insufficiency, bladder infection and cancer in the asymptomatic individual). If she has not had a normal level of thyroid-stimulating hormone (TSH) documented in the past, you would obtain a TSH level, screening for hypothyroidism, especially if Mrs. Shipper had related non-specific symptoms such as decreased energy or constipation.

Were Mrs. Shipper to have additional health risks as determined by her history, you would consider the other preventive measures listed in Table 18.1 under 'high-risk older adults'.

Health promotion

During the history of previous preventive services, you have actually already obtained some of the information that is important in assessing Mrs. Shipper's needs for counseling about a healthy lifestyle. A program of health promotion should include sponsorship of increased physical activity and exercise, proper nutrition, reduced use of tobacco and alcohol, and efforts at improving mental health and reducing adverse effects of stress and mental disorders.

Physical activity

This increases healthful life span (Wagner et al. 1994). It is important for maintaining functional independence and quality of life. It can prevent falls and major diseases while building strength, endurance, and self-confidence (Tinetti et al. 1994). The physical activity should be habitual, light or moderate in intensity, and last about 20–30 minutes a day. Walking is an excellent form of exercise. Advancing age, even in the presence of cardiopulmonary and musculoskeletal disease and disability, is no absolute barrier to improving endurance and strength (Fiatarone et al. 1994).

Mrs. Shipper is interested in improving her health. Regular exercise, together with a balanced diet, will have important benefits.

Nutrition and dietary habits

You ask Mrs. Shipper to describe how she nourishes herself. Does she shop for herself and cook her own food? How often does she eat out? Does she rely on meals prepared by others? You attempt to assess the quality of her nutrition by reviewing her three–day recall of all dietary intake. Recall can be facilitated by having Mrs. Shipper prepare this information prior to the next visit. Tables of nutritional content of foods are available (Pennington 1994), or you may consider a referral to a registered dietitian.

Nutrition and dietary habits are often related to patterns of physical activity. Excesses or imbalances of food components should be avoided. Physicians should encourage consumption of a variety of foods and recommend a diet low in fat and cholesterol. The diet should be high in fruit and vegetables (including grain), modest in sugar and salt (sodium), and low in alcohol. Caloric intake should be adjusted to maintain a healthy weight. Dietary fat intake should be less than 30% of the total calories. The target intakes of fat, cholesterol, fiber, sodium, and calcium are included in Table 18.2.

Use of tobacco

This is the most important single preventable risk for death in developed countries around the world. Although Mrs. Shipper has never smoked tobacco, promotion of tobacco-free environments would benefit her. Restrictions on smoking in public places should be strengthened and enforced. More prevention education and smoking cessation programs should be offered to patients (Rimer and Orleans 1994).

Alcohol

You inquire as to Mrs. Shipper's usual alcohol intake. Has she ever had problems with excessive drinking of alcohol? You recommend that she limit her alcohol intake to less than the equivalent of one small glass of wine per day. Excess alcohol intake could be a remediable cause of falls and accidental injury, cardiac arrhythmias, stroke, urinary incontinence, and delirium.

Alcohol excess is associated with well-known health problems and diseases. Educational programs to reduce alcohol and other drug consumption, together with raising awareness of the harmful effects of addictive substances, are important. Increased access to treatment programs for drug dependence and increased social support programs should also be beneficial.

Mental health

To improve mental health and reduce mental disorders in the elderly, it is necessary to maintain or strengthen the older person's ability to negotiate the day-to-day challenges and interactions of life without sustaining emotional or behavioral decompensation. You ask Mrs. Shipper if she lives alone. Two of every three women her age live alone (Magaziner and Cadigan 1988). You ask her to describe her social network: does she have someone to depend on in case of need? Mrs. Shipper may have recently lost her husband, or a close friend, or stopped working. Such personal losses or losses in function through illness or withdrawal from social engagement and support are major risk factors for depression. A family history of depression or any previous personal bout with depression are also risk factors. Any difficulty with sleeping at night, or the use of sedatives or alcohol, would heighten your suspicion of new-onset depression and prompt further evaluation and the possibility of treatment.

Mental stress due to life events, chronic strain, or environmental pressures may result in cognitive, emotional, and behavioral problems, as well as physiologic changes. The promotion of healthful habits, such as proper nutrition and ample physical activity, are also important in the reduction of stress.

Equally important in reducing mental stress would be to increase one's self-resourcefulness and sense of control over the environment. Possible medical interventions, when needed, include antidepressant and other psychotherapeutic agents, and biofeedback. Treating behavioral problems in elderly patients with dementia would require additional efforts at designing a more supportive living environment and other therapeutic activities, such as art and music therapy.

Health protection

Important elements of an educational program on health protection include injury-prevention instruction, environmental safety information, food and drug education, and oral health instruction.

Accidental injuries

These are a leading cause of morbidity and mortality in older persons. Recent research has shown that strength training can improve gait in very old individuals at all levels of health (Fiatarone et al. 1994). This benefit is important as you counsel Mrs. Shipper on preventing falls, the most common cause of injury among older persons. Strength training can reduce the risk of falls by improving the effector component (muscles, tendons, bones) of balance and gait. Other strategies for the prevention of falls are necessary in order to prevent decline in the sensory component (vision, hearing, vibration, and position sense) and in the central integrative component, the central nervous system. You also ask Mrs. Shipper about potential hazards in her living environment, such as scatter or throw rugs (which should be removed), telephone wires, and inadequate lighting. Such hazards are included in home safety checklists available from many agencies that serve the aging population. A home visit is the most direct source of information that is important in the counseling for injury prevention.

Motor vehicle accidents and falls are the major causes of injuries. To help the elderly reduce injuries, there is a need for societies to develop an effective injury-preventing instruction program with the co-operation of leaders in transportation and highway safety, law, engineering, architecture, and safety sciences.

Another important aspect of injury prevention is automobile driving. Does Mrs. Shipper drive a car? Does she have concerns about her driving? Has she had any vehicular accidents? The American Association of Retired Persons and the American Automobile Association sponsor safe driving programs aimed at improving skills of older drivers. If she still drives, she might consider attending such a program. Does she wear seat belts? If she rides a bicycle or motorcycle, does she wear a safety helmet?

Does Mrs. Shipper live close to any chemical plants or factories? Does she live close to a major street or highway? Does she live close to a train station? How many flights of stairs does she climb to get to her apartment?

Environmental safety

Health programs designed to promote environmental safety should include efforts to minimize exposure to pollutants and carcinogens, such as smoke, lead and radon, reduce cumulative trauma from repetitive motion, pressure or noise, and reduction of back injury through exercise and prevention programs. Improved household management of toxic waste materials and recyclable materials would also be important.

Food and drug hazards

Has Mrs. Shipper been ill with gastrointestinal symptoms of vomiting or diarrhea? Does she prepare her own food or does someone else do the cooking? Does she take home remedies or over-the-counter preparations? Does she share or trade medications with her neighbors or friends? Adverse food and drug outcomes may be reduced through efforts at reducing food-borne diseases and improving pharmacy-based information systems. Better food preparation, handling, and storage techniques can reduce food-borne disease. Increased patient education regarding drug interactions, drug side-effects and food–drug interactions (Roe 1989) would be extremely important. Equally important, a reduction in the number of drugs per person can significantly lower drug toxicities.

Oral health

Does Mrs. Shipper have her native teeth, or partial or complete dental prostheses (plates or dentures)? Does she see a dentist every six to twelve months? For what oral conditions is she being treated? Has she noticed any changes in her chewing ability or in the fit of her dental prostheses, or any discomfort or lumps in her mouth? Tobacco and alcohol are synergistic risk factors for cancers of the oral cavity and elsewhere.

Oral health is an important part of screening at all ages. Regular care for dental caries, periodontal disease, and dentures are of primary importance. Instructions for improved oral self-care and avoidance of foods that increase caries, tobacco, and excessive alcohol are equally important. Improving the use of oral health screening and follow-up services would also be helpful.

Preferences for future medical care

You conclude your counseling efforts with a determination of Mrs. Shipper's preferences for life-prolonging or life-sustaining care in the event of serious illness. This process involves three steps: discussion, decision, and documentation. The conversation can proceed over several visits. First, you raise the issue: elicit her previous experiences, concerns, and values about life-prolonging therapy, death, and dying. You ask if Mrs. Shipper has considered a surrogate or proxy who knows her values and preferences and with whom you and other care providers could interact if Mrs. Shipper were to become unable to communicate her wishes. You explore her reasons for deciding for or against a future attempt of cardiopulmonary resuscitation, considering both her current state of health and potential future states of deterioration. Does she understand the consequences of her decisions?

You ask Mrs. Shipper to discuss the matter further with those close to her, and you offer to provide her with additional written or audiovisual information if desired. On return visit, you will ask if she has decided her preferences and if she wishes to document these decisions.

A growing number of countries have provisions for the appointment of surrogate decision makers who can decide for the patient in the event of an illness or accident that would render the patient unable to communicate his or her desires. Conversation and discussion before a crisis can help to give guidance to the surrogate decision maker and can maximize the chance that Mrs. Shipper's wishes will be respected if she became seriously ill and could not speak for herself. You also make it clear to Mrs. Shipper that she can always rediscuss these decisions with you and that you will help her with any changes that she might wish to make.

Summary

Mrs. Shipper has hypertension but is an otherwise healthy 80-year-old woman. Priority concerns in prevention and promotion are described in this chapter: heart disease and stroke, cancer screening, sensory loss, oral health, depression, falls, misuse of medications (she takes only her antihypertensive medication), nutrition, physical activity, infectious diseases, osteoporosis, smoking (not relevant in the care of Mrs. Shipper), and social isolation. Recent and ongoing research on these issues supports your recommendations for a comprehensive and continuing program of disease prevention, health promotion, and health protection, with the goals of maintaining independence and enhancing quality of life.

Questions for further reflection

1. What immunizations are appropriate for an older person?
2. In considering the primary care of older women, what preventive measures, if any, would you take toward early detection of breast cancer and in limiting osteoporosis?
3. If a healthy 75-year-old man with little significant past medical history came to your office and asked for your advice on what should be done to live longer and healthier, what would you tell him?

References

Albers, G.W. (1995). Antithrombotic agents in cerebral ischemia. *American Journal of Cardiology*, 75, 34B-8.

ATC (Antiplatelet Trialists' Collaboration) (1994). Collaborative overview of randomized trials of antiplatelet therapy. *British Medical Journal*, 308, 81–106.

Austoker, J. (1994). Melanoma: prevention and early diagnosis. *British Medical Journal*, 308, 1682–6.

Block G. (1992). The data support a role for antioxidants in reducing cancer risk. *Nutrition Reviews*, 50, 207–13.

Butler, J.C., Breiman, R.F., Campbell, J.F., Lipman, H.B., Broome, C.V., and Facklam, R.R. (1993). Pneumococcal polysaccharide vaccine efficacy. An evaluation of current recommendations. *Journal of the American Medical Association*, 270, 1826–31.

Eisdorfer, C. (1994). Community resources and the management of dementia patients. *Medical Clinics of North America*, 78, 869–75.

Fabacher, D., Josephson, K., Pietruszka, F., Linderborn, K., Morley, J.E., and Rubenstein, L.Z. (1994). An in-home preventive assessment program for independent older adults: a randomized controlled trial. *Journal of the American Geriatrics Society*, 42, 630–8.

Fiatarone, M.A., O'Neill, E.F., Ryan, N.D., Clements, K.M., Solares, G.R., Nelson, M.E., et al. (1994). Exercise training and nutritional supplementation for physical frailty in very elderly people. *New England Journal of Medicine*, 330, 1769–75.

Fitten, L.J. (1994). Evolving trends in Alzheimer research. In *Dementia and cognitive impairments. Facts and research in gerontology 1994 (supplement)*, (ed. J. Vellas, J.L. Albarede, and P.J. Garry), pp. 7–9, Serdi, Paris.

Geelhoed, E., Harris, A., and Prince, R. (1994). Cost-effectiveness analysis of hormone replacement therapy and lifestyle intervention for hip fracture. *Australian Journal of Public Health*, 18, 153–60.

HCFAOA (Health Care Financing Administration, Office of the Actuary) (1988). Expenditures and percent of gross national product for national health expenditures, by private and public funds, hospital care, and physician services; calendar years 1960–87. *Health Care Financing Review*, 10, 2.

Henderson, I.C. (1995). Paradigmatic shifts in the management of breast cancer. *New England Journal of Medicine*, 332, 951–2.

Herman, C.J. and Robertson, J.M. (1993). Issues in Preventive Geriatrics. In *Geriatrics review syllabus supplement: a core curriculum in geriatric medicine*, (ed. D.B. Reuben, T.T. Yoshikawa, and R.W. Besdine), pp. 33S-38. American Geriatrics Society, New York.

Kupersmith, J., Homes-Rovner, M., Hogan, A., Rovner, D., and Gardiner, J. (1995). Cost-effectiveness analysis in heart disease, Part II: Preventive therapies. *Progress in Cardiovascular Diseases*, 37, 243–71.

Lemley, K.B., O'Grady, E.T., Rauckhorst, L., Russell, D.D., and Small, N. (1994). Baseline data on the delivery of clinical preventive services provided by nurse practitioners. *Nurse Practitioner*, 19, 57–63.

Magaziner, J. and Cadigan, D.A. (1988). Community care of older women living alone. *Women's Health*, 14, 121–38.

Manson, J.E. (1994). Postmenopausal hormone therapy and atherosclerotic disease. *American Heart Journal*, 128, 1337–43.

Massie, B.M. (1994). First-line therapy for hypertension: different patients, different needs. *Geriatrics*, 49, 22–30.

McCormick, W.C. and Inui, T.S. (1992). Geriatric preventive care: Counseling techniques in practice settings. *Clinics in Geriatric Medicine*, 8, 215–28.

Mufson, M.A., Hughey, D.A., Turner, C.E., and Schiffman, G. (1991). Revaccination with pneumococcal vaccine of elderly persons 6 years after primary vaccination. *Vaccine*, 9, 403–7.

NINDS (National Institute of Neurological Disorders and Stroke) (1994). *Carotid endarectomy for patients with asymptomatic internal carotid artery stenosis*. (Clinical Advisory). National Institutes of Health, Bethesda, MD.

Pennington, J.A. (1994). *Bowes and Church's: Food values of portions commonly used*, (16th edn). Lippincott, Philadelphia, PA.

Rimer, B.K. and Orleans, C.T. (1994). Tailoring smoking cessation for older adults. *Cancer*, 74, 2051–4.

Roe, D.A. (1989). *Handbook: Interactions of selected drugs and nutrients in patients*, (4th edn). American Diatetic Association, Chicago.

Singh, B.M., Prescott, J.J., Guy, R., Walford, S., Murphy, M., and Wise, P.H. (1994). Effect of advertising on awareness of symptoms of diabetes among the general public: the British Diabetic Association Study. *British Medical Journal*, 308, 632–6.

Sox, H.C. (1994). Preventive health services in adults. *New England Journal of Medicine*, **330**, 1589–95.

Sutton, S.M., Eisner, E.J., and Burklow, J. (1994). Health communications to older Americans as a special population. The National Cancer Institute's consumer-based approach. *Cancer*, **74**, 2194–9.

Taplin, S.H., Anderman, C., Grothaus, L., Curry, S., and Montano, D. (1994). Using physician correspondence and postcard reminders to promote mammography use. *American Journal of Public Health*, **84**, 571–4.

Tinetti, M.E., Baker, D.I., McAvay, G., Claus, E.B., Garrett, P., Gottschalk, M., *et al.* (1994). A multifactorial intervention to reduce the risk of falling among elderly people living in the community. *New England Journal of Medicine*, **331**, 821–7.

Wagner, E.H., LaCroix, A.Z., Grothaus, L., Leveille, S.G., Hecht, J.A., Artz, K., *et al.* (1994). Preventing disability and falls in older adults: a population-based randomized trial. *American Journal of Public Health*, **84**, 1800–6.

Wei, J.Y. (1989). Use of calcium entry blockers in elderly patients. *Circulation*, **80**, IV171–7.

Index